Springer
Berlin
Heidelberg
New York
Barcelona
Budapest
Hongkong
London
Mailand
Paris
Santa Clara
Singapur
Tokio

Springer
Berlin
Heidelberg
New York
Barcelona
Budapest
Hongkong
London
Mailand
Paris
Santa Clara
Singapur
Tokio

Martin C. Woodle, Ph.D. • Gerrit Storm, Ph.D. (Eds.)

Long Circulating Liposomes:
Old Drugs, New Therapeutics

Springer

Martin C. Woodle, Ph.D.
Genetic Therapy, Inc.
Gaithersburg, Maryland
U.S.A.

Gerrit Storm, Ph.D.
Department of Pharmaceutics
University of Utrecht
Utrecht, The Netherlands

ISBN: 3-540-63082-1 Springer-Verlag Berlin Heidelberg New York
Biotechnology Intelligence Unit

Library of Congress Cataloging-in-Publication data

Long circulating liposomes: old drugs, new therapeutics / [edited by] Martin C. Woodle, Gerrit Storm.
p. cm. — (Biotechnology intelligence unit)
Includes bibliographical references and index.
ISBN 1-57059-508-9 (alk. paper)
1. Liposomes—Therapeutic use. 2. Drug carriers. 3. Drugs—Vehicles. I. Woodle, Martin C. II. Storm, Gerrit. III. Series.
[DNLM: 1. Liposomes—therapeutic use. 2. Liposomes—Diagnostic use. 3. Drug Carriers. 4. Antineoplastic Agents—administration & dosage. QU 93 L848 1997]
RS201.L55L66 1997
615'.7—dc21
DNLM/DLC 97-30426
for Library of Congress CIP

Typesetting: Landes Bioscience Georgetown, TX, U.S.A.

SPIN: 10630491 31/3111 - 5 4 3 2 1 0 - Printed on acid-free paper

PREFACE

The overall theme of this book is that new formulations of existing drugs can create entirely new therapeutic agents. The changes that can occur impact important clinical issues such as therapeutic index, schedule, route, side effects and—most importantly—indications. Thus, the principle of using advanced formulation technology with existing or old agents to create effectively new therapeutic agents appears as one of the most promising to address the continuing need for better therapeutic agents for clinical practice. The Doxil® and DaunoXome® long circulating liposome formulations of well known drugs now available to clinicians are good examples of application of this principle. This book focuses on these and long circulating liposome formulations in general because they are just beginning to appear in approved products. In fact, the development of long circulating liposome formulations of other anti-tumor therapeutic agents is progressing with such rapidity that practicing oncologists in the U.S. can expect to encounter other long circulating formulations in the near future. However, rather than a treatise in support of one product or the another, the goal of this book is to provide an understanding of the basis for these liposome products and thus how they alter the way encapsulated drugs should be used clinically. This is especially important given the large body of knowledge on use of these drugs in their original formulation, usually referred to in this volume as the "free" drug, which often is contrary to the use of these new formulations. Thus, such understanding of differences between use of the original product and these new ones is critically important for their best use in clinical practice. The book describes the resulting changes in properties of a few well-known drugs when formulated with long circulating liposomes, including the most recent clinical experience.

About 30 years ago materials now commonly called liposomes, meaning "fat bodies", were identified as self-organized particles formed when natural phospholipids were dispersed in water, where the lipids form a shell surrounding an internal "entrapped" aqueous compartment. Almost immediately afterwards, the possibility that they could be used to improve drugs was suggested and appropriate investigations launched. Some of their properties leading to this proposal included:

1) The ability to solubilize aqueous insoluble drugs.

2) The ability to encapsulate aqueous drugs in their internal compartment, thus possibly protecting and delivering them.

3) Biocompatibility due to composition of natural lipid components of cell membranes.

4) Presumed lack of toxicity and recognition by host defense mechanisms.

Within a few years, liposomes became very popular with both medical and basic biochemical researchers, something which persists to this day. However, despite great interest and considerable efforts for many years, commercial liposome formulations of drugs are only just beginning to appear. The first therapeutic liposome products appear to be formulations of the effective but insoluble antifungal drug amphotericin B, which provide reduced toxicity and at least equivalent efficacy. Thus these formulations are significantly better than the original, highly toxic detergent formulation. The first of these has been available in several Western European countries for several years now. Most recently, special long circulating liposome formulations of oncology agents are achieving marketing approval, again due to reduced toxicity and improved efficacy. Consequently, it seems as if these new formulations of old drugs will become increasingly common alternatives for clinicians and their patients.

The protracted time between initial tests of the therapeutic potential of liposomes, suggesting benefits, and their appearance in the marketplace is due to the existence of several difficulties in their commercial development, many not originally foreseen. Of these, perhaps the most troubling has been the realization that despite their composition of natural lipids typical of cell membranes and their lack of protein or glycosylation typically responsible for an immune response, i.e., recognition by the host defense, they are rapidly removed from the blood and digested by cells of the mononuclear phagocytic system (MPS), with only limited exception. Nonetheless, a few drugs were identified that can retain activity even after macrophage uptake in liposome-encapsulated form, doxorubicin being a clear example.

One exception to rapid liposome uptake in vivo was identified by the use of very small, neutral liposomes composed of so-called "rigid" lipids. These liposomes have been developed as a formulation for daunorubicin with the trade name DaunoXome®, marketed by NeXstar Pharmaceuticals, which has recently been given U.S. FDA marketing approval for first line therapy of Kaposi's sarcoma patients. The fact that daunorubicin is not a major part of the first line therapy in the U.S. demonstrates that long circulating liposomes can convert old drugs into different, hopefully better, therapeutic agents. Subsequently, it was found that liposomes with the water soluble polymer polyethylene glycol conjugated to their surface have the ability to partially evade recognition largely independent of particle size, surface charge, and lipid composition. The phenomenon appears to depend upon a property of the polymer to act as a "fuzzy" coating, or steric barrier, which impairs the ability of blood proteins to bind to the liposomes, thereby reducing their recognition by phagocytic cells. The result is a significantly prolonged residence in the blood, making them much more effective for the delivery of drugs via the systemic compartment. Development of PEG grafted liposomes has progressed fastest and farther as a formulation for the well known drug doxorubicin. Their use for doxorubicin has culminated in a formulation called Stealth® Liposomal Doxorubicin or Doxil®, marketed by SEQUUS Pharmaceuticals, which is approved for marketing by the U.S. FDA for the treatment of patients with Kaposi's sarcoma refractory to first line therapy.

The purpose of this book is to give a general introduction to the physicochemical and biological properties of long circulating liposomes, some with advanced surface properties, followed by detailed descriptions of their therapeutic properties as formulations of drugs. Specifically, the book is intended to assist oncologists and other physicians facing the option to use Doxil® and DaunoXome®, and eventually many long circulating liposome drug formulations in their clinical practice. In addition, one goal is to give physicians a feel for the future developments that can be expected, hopefully soon, from advances in drug delivery technologies and how those technologies will impact treatment of patients.

The book is broken up into the following three sections. First, the introduction simply provides a quick overview of the physicochemical and biological properties of long circulating liposomes, providing a general understanding of the basic biological phenomenon and current understanding of the underlying mechanism. This section will facilitate understanding of the capabilities and limitations of these novel liposomes.

The second section describes the achievements that have been made in developing long circulating liposome formulations of existing or conventional small molecule therapeutic or diagnostic agents, relying on the ability of these liposomes to selectively localize into sites of pathology by a passive means. This represents the major part of the book since it is where the first commercial products based on long circulating liposomes, as exemplified by Doxil and DaunoXome, can be expected. Long circulating liposomes can be expected to give rise to several commercially available products in the very near future. Development of long circulating liposome systems will require focusing upon essentially only two issues: achieving adequate levels of each new drug stably entrapped in the liposome, and achieving adequate release at the site of pathology. This section is further divided into three subsections: malignant diseases, other pathological therapeutic realms, and diagnostics.

The third section describes advances that have been made in efforts to improve upon passive targeting by adding mechanisms for active targeting. Clearly, antibodies are the first place to start, and considerable activity is currently ongoing for their use in this manner. Nevertheless, while these and small molecule alternatives are being evaluated, products based upon active targeting of long circulating liposome formulations are considerably further away from commercial reality.

It is the hope of the editors that this book presents an adequate perspective of where commercial drug formulations based upon long circulating liposomes are today and where they can be expected to appear in the future.

CONTENTS

EDITORS

Martin C. Woodle, Ph.D.
Genetic Therapy, Inc.
Gaithersburg, Maryland, U.S.A.
Chapters 1, 12, 20

Gerrit Storm, Ph.D.
Department of Pharmaceutics
Utrecht Institute for Pharmaceutical Sciences
University of Utrecht
Utrecht, The Netherlands
Chapters 1, 12, 16, 20

CONTRIBUTORS

Jill P. Adler-Moore
NeXstar Pharmaceuticals
San Dimas, California, U.S.A.
Chapter 13

Theresa M. Allen
Department of Pharmacology
University of Alberta
Edmonton, Alberta, Canada
Chapter 2

Irma A.J.M. Bakker-Woudenberg
Dept. Medical Microbiology &
Infectious Diseases
Erasmus University
Rotterdam, The Netherlands
Chapters 12, 14

Marcel B. Bally
Division of Medical Oncology
Section of Advanced Therapeutics
British Columbia Cancer Agency
and
Liposome Research Unit
Department of Biochemistry
The University of British Columbia
Vancouver, B.C., Canada
Chapter 3

Otto C. Boerman
Department of Nuclear Medicine
University Hospital Nijmegen
The Netherlands
Chapter 16

Nancy L. Boman
Inex Pharmaceuticals Corporation
Vancouver, B.C., Canada
Chapter 3

Frans H.M. Corstens
Department of Nuclear Medicine
University Hospital Nijmegen
The Netherlands
Chapter 16

Pieter R. Cullis
Inex Pharmaceuticals Corporation
and
Liposome Research Unit
Department of Biochemistry
The University of British
Columbia
Vancouver, B.C., Canada
Chapter 3

Alexander M. M. Eggermont
Department of Surgical Oncology
University Hospital Rotterdam
Daniel den Hoed Cancer Center
Rotterdam, The Netherlands
Chapter 4

Robert M. Fielding
NeXstar Pharmaceuticals, Inc.
Boulder, Colorado, U.S.A.
Chapter 15

Eric A. Forssen
NeXstar Pharmaceuticals
San Dimas, California, U.S.A.
Chapters 6, 10

Alberto A. Gabizon
Hadassah Hebrew University Medical Center
Sharet Institute of Oncology
Jerusalem, Israel
Chapter 11

Mark Harrison, MRCP FRCR
Imperial Cancer Research Fund
Clare Hall Laboratories
Potters Bar Herts, U.K.
Chapter 8

K.J. Harrington, MRCP FRCR
Imperial Cancer Research Fund
Royal Postgraduate Medical School
Hammersmith Hospital
London, England
Chapter 8

Alexander L. Klibanov
Mallinckrodt Inc.
Medical Imaging Division
Hazelwood, Missouri, U.S.A.
Chapter 19

Danilo D. Lasic
Liposome Consultations
Newark, California, U.S.A.
Chapter 20

Kazuo Maruyama
Faculty of Pharmaceutical Sciences
Teikyo University
Sagamiko, Tsukui-gun, Japan
Chapter 7

Lawrence D. Mayer
Division of Medical Oncology
Section of Advanced Therapeutics
British Columbia Cancer Agency
and
Liposome Research Unit
Department of Biochemistry
The University of British Columbia
Vancouver, B.C., Canada
Chapter 3

Franco M. Muggia
Kaplan Comprehensive Cancer Center
New York University Medical Center
New York, New York, U.S.A.
Chapter 11

G. Mukwaya
NeXstar Pharmaceuticals
San Dimas, California, U.S.A.
Chapters 10, 15

Donald W. Northfelt, MD, FACP
Pacific Oaks Medical Group
Palm Springs, California, U.S.A.
and
University of California
San Diego, California, U.S.A.
Chapter 9

Naoto Oku
Department of Radiobiochemistry
School of Pharmaceutical Sciences
University of Shizuoka
Yada, Shizuoka, Japan
Chapter 18

Wim J.G. Oyen
Department of Nuclear Medicine
University Hospital Nijmegen
The Netherlands
Chapter 16

Richard T. Proffitt
NeXstar Pharmaceuticals
San Dimas, California, U.S.A.
Chapters 6, 13

M. Ross
NeXstar Pharmaceuticals
San Dimas, California, U.S.A.
Chapter 10

Robert A. Sandhaus
NeXstar Pharmaceuticals, Inc.
Boulder, Colorado, U.S.A.
Chapter 15

Paul Schmidt
NeXstar Pharmaceuticals, Inc.
Boulder, Colorado, U.S.A.
Chapter 10

J.S.W. Stewart, FRCP, FRCR
St. Mary's Hospital
London, England
Chapter 8

Timo L.M. ten Hagen
Department of Surgical Oncology
University Hospital Rotterdam
Daniel den Hoed Cancer Center
Rotterdam, The Netherlands
Chapter 4

Vladimir P. Torchilin
Center for Imaging and
Pharmaceutical Research
Massachusetts General Hospital
Harvard Medical School
Charlestown, Massachusetts, U.S.A.
Chapter 17

Vladimir S. Trubetskoy
Center for Imaging and
Pharmaceutical Research
Massachusetts General Hospital
Harvard Medical School
Charlestown, Massachusetts, U.S.A.
Chapter 17

Paul S. Uster
SEQUUS Pharmaceuticals Inc.
Menlo Park, California, U.S.A.
Chapter 5

Jan Vaage
Dept. Experimental Pathology
Roswell Park Cancer Institute
Buffalo, New York, U.S.A.
Chapter 5

Alexander H. van der Veen
Department of Surgical Oncology
University Hospital Rotterdam
Daniel den Hoed Cancer Center
Rotterdam, The Netherlands
Chapter 4

Els W.M. van Etten
Dept. Medical Microbiology &
Infectious Diseases
Erasmus University
Rotterdam, The Netherlands
Chapter 14

Murray S. Webb
Inex Pharmaceuticals Corporation
Vancouver, B.C., Canada
Chapter 3

Peter K. Working
SEQUUS Pharmaceuticals Inc.
Menlo Park, California, U.S.A.
Chapter 5

ABBREVIATIONS

ABV	Adriamycin®, bleomycin and vincristine
ACTG	AIDS Clinical Trials Group
ampB	amphotericin B
AUC	area under concentration-time curve
AUC	area under the curve
BV	bleomycin + vincristine
CDC	Centers for Disease Control
Chol	cholesterol
CI	confidence interval
CR	complete response
CSF	cerebrospinal fluid
CT	computed tomography
DAPC	diarachidoyl phosphatidylcholine
DBPC	dibehenoyl phosphatidylcholine
DLPG	dilauryl phosphatidylglycerol
DMPC	dimyristoyl phosphatidylcholine
DMPG	dimiristoyl phosphatidylglycerol
DOPE	dioleoyl phosphatidylethanolamine
DPPC	dipalmitoyl phosphatidylcholine
DPPE-DTTA	dipalmitoyl phosphatidylethanolamine-diethylenetriaminetetraacetic acid
DSPC	distearoyl phosphatidylcholine
DSPE	distearoyl phosphatidylethanolamine
DSPG	distearoyl phosphatidylglycerol
DTPA	diethylenetriaminepentaacetic acid
DTTA	diethylenetriaminetetraacetic acid
DXR	doxorubicin
EPC	egg phosphatidylcholine
EPG	egg phosphatidylglycerol
FDA	U.S. food and drug administration
Gd-BOPTA	gadolinium benzoyloxypropionic tetraacetate
GI	gastrointestinal
G_{M1}	ganglioside G_{M1}
HEPC	hydrogenated egg phosphatidylcholine
HIV	human immunodeficiency virus
HMPAO	hexamethylpropyleneamine oxime
HSPC	hydrogenated soy phosphatidylcholine
ID	injected dose
ip	intraperitoneal
iv	intravenous

LVEF	left ventricular ejection fraction
MBC	minimum bactericidal concentration
MIC	minimum inhibitory concentration
MLV	multilamellar vesicle
MPB-PE	maleimidophenylbutryl phosphatidylethanolamine
MPEG	monomethoxy PEG
MPEG-PE	methoxy PEG-PE
MPS	mononuclear phagocyte system
MRI	magnetic resonance imaging
NGPE	N-glutaryl phosphatidylethanolamine
PC	phosphatidylcholine
PE	phosphatidylethanolamine
PEG	polyethylene glycol
PEG-DOPE	PEG-doleoyl phosphatidylethanolamine
PEG-PE	PEG-phosphatidylethanolamine
PEG-OSu	monomethoxy polyethylene glycol succinimidyl succinate
PG	phosphatidylglycerol
PR	partial response
PS	phosphatidylserine
RES	reticuloendothelial system
sc	subcutaneous
SD	stable disease
SM	sphingomyelin
TSL	thermosensitive liposome
[^{67}Ga]DF	[^{67}Ga]deferoxamine

Section I

Introduction

CHAPTER 1

Long Circulating Liposome Therapeutics: From Concept to Clinical Reality

Gerrit Storm, Martin C. Woodle

Introduction: Long Circulating Liposomes for Drug Delivery

There is a strong need for improved drug delivery in clinical practice. The performance of both existing and promising new drugs, especially the new biotechnology derived agents based on biomacromolecules such as proteins, would benefit greatly from delivery strategies which ensure the delivery of the drug at the right site, at the right time and in the right concentration. Over the past few decades, there has been a multitude of efforts aimed at developing new drug delivery systems in order to improve the poor benefit/risk ratio associated with many drugs used for the treatment of life-threatening diseases, such as cancer and certain infectious diseases. Among a variety of delivery systems that have been devised are many particulate carrier systems, for example microspheres, nanoparticles, lipoproteins, micellular systems, and liposomes.[1-6] Liposomes have been investigated extensively for over 20 years and the findings of mostly preclinical studies have demonstrated their versatility to accommodate a large variety of drugs for a wide range of therapies. These efforts have now culminated in several liposomal formulations in clinical trials. Initial introduction of approved products is primarily based on long circulating forms of liposomes described here and in other chapters of this volume.

The therapeutic goals of drug delivery applications of liposomes are numerous. In general, for parenteral routes of administration, liposomal incorporation aims to achieve one or more of the following effects on drug behavior:

1) Favorably alter the pharmacokinetics.
2) Protect against degradation.
3) Reduce distribution to sensitive tissues and hence toxicity.
4) Enhance distribution to target tissues and hence efficacy.

Recent approval of several liposome-based pharmaceutical products intended for parenteral administration illustrates that liposomes can provide at least some of these services. In the past, though, the value of liposomes for drug delivery has not always been clear nor certain. Studies revealed many hurdles on the road from concept to pharmaceutical products which converted much of the original enthusiasm into pessimism. Nonetheless, a multitude of efforts have now addressed many of the pharmaceutical issues. Many contributions to address these pharmaceutical issues have been made by venture capital financed companies such as Liposome Technology, Inc. (now Sequus Pharmaceuticals, Inc.), The Liposome Company and Vestar Inc. (now Nexstar Pharmaceuticals, Inc.) which have now introduced liposome-based pharmaceutical products. As a consequence, there is a growing acceptance of liposome formulations as an important member of parenteral drug delivery systems.

The attraction of liposomes for drug delivery was based on expectations of good biocompatibility, low toxicity and a lack of immune system activation or suppression. These assumptions arose because liposomes are typically composed of natural biological lipids forming bilayers that bear a structural resemblance to cell membranes but without the proteins thought necessary for recognition. In addition, their internal aqueous phase surrounded by the lipid bilayer suggested they would be applicable to both aqueous and organic soluble drugs. Of course, reality turned out to be more complex and the surprises resulted in protracted development of liposomes as drug formulations. Now, though, liposomes have developed into a viable pharmaceutical dosage form and the techniques required for pharmaceutical-scale production are well established.[1,4,5,7]

One of the important barriers limiting the usefulness of liposomes for drug delivery has been a short blood circulation time resulting from rapid recognition and removal from blood by cells of the mononuclear phagocytic system (MPS), particularly those in the liver and spleen.[1,8-10] The rapid sequestration of encapsulated drugs by these macrophages often eliminates the intended beneficial effects and moreover, can pose considerable risk of toxicity to these cells.[11] Thus an important challenge has been the development of liposomes which can persist for prolonged times in the bloodstream, giving them a greater chance to reach target sites in other tissues.

The development of long circulating liposomes (LCL) represents a milestone in liposomal delivery research. These newer forms of liposomes are actively being investigated worldwide and the results have substantially expanded the role of liposomes in developing new therapeutics. The results obtained with LCL so far indicate that incorporation of existing therapeutic agents into LCL can greatly alter their in vivo fate, providing—at least in a couple of cases—a significantly enhanced therapeutic index. In fact, LCL have opened a new realm of therapeutic opportunities, and a multitude of novel applications can be expected through use of LCL formulations.

Another significant obstacle, especially for iv-injected particulate drug delivery systems, is a need to cross the vascular wall in order to reach extravascular targets. In most tissues the vascular system is lined with a continuous layer of endothelial cells often supported by a basement membrane. This barrier virtually excludes extravasation of particles such as liposomes except for a few selected sites where the endothelial lining is discontinuous, for example, the fenestrations in the endothelium of the liver allow extravasation of small liposomes (see Fig. 6.2). Liposomes and other particulates able to persist in the blood for prolonged periods have been found to extravasate but this occurs only in areas where the permeability of the endothelial barrier is increased.[12-16] Fortunately, regions of increased capillary permeability include pathological tissues such as tumors and sites of infection and inflammation. Thus prolonged circulation of liposomes increases the probability that they extravasate into these pathological areas and therefore could enhance the efficacy of encapsulated drugs. The ability to target drugs to extravascular sites via liposomes appears dependent on a combination of:

1) Prolonged blood circulation, providing ample opportunities to encounter the region of disease.
2) Adequate access to the pathological tissue and target cells therein.
3) Ability of the liposomes to arrive at the target site with the encapsulated drug and to release it in an active form.[13,17,18]

For most liposomal drugs improved efficacy requires a composition capable of retaining the drug in the liposome during prolonged circulation but releasing it once the liposomes have accumulated in the site of pathology. This is illustrated for vincristine in chapter 3 and using hyperthermia to enhance localization and release in chapter 7. The achievement of reduced MPS uptake and prolonged blood circulation of liposomes has provided a significant level of improved drug delivery to certain pathological tissues and therefore has restored much of the original promise of liposomes as a drug delivery system.

The aim of this introductory chapter is to describe the general physicochemical and biological properties of long circulating liposomes as relevant to their application for drug delivery. Specific applications of LCL are discussed in subsequent chapters.

Biopharmaceutical Aspects

Liposomes are self-organized dispersions of phospholipids in water where the lipids orient themselves into bilayer sheets, sequestering their hydrophobic acyl chains into the interior of the sheet with their aqueous soluble "head groups" at the surface in contact with the water. These sheets curve to form a sphere surrounding one or more internal aqueous compartments (see Fig. 1.1). Below we will briefly describe pharmaceutical aspects of liposomes (preparation, characterization and stability). For more detailed information, the reader is referred to some recent reviews and books.[1,4,5,8,19]

Basically, the backbone of the liposomal bilayers consist of phospholipids. Several classes of phospholipids are being used for liposome production:

1) Phospholipids purified from natural sources [e.g., phosphatidylcholine (PC), phosphatidylethanolamine (PE), and sphingomyelin, which are obtained from egg yolk or soy beans].
2) Modified natural phospholipids [e.g., by partial or complete hydrogenation of the acyl chains, yielding (partially) saturated phospholipids which are less sensitive to oxidation and have an increased main transition temperature].
3) Semi- or fully synthetic phospholipids which have the advantage of a defined fatty acid composition and can be tailored to specific needs.

A large variety of preparation techniques for preparing different types of liposomes on a laboratory scale have been described. However, techniques allowing the large-scale production of liposome dispersions are required for use in clinical practice. Several steps may be discerned in the preparation process of liposomes:

1) Hydration of the lipids, resulting in vesicle formation.
2) Drug encapsulation.
3) Sizing of the vesicles.

In some cases these steps are linked together in the production process. In general, hydrophilic drugs are located in the aqueous phase inside the liposomes. More

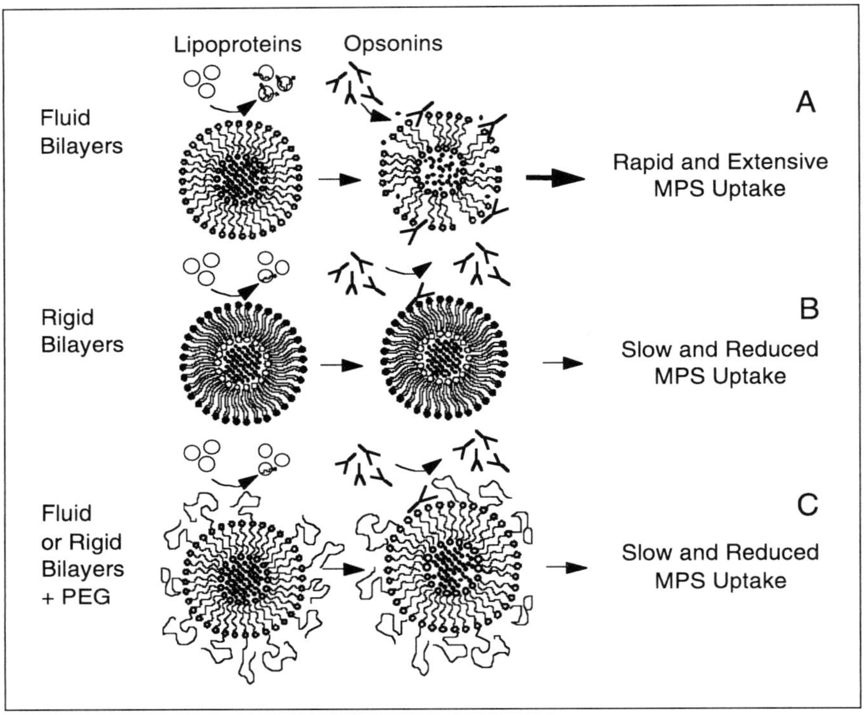

Fig. 1.1. Diagram of liposomes showing drug (as dots) in the aqueous compartment surrounded by the lipid bilayer and interactions leading to MPS uptake or prolonged circulation. The nature of the liposome cohesion and surface polymer coating are correlated with enhanced in vivo stability and reduced opsonin binding, providing reduced MPS uptake and prolonged blood clearance. Either strong lipid cohesion (rigid bilayers) or PEG coating permit prolonged circulation whereas drug leakage is determined by lipid cohesion (adapted from ref. 28).

lipophilic drugs tend to associate with the bilayers. The product must be sterile, well characterized and stable.

Liposome characteristics of major importance are:

1) Vesicle size.
2) Number of bilayers and morphology.
3) Bilayer fluidity (see Fig. 1.2).
4) Surface characteristics such as charge and hydrophilicity.

Vesicle size can range between 0.03 and 10 µm. The larger vesicles [referred to as multilamellar vesicles (MLV)] may have more than 10 bilayers, but this number can be manipulated by the chosen preparation method. Vesicles smaller than 0.1 µm are mainly unilamellar, independent of the preparation technique. Depending on temperature, a phospholipid bilayer may exist in several physical phases which differ in molecular packing of the phospholipids. Two phases most often discerned in studies on the biological effect of liposomes are the rigid, gel phase, with ordered acyl chains, and the more disordered and fluid, liquid-crystalline phase (see Fig. 1.2). At

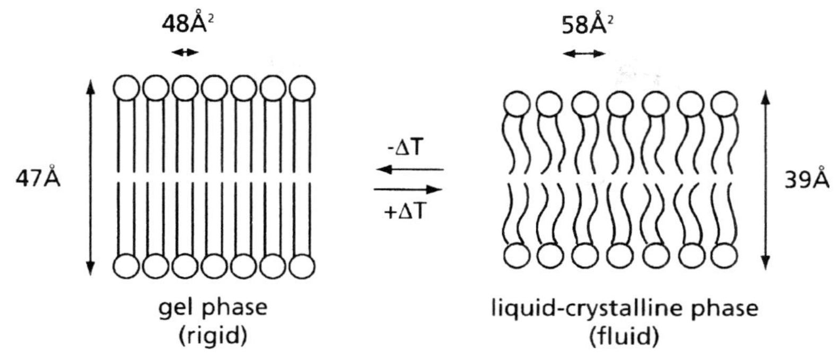

Fig. 1.2. Two phases most often discerned in studies on the biological effect of liposomes are the rigid, gel phase, with ordered acyl chains, and the more disordered and fluid, liquid-crystalline phase. The dimensions shown are those of DPPC (taken from ref. 20).

the transition temperature (T_m) the bilayer passes from one phase to the other.[20] The T_m depends mainly on the (phospho)lipid composition, but can also be affected by the presence of solutes, or dehydration. In general, fluid bilayers are more permeable to solutes than rigid bilayers, but the permeability shows a maximum around T_m, where phase separation occurs. Compounds that interact with the bilayer surface or that are located between the phospholipids (e.g., cholesterol) can affect or abolish phase transitions. In addition, they may change the molecular packing of these phospholipids and the bilayer fluidity. The surface charge of the bilayers can be tailored by adding negatively or positively charged (phospho)lipids. The surface characteristics of liposomes may also be altered by attachment of hydrophilic moieties to phospholipids or cholesterol, such as polyethylene glycol (PEG).

Liposomes to be used as delivery systems for drugs must be stable over prolonged periods of time. A shelf life of at least 2 years is normally requested for pharmaceuticals. Liposome stability studies should address physical, chemical and biological stability issues. Physical processes that affect shelf life include loss of liposome-associated drug and changes in size: aggregation and/or fusion. Drug molecules can leak from the liposomes; rapid loss of encapsulated drug in vivo is usually highly undesirable. The leakage rate strongly depends on the bilayer composition and the physicochemical properties of the drug. Cholesterol is often included in the liposomal bilayers to increase stability against drug leakage. Very cohesive bilayers, i.e., those in the so-called gel phase (see Fig. 1.2), tend to lose the associated drug slowly. Aggregation is the formation of larger units of liposomal material; these units are still composed of individual liposomes. In the case of fusion processes, new particle structures are formed. The chemical stability of liposomes is also a concern since phospholipids can be affected by two types of chemical degradation reactions:

1) Hydrolysis of the ester bonds, leading to the formation of lysophospholipids and free fatty acids.

2) Peroxidation of double bonds in unsaturated acyl chains (if present) which are sensitive to oxidation via free radical reactions. The consequence of lipid degradation is often a change in physical properties. Loss of drug and changes in particle size will change the fate of the drug upon administration. Therefore the stability should be carefully monitored and, if necessary, measures should be taken to make the liposome formulations meet the demands of the pharmaceutical industry. In this regard, important progress has been made over the last decade. However, discussion of the approaches that can be chosen to ensure the chemical and physical stability of liposomes are outside the scope of this chapter.

The form of stability with which we are most concerned here is that of biological stability. In this case, stability means the ability to resist interaction with molecular and cellular components in vivo. Many plasma components are known to interact with liposomes and such interactions can lead to liposome destablization or subsequent recognition and uptake by the MPS. Consequently, to maintain stability upon iv administration, liposomes must be able to resist interactions with the plasma proteins and cells with which they come into contact (see Fig. 1.1).

A key achievement leading to renewed interest in liposome drug delivery has been the introduction of liposomes exhibiting prolonged circulation (LCL). Qualitatively, the mechanism behind the approaches taken to enhance the residence time of liposomes in the blood compartment can be explained by reduction of the adsorption of various blood components onto the liposomal surfaces (e.g., opsonin ligands interacting with one or more receptors on the macrophage cell surface). Thus liposome types able to resist rapid opsonization are likely to show prolonged blood circulation times. One of the first major advances in prolonging blood residence was made possible through careful studies of the dependence of uptake on lipid components as well as physical properties of liposomes, described in a comprehensive review.[8] These studies led to findings that small (i.e., less than or equal to 100 nm in diameter), neutral and rigid (i.e., composed of fully saturated lipids below their phase transition and a high cholesterol content) liposomes can exhibit prolonged circulation, but only at relatively high lipid doses.[1,8,21] This success in generating LCL has been exploited for development of liposome drug delivery formulations, exemplified with daunorubicin and amphotericin B (see chapter 6 and 13).[16,22,23] It is believed that the use of highly cohesive bilayers inhibits interaction of plasma proteins with the hydrophobic interior of the bilayer (illustrated in Fig. 1.1). In this approach, opsonic proteins may interact with the liposome exterior but they are inhibited from penetration into the bilayer and consequently, not able to induce the rearrangements which otherwise would 'mark' the liposome for MPS uptake. An interesting extension of these findings is work with fluorocarbon-based phospholipids which seems to further increase the cohesive nature of the bilayer.[24,25] Another approach utilized inclusion of specific glycolipids such as monosialoganglioside G_{M1} or phosphatidylinositol (PI). It was hypothesized that these glycolipids act through creating a carbohydrate "shield" over negatively charged groups located underneath it.[26,27] Overall, these efforts achieved some success, but all are dependent on rigid liposome bilayers which can impose a limitation when fluid bilayers are needed to achieve appropriate drug release rate profiles in vivo.

More recently, reduced liposome MPS uptake without restriction to rigid bilayers was realized by modification of the surface with polyethylene glycol (PEG) (illustrated in Fig. 1.1).[28] The PEG surface modification is commonly obtained simply by addition of a hydrophilic polymer-lipid conjugate, polyethylene glycol-phosphatidylethanolamine (PEG-PE), to the organic solution of lipids used to prepare the liposomes. Many reports have appeared in the last 5 years describing the ability of PEG polymers, in particular those with molecular weights of 1900 and 5000 Da, to extend liposome circulation times and reduce uptake by the liver and the spleen macrophages; typical blood clearance profiles in rats are shown in Figure 1.3. Pharmacokinetic studies in humans confirmed the PEG effect as demonstrated by a 40-50 half-life for doxorubicin when encapsulated in these liposomes (described in chapter 11). The mechanism by which PEG polymers exert their effects has been suggested to rely on the physicochemical concept of steric stabilization.[6,29] The steric stabilization mechanism is thought to entail the polymer coating operating as a repulsive steric barrier opposing interactions with biological components, including opsonins. This proposed mechanism has resulted in the name 'sterically stabilized liposomes' to denote PEG-liposomes.

PEG surface modification has been shown to have important advantages over the other methods to obtain prolonged circulation identified to date.[6,30] One of the principle advantages is that while classical liposomes have nonlinear, saturable pharmacokinetics, PEG-liposomes possess dose-independent, nonsaturable, log-linear pharmacokinetics.[26] This permits dose escalation without complications arising from changes in pharmacokinetics. Another advantage is a versatility in choice of lipid composition which provides an ability to optimize the liposome physicochemical properties for drug loading and leakage without loss of prolonged blood circulation and tissue distribution characteristics. These properties seem to be a direct consequence of the physicochemical effects of the surface-bound polymer layer on interactions with plasma proteins and cells, interactions which are only just beginning to be understood. Studies with other surface-bound polymers (e.g., polyoxazolines) have reinforced the notion of a physical rather than chemical mechanism. However, PEG remains the most popular material for surface modification to obtain LCL. While these surface modified LCL have certain clear advantages, the earlier types of LCL also can provide therapeutic enhancements if the physicochemical properties of the drug allow appropriate incorporation, retention, and release in vivo.

Localization at Pathological Sites

One of the most important tasks a targeted drug delivery system should fulfill is preferential localization at pathological sites, the "magic bullet" concept. In this regard, LCL are promising since they exhibit accumulation into several important sites of pathology including solid tumors, infections, and inflammations, described in recent reviews.[29,31] For example, concentrations of Doxorubicin and Daunorubicin in an implanted tumor increase substantially when these drugs are administered as LCL formulations (see chapters 5 and 6). Also, the time over which accumulation occurs can be prolonged, as shown in Figure 1.4.[32] Tumor studies suggest that the angiogenesis process can result in increased vascular leakage which may help to explain this phenomenon,[13] as shown in Figure 6.2. Moreover, despite significant differences in the types of pathological tissues into which LCL can localize, the results may be indicative of a common mechanism: increased extravasation into areas characterized by increased permeability of the vasculature by virtue of the prolonged circulation capability.[33] Up to 10% of the injected LCL dose has been shown to localize in several important sites of pathology, suggesting potential for substantial improvements in efficacy of encapsulated therapeutic agents active towards these pathologies.

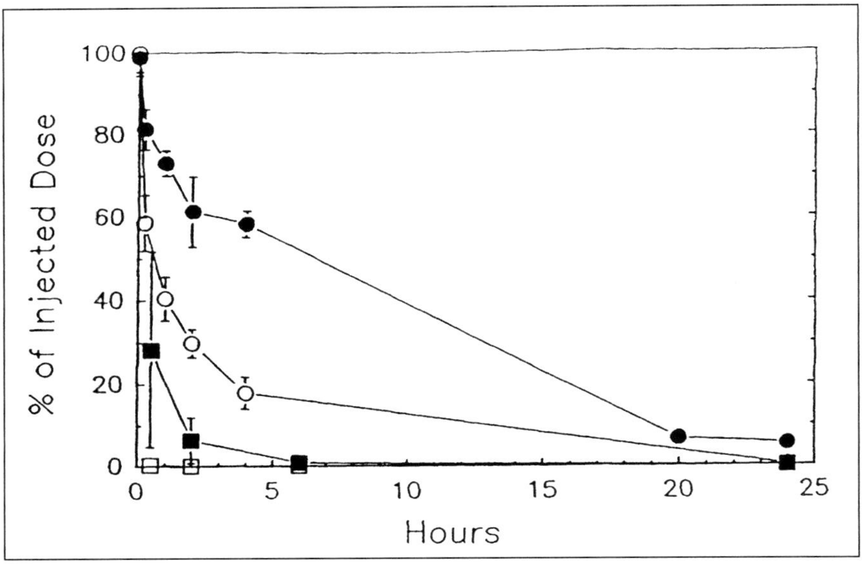

Fig. 1.3. Blood circulation of radiolabeled PEG-grafted liposomes following iv administration in rats. Closed circles are PEG-PE:PC:Cholesterol, open circles are PC:Cholesterol, closed squares are PG:PC:Cholesterol, and open squares are free label. (taken from reference 30). Copyright CRC Press, used with permission.

Clinical Applications of Long Circulating Liposomes

Doxorubicin and Daunorubicin formulated in LCL currently represent the best understood clinical applications of LCL (see chapters 8-11). Doxorubicin was one of the first oncology agents considered for encapsulation in liposomes (see refs. 34,35). Also, pH and ion gradient "after loading" methods were applied to these two agents, achieving high encapsulation efficiencies largely independent of lipid composition; this has permitted new formulations to be easily prepared and tested.[23,32,36-38] Doxorubicin formulations based on PEG-grafted liposomes have resulted in numerous investigations demonstrating enhanced efficacy in preclinical animal tumor models.[29] In fact, commercially developed formulations of both Doxorubicin and Daunorubicin—Doxil and DaunoXome, respectively—have been approved for treatment of Kaposi's Sarcoma (KS) patients, and are described in chapters 8-11.

Studies on the plasma pharmacokinetics of Doxil in humans show a dose independent half-life of 40-60 h.[39,40] Note that such prolonged circulation times means that about 10% of the iv injected dose is still in the blood after one week. As would be expected for dose independent kinetics, the initial plasma concentrations and area-under-the-curve (AUC) are linearly dependent on dose. The clinical testing of this well known anticancer drug reformulated into PEG-grafted liposomes confirms the applicability of the preclinical findings, at least in terms of plasma pharmacokinetics (see chapter 11).

This achievement—the development of liposomes which circulate for up to one week in humans—has created new challenges and requirements for their use in de-

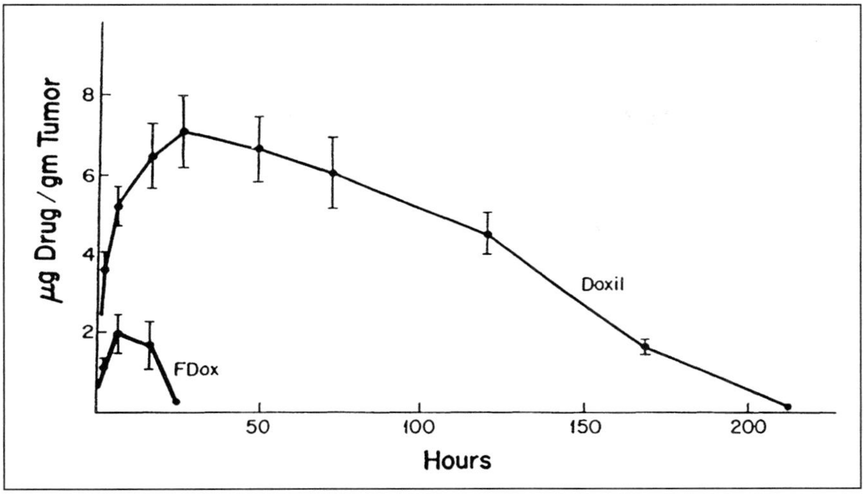

Fig. 1.4. Localization of doxorubicin in implanted tumors in mice determined by microfluorimetry following i.v. dose of 0.9 mg/kg doxorubicin given as either in PEG-grafted liposomal doxorubicin (Doxil) or as unencapsulated "free" doxorubicin (FDox).[32] Copyright CRC Press, used with permission.

livering therapeutic agents. If these liposomes are to prove successful for drug delivery then they must have the ability to hold onto the drug for the extended period of time while circulating and they must be able to release the drug at the appropriate time within the target site. These requirements appear to have been attained to some extent with the commercial formulations of Doxorubicin and Daunorubicin, as they show striking tumor growth inhibition, and even regression, both in animal models and, most recently, in humans (see chapters 8-11). This 'stability/instability' issue and approaches to optimize LCL accordingly have been evaluated with studies of vincristine, described in chapter 3.

An interesting question is whether additional optimization of the commercial formulations is required to further improve the release after extravasation into the target tissue. Attempts have been made to improve efficacy through the use of temperature-sensitive liposomes and application of heat to the anatomical region of the tumor.[41-46] The concept is to selectively enhance drug release within the tumor. In practice it turns out that extravasation is also enhanced at the heated site and thus both localization and release are positively influenced. The actual extent of efficacy improvement which can be obtained by this approach has yet to be fully determined.

The achievement of liposomes which persist in the blood and localize in vascularized tumors promises general applicability in the delivery of many anticancer agents. Conceptually, most current oncology agents could benefit from enhanced delivery to solid tumors, or even "just" reduced distribution to healthy tissues, and a number have been considered for encapsulation in LCL. Improved efficacy has been reported with long circulating liposome formulations containing araC, epirubicin, vincristine or mitoxantrone.[47-53] These results suggest broad applicability of LCL for drug delivery in oncology.

The ability of LCL to persist in the blood and localize in sites of pathology also has applications for drug delivery outside of oncology. Other examples can be found in recent evaluations of their usefulness for delivery of antibiotics,[54-56] cytokines[57] and radiopharmaceuticals.[58-66] Their persistence in blood suggests that they also may prove effective in providing slow release of therapeutics such as biomacromolecules into the systemic circulation.[67] One study of this approach has been reported using a peptide hormone, vasopressin.[68]

An area of research currently very active is addition of specific targeting ligands to long circulating liposomes (discussed in chapters 15, 17 and 18).[69-73] Nonetheless, the added clinical value for localization beyond that achieved with simple LCL for tumors, infections, and inflammations remains to be seen. An important alternative use of "targeted" LCL is to act as a platform for prolonged display of ligands to blood elements.[74] However, the latter area of investigation is just beginning to be addressed.

Finally, an exciting future application being considered for LCL is in nonviral gene delivery. Unfortunately, intracellular delivery with PEG-grafted liposomes is likely a significant problem because the surface polymer seems to inhibit protein and cellular interactions, exactly the contrary of what is required for intracellular delivery. However, this problem may not be as troubling given recent in vitro data showing that surface PEG was not prohibitive of endocytosis and cellular uptake of PEG-liposomes bearing the folate ligand.[75] Even if the PEG is prohibitive of intracellular delivery, approaches to obtain PEG detachment from the liposome surface after arrival in the target tissue may be successful. The possibility to introduce fusogenic capabilities into the PEG-liposome system is also being explored, which represents the intriguing but formidable challenge to create a 'synthetic virus' for cytosolic delivery of genetic material.

Conclusions

LCL appear to provide multifaceted capabilities applicable to many indications and agents, but their full impact on clinical practice will only be realized through further exploration and development. The LCL can improve the therapeutic index of encapsulated drugs, as seen with doxorubicin and daunorubicin formulations now approved for the treatment of KS patients. Other drug and diagnostic formulations are currently in development and being evaluated in clinical studies. These studies can be expected to result in additional approved LCL formulations in the near future. Consequently, practicing physicians will need to appreciate the properties of the LCL formulations and understand the striking differences they make in the in vivo behavior of drugs when they are encapsulated. Importantly, drug encapsulation in LCL can result in such differences that administration schedule, potential toxicities, and other issues in clinical practice must be completely re-evaluated. The best approach may be to consider LCL formulations of existing agents as entirely new agents. We can look forward to the results of current and future efforts defining the requirements and strategies to expand the utility of LCL via surface ligands and new functionalities for intracellular delivery.

References

1. Lasic DD, ed. Liposomes: from Physics to Applications. Amsterdam: Elsevier, 1993:575.
2. Storm G, Vingerhoeds MH, Bakker-Woudenberg IAMJ et al. Biodistribution and therapeutic utility of liposomal drug carrier systems. J Lipo Res 1993; 3:551-562.

3. Domb AJ. Polymeric Site-Specific Pharmacotherapy. New York: John Wiley & Sons, 1994.
4. Crommelin DJA, Schreier H. Liposomes. In: Kreuter J, ed. Colloidal Drug Delivery Systems. New York: Marcel Dekker, Inc., 1994:73-190.
5. Barenholz Y, Crommelin DJA. Liposomes as pharmaceutical dosage forms. In: Boylan JSaJC, ed. Encyclopedia of Pharmaceutical Technology. New York: Marcel Dekker Inc., 1994:1-39.
6. Storm G, Belliot SO, Daemen T et al. Surface modification of nanoparticles to oppose uptake by the mononuclear phagocyte system. Adv Drug Delivery Rev 1995; 17:31-48.
7. Amselem S, Cohen R, Barenholz Y. In vitro tests to predict in vivo performance of liposomal dosage forms. Chemistry and Physics of Lipids 1993; 64:219-237.
8. Senior JH. Fate and behavior of liposomes in vivo: a review of controlling factors. Critical Reviews in Therapeutics and Drug Carrier Systems 1987; 3:123-193.
9. Gregoriadis G. Liposomes as Drug Carriers. Recent Trends and Progress. In: New York: John Wiley & Sons, 1988.
10. Porter CJ, Davies MC, Davis SS et al. Microparticulate systems for site-specific therapy—bone marrow targeting. In: Domb AJ, ed. Polymeric Site-Specific Pharmacotherapy. New York: John Wiley & Sons, 1994:157-204.
11. Storm G, Oussoren C, Peeters PAM et al. Tolerability of liposomes in vivo. In: Gregoriadis G, ed. Liposome Technology. Boca Raton: CRC Press, 1993:345-383.
12. Thorpe PE, Burrows FJ. Antibody-directed targeting of the vasculature of solid tumors. Breast Cancer Research Treatment 1995; 36:237-251.
13. Jain RK. Delivery of molecular medicine to solid tumors. Science 1996; 271:1079-1080.
14. Pimm MV, Hudecz F. Biodistribution in tumour-bearing mice of polycationic, amphoteric and plyanionic branched polypeptides with a poly(L-lysine) backbone labelled with 125I and 111In: tumour accumlation less than that of labelled serum proteins. J Cancer Res Clin Oncol 1996; 122:45-54.
15. Gabizon AA. Selective tumor localization and improved therapeutic index of antracyclines encapsulated in long circulating liposomes. Cancer Research 1992; 52:891-.
16. Forssen EA, Male-Brune R, Adler-Moore JP et al. Fluorescence imaging studies for the disposition of daunorubicin liposomes (DaunoXome) within tumor tissue. Cancer Research 1996; 56:2066-2075.
17. Storm G, Nassander UK, Vingerhoeds MH et al. Antibody-targeted liposomes to deliver doxorubicin to ovarian cancer cells. J Lipo Res 1994; 4:641-666.
18. Woodle MC. Sterically stabilized liposome therapeutics. Adv Drug Del Rev 1995; 16:249-265.
19. Crommelin DJA, Storm G. Liposomes, Quo Vadis? A personal viewpoint. In: F. Puisieux PC, J. Delattre, Devissaguet J-Ph, eds. Liposomes, New Systems and New Trends in the Applications. Paris: Editions de Sante, 1995:767-781.
20. Wilschut J. In: Leserman LD, Barbet J, eds. Methologies des Liposomes. Paris: INSERM, 1982:9-24.
21. Hwang KJ. In: Ostro M, ed. Liposomes from Biophysics to Therapeutics. New York: Marcel Dekker, 1987:109-156.
22. Forssen EA, Coulter DM, Proffitt RT. Selective in vivo localization of daunorubicin small unilamellar vesicles in solid tumors. Cancer Research 1992; 52:3255-3261.
23. Forssen EA, Ross ME. Daunoxome treatment of solid tumors: preclinical and clinical investigations. J Lipo Res 1994; 4:481-512.
24. Riess JG. Fluorinated vesicles from self-aggregated perfluoroalkylated amphipliles. J Drug Targ 1994.

25. Frezard F, Santaella C, Montisci MJ et al. Fluorinated phosphatidylcholine-based liposomes: H+/Na+ permeability, active doxorubicin encapsulation and stability, in human serum. Biochimica et Biophysica Acta 1994; 1194:61-68.
26. Allen T. The use of glycolipids and hydrophilic polymers in avoiding rapid uptake of liposomes by the mononuclear phagocyte system. Adv Drug Deliv Rev 1994; 13:285-309.
27. Gabizon AA. Liposome circulation time and tumor targeting: implications for cancer chemotherapy. Adv Drug Delivery Reviews 1995; 16:285-294.
28. Woodle MC, Lasic DD. Sterically stabilized liposomes. Biochim Biophys Acta 1992; 1113:171-199.
29. Lasic DD, Martin FJ, eds. Stealth Liposomes. Boca Raton: CRC Press, 1995:289.
30. Woodle MC, Newman MS, Working PK. Biological properties of sterically stabilized liposomes. In: Lasic D, Martin F, eds. Stealth Liposomes. Boca Raton: CRC Press, 1995:103-118.
31. Oku N, Tokudome Y, Tsukada H et al. In vivo trafficking of long circulating liposomes in tumour-bearing mice determined by positron emission tomography. Biopharm Drug Dispos 1996; 17:435-441.
32. Vaage J, Barbera E. Tissue uptake and therapeutic effects of Stealth doxorubicin. In: Lasic D, Martin F, eds. Stealth Liposomes. Boca Raton: CRC Press, 1995:149-171.
33. Huang SK, Martin FJ, Firend DS et al. Mechanism of Stealth liposomes accumulation in some pathological tissues. In: Lasic D, Martin F, eds. Stealth Liposomes. Boca Raton: CRC Press, 1995:119-126.
34. Gabizon A, Price DC, Huberty J et al. Effect of liposome composition and other factors on the targeting of liposomes to experimental tumors: biodistribution and imaging studies. Cancer Res 1990; 50:6371-6378.
35. Oku N, Doi K, Namba Y et al. Therapeutic effect of adriamycin encapsulated in long circulating liposomes on Meth-A-sarcoma-bearing mice. Int J Cancer 1994; 58:415-419.
36. Mayer LD, Tai LC, Bally MB et al. Characterization of liposomal systems containing Doxorubicin entrapped in response to pH gradients. Biochimica et Biophysica Acta 1990; 1025:143-151.
37. Haran G, Cohen R, Bar LK et al. Transmembrane ammonium sulfate gradients in liposomes produce efficient and stable entrapment of amphipathic weak bases. Biochimica et Biophysica Acta 1993; 1151:201-215.
38. Lasic DD, Frederik P M, Stuart MCA et al. Gelation of liposome interior: a novel method for drug encapsulation. FEBS Letters 1992; 312:255-258.
39. Northfelt, DW, Martin FJ, Kaplan LD et al. Pharmacokinetics, tumor localization and safety of Doxil (liposomal doxorubicin) in AIDS patients with Kaposi's sarcoma. Proc Am Soc Clin Oncol 1993; 12:51.
40. Gabizon A, Catane R, Uziely B et al. Prolonged circulation time and enhanced accumulation in malignant exudates of doxorubicin encapsulated in polyethylene-glycol coated liposomes. Cancer Research 1994; 54:987-992.
41. Iga K, Ogawa Y, Toguchi H. Heat-induced drug release rate and maximal targeting index of thermosensitive liposome in tumor-bearing mice. Pharm Res 1992; 9:658-662.
42. Maruyama K, Unezaki S, Takahashi N et al. Enhanced delivery of doxorubicin to tumor by long circulating thermosensitive liposomes and local hyperthermia. Biochim Biophys Acta 1993; 1149:209-216.
43. Huang SK, Stauffer PR, Hong K et al. Liposomes and hyperthermia in mice: increased tumor uptake and therapeutic efficacy of doxorubicin in sterically stabilized liposomes. Cancer Research 1994; 54:2186-2191.

44. Ning S, MacLeod K, Abra RM et al. Hyperthermia induces doxorubicin release from long circulating liposomes and enhances their anti-tumor efficacy. Int J Radiat Oncol Biol Phys 1994; 29:827-834.
45. Unezaki S, Maruyama K, Takahashi N et al. Enhanced delivery and antitumor activity of doxorubicin using long circulating thermosensitive liposomes containing amphipathic polyethylene glyco in combination with local hyperthemia. Pharm Res 1994; 11:1180-1185.
46. Oku N, Naruse R, Doi K et al. Potential usage of thermosensitive liposomes for macromolecule delivery. Biochim Biophys Acta 1994; 1191:389-391.
47. Allen TM, Mehra T, Hansen C et al. Stealth liposomes: an improved sustained release system for 1-b-D-arabinofuranosylcytosine. Cancer Research 1992; 52:2431-2439.
48. Mayhew EG, Lasic DD, Babbar S et al. Pharmacokinetics and antitumor activity of epirubicin encapsulated in long circulating liposomes incorporating a polyethylene glycol-derivatized phospholipid. Int J Cancer 1992; 51:302-309.
49. Mayer LD, Bally MB, Loughrey H et al. Liposomal vincristine preparations which exhibit decreased drug toxicity and increased activity against murine L1210 and P388 tumors. Cancer Research 1990; 50:575-579.
50. Vaage J, Donovan D, Mayhew E et al. Therapy of mouse mammary carcinomas with vincristine and doxorubicin encapsulated in sterically stabilized liposomes. Int J Cancer 1993; 54:959-964.
51. Beck P, Kreuter J, Reszka R et al. Influence of polybutylcyanoacrylate nanoparticles and liposomes on the efficacy and toxidity of the anticancer drug mitoxantrone. J Microencapsulation 1993; 10:101-114.
52. Genne P, Olsson NO, Gutierrez G et al. Liposomal mitozantrone for the local treatment of peritoneal carcinomatosis induced by colon cancer cells in mice. Anticancer Drug Des 1994; 9:73-84.
53. Tokudome Y, Oku N, Doi K et al. Antitumor activity of vincristine encapsulated in glucuronide-modified long circulating liposomes in mice bearing Meth A sarcoma. Biochim Biophys Acta 1996; 1279:70-74.
54. Bakker-Woudenberg IAJM, Lokerse AF, ten Kate MT et al. Liposomes with prolonged blood circulation and selective localization in Klebsiella pneumoniae-infected lung tissue. J Infect Dis 1993; 168:164-171.
55. Bakker-Woudenberg IAJM, ten Kate MT, Stearne-Cullen LET et al. Efficacy of gentamicin or ceftazidime entrapped in liposomes with prolonged blood circulation and enhanced localization in Klebsiella Pneumoniae infected lung tissue. J Infect Dis 1995.
56. van Etten EWM, van Vianen W, Tijhuis RHG et al. Sterically stabilized amphotericin B-liposomes: toxicity and biodistribution in mice. J Controlled Rel 1995; 37:123-129.
57. Kedar E, Rutkowsky Y, Braun E et al. Delivery of cytokines by liposomes. I. Preparation and characterization of interleukin-2 encapsulated in long circulating sterically stabilized liposomes. J Immunther. Emphasis Tumor Immunol 1994; 16:47-59.
58. Woodle MC. ^{67}Gallium-labeled liposomes with prolonged circulation: Preparation and potential as nuclear imaging agents. Nucl Med Biol 1993; 20:149-155.
59. Tilcock C, Ahkong QF, Fisher D. Polymer-derivatized technetium 99mTc-labeled liposomal blood pool agents for nuclear medicine applications. Biochimica et Biophysica Acta 1993; 1148:77-84.
60. Oku N, Namba Y, Takeda A et al. Tumor imaging with technetium-99m-DTPA encapsulated in RES-avoiding liposomes. Nucl Med Biol 1993; 20:407-412.
61. Kumar S, Singh T, Khar RK et al. Evaluation of sterically stabilized liposomes as a vehicle for targeting technetium-99m labelled radiopharmaceuticals. Pharmazie 1993; 48:613-616.

62. Tilcock C. Imaging tools: liposomal agents for nuclear medicine, computed tomography, magnetic resonance, and ultrasound. In: Philippot JR, Schuber F, eds. Liposomes as Tools in Basic Research and Industry. Boca Raton: CRC Press, 1994:225-240.
63. Boerman OC, Storm G, Van Oyen W et al. Sterically stabilized liposomes labeled with indium-111 to image focal infection. J Nucl Med 1995; 36:1639-1644.
64. Boerman OC, Oyen WJG, Van Bloois L et al. Optimization of 99m Tc-labeled PEG-liposomes to image focal infection: effects of particle size and circulation time. J Nucl Med 1997: (in press).
65. Oyen WJG, Boerman OC, Storm G et al. Labeled StealthTM liposomes in experimental infection: an alternative for leukocyte scintigraphy? Nucl Med Commun 1996; 17:742-748.
66. Oyen WJG, Boerman OC, Storm G et al. Detecting infection and inflammation with technetium-99m-labeled StealthTM liposomes. J Nucl Med 1996; 37:1392-1397.
67. Storm G, Koppenhagen FJ, Heeremans ALM et al. Novel developments in liposomal delivery of peptides and proteins. J Controlled Rel 1995; 36:19-24.
68. Woodle MC, Storm G, Newman MS et al. Prolonged Systemic delivery of peptide drugs by long circulating liposomes: Illustration with vasopressin in the Brattleboro rat. Pharm Res 1992; 9:260-265.
69. Vingerhoeds MH, Storm G, Crommelin DJA. Immunoliposomes in vivo. Immuno Methods 1994; 4:259-272.
70. Crommelin DJA, Herron J, Storm G. (Protein)-targeted delivery with particulate systems. In: Lee VHL, H, M, Mizushima YM, eds. Trends and future perspectives in peptide and protein delivery. Harwood Academic Publishers, GmbH, 1994:207-239.
71. Blume G, Cevc G, Crommelin DJA et al. Specific targeting with (polyethylene glycol)-modified liposomes: coupling of homing devices to the ends of the polymeric chains combines effective target binding with long circulation times. Biochim Biophys Acta 1993; 1149:180-184.
72. Allen TM, Agrawal AK, Ahmad I et al. Antibody-mediated targeting of long circulating (StealthTM) liposomes. J Lipo Res 1994; 4:1-25.
73. Maruyama K, Takizawa T, Yuda T et al. Targetability of novel immunoliposomes modified with amphipathic poly(ethylene glycol)s conjugated at their distal terminals to monoclonal antibodies. Biochimica et Biophysica Acta 1995; 1234:74-80.
74. Zalipsky S, Poulikas P, Mullah N et al. Increased systemic exposure of peptides by covalent fixation to extremities of PEG-grafted liposomes. In: Western Biotech Conference. San Diego: American Chemical Society, 1995.
75. Lee RJ, Low PS. Delivery of liposomes into cultured KB cells via folate receptor-mediated endocytosis. J Bio Chem 1994; 269:3198-3204.

Section II

Current Medical Applications

A. Malignant Diseases

CHAPTER 2

Oncologic Agents in Sterically Stabilized Liposomes: Basic Considerations

Theresa M. Allen

Why Use Liposomes for Oncologic Agents?

Drug carriers such as liposomes are employed in an effort to improve upon a number of properties of oncologic agents, as outlined in Table 2.1. Liposomes have the ability to function as carriers for either water-soluble drugs or lipid-soluble drugs. Thus hydrophilic drugs such as 1-β-D-arabinofuranosylcytosine (cytosine arabinoside, ara-C) can be dissolved in the aqueous interior of the liposomes,[1,2] and lipid soluble drugs, such as the lipophilic derivatives of cis-platin,[3] can be associated with the hydrophobic core of the phospholipid bilayer. Even difficult to formulate drugs such as taxol can be formulated into liposomes.[4] Because liposomes themselves are formed from substances which have low intrinsic toxicity, such as neutral phospholipids and cholesterol, use of liposomes as a solubilizing agent for hydrophobic antineoplastic drugs can prevent the toxicity associated with Cremaphor or dimethylsulfoxide (DMSO) which have traditionally been used to formulate such drugs.

Table 2.1. Objectives for the use of drug carriers with oncologic agents

1. Enhance drug solubility
2. Retard drug degradation
3. Achieve drug sustained release
4. Decrease drug toxicity
5. Alter drug biodistribution
6. Improve drug bioavailability
7. Overcome multidrug resistance
8. Increase drug localization in tumor cells through:
    a. passive targeting
    b. ligand-mediated targeting

Long Circulating Liposomes: Old Drugs, New Therapeutics, edited by Martin C. Woodle and Gerrit Storm.

Liposomes have another useful property which can help to improve the therapeutic efficacy of oncologic agents. Drugs associated with liposomes, in particular those drugs entrapped in the liposome aqueous interior, are protected by the phospholipid bilayer of the liposomes from interaction with degradative enzymes which can result in rapid breakdown of some antineoplastic drugs. This occurs because large protein molecules such as enzymes cannot gain access to the liposome interior. For example, ara-C, as a result of the action of the enzyme cytidine deaminase, has a half-life in the blood of less than one-half hour when the free drug is given by bolus injection into humans.[5,6] Encapsulation of rapidly degraded drugs like ara-C into liposomes will protect the drug from degradation following iv administration, and dramatically extend the half-life of the drug in vivo.

Not only are liposomes impermeant to large molecules such as enzymes, but they can have low permeability to protons and other cations. This enables one to sustain a pH or other ion gradient across the liposome bilayer for considerable periods of time. Taking advantage of this fact, another example of the use of liposomes in the protection and stabilization of anticancer drugs is liposomal topotecan and related structures such as camptothecin, 9-aminocamptothecin and CPT-11. For these promising antineoplastic agents, the biological activity depends on the existence of a closed α-hydroxy lactone ring moiety, which unfortunately hydrolyzes rapidly at physiological pH, inactivating the drugs. Encapsulation of topotecan in liposomes having an internal pH of 5 can result in a significant extension of the half-life of the drug in human plasma.[7]

Hydrophilic drugs, which have a very low octanol to water partition coefficient ($\log P_{oct} < 1.7$), e.g., ara-C, are readily retained in the liposome interior and are released from the liposome over an extended period of time,[8] resulting in a sustained release of cytotoxic levels of the drug over a period of several hours to several days.[1] The rate of drug release can be manipulated over a wide range depending on the chosen liposome composition and the manner of drug loading.[9,10] In general, using phospholipids which form a rigid liposomal bilayer will tend to retard the rate of release of drugs relative to liposomes with a fluid bilayer. Liposome contents can be rapidly released at a temperature corresponding to the phase transition of the phospholipid or phospholipid mixture in the liposomes. We can take advantage of this by formulating oncologic agents in liposomes with a phase transition of approximately 42°C so that their contents are released in response to local hyperthermia, for example, following localization of the liposomes in a solid tumor nodule.[11] Plasma drug levels will also depend on the circulation half-lives of the liposomes, which can be similarly influenced by manipulation of liposome size, surface charge, and the attachment of sterically stabilizing molecules such as polyethylene glycol (PEG) to the liposome surface.[12]

Drugs which have intermediate solubility ($1.7 < \log P_{oct} < 4$) do not readily associate with liposomes in a stable manner and partition rapidly between the liposome bilayer and the aqueous phase.[8] Liposomes, in these cases, are acting more as a solubilizing agent than as drug carriers. However, methods have been devised for this situation such as manipulation of the liposomal internal pH to form molecular complexes in the liposome interior which results in excellent liposomal preparations for some compounds, e.g., remote-loaded doxorubicin (DXR)[13,14] and daunorubicin or Doxil® and DaunoXome®, respectively (discussed in chapters 5-11).

For highly lipid-soluble drugs ($\log P_{oct} > 5$), the degree of incorporation of the compound into the phospholipid bilayer of the liposomes can be virtually complete, with excellent retention of the drug by the liposomes.[8] To date, though, very few highly hydrophobic oncologic agents which could take advantage of these liposome properties are in clinical development. Two hydrophobic oncologic agents in clinical development include annamycin, a hydrophobic derivative of DXR,[15] and lipid derivatives of cisplatin.[3]

Alterations in Drug Pharmacokinetics Upon Liposome Association

Association of antineoplastic drugs with liposomes can have significant effects on their pharmacokinetics, i.e., adsorption, biodistribution, metabolism and elimination.[12,16] As a general rule-of-thumb, the biological activity of a liposome-associated drug will be delayed as the drug will remain within the liposome for a certain period of time. For drugs in long circulating liposomes given iv, they will circulate in blood with a low rate of clearance and follow a biodistribution pattern restricted to the vascular compartment and those few tissues accumulating the liposomes. This can delay or alter tissue exposure, metabolism and elimination compared to the same amount of drug administered as an aqueous solution (i.e., "free" drug). These changes can have dramatic effects on both the efficacy and the toxicity of the entrapped compound. The degree to which the pharmacokinetics of the drug-liposome package is different from that of the free drug depends on the liposome size, composition, surface charge, the administered dose and the route of administration, as well as the release rate of the drug from the liposomes, and of course the pharmacokinetic properties of the free drug.[12,16] To consider the two extremes, if the drug is released very rapidly from the liposomes following administration, the pharmacokinetics will approach that of the free drug. On the other hand, if the drug has a very slow rate of release from the liposomes, the pharmacokinetics will approach that of the carrier.

Sterically stabilized liposomes (SL), with PEG-grafted surfaces, have two important pharmacokinetic advantages over classical liposomes (CL). SL have decreased rates of clearance (i.e., long circulation half-lives) and dose-independent pharmacokinetics in contrast to dose-dependent, saturation-type pharmacokinetics exhibited by classical liposomes. This difference is illustrated by results obtained in mice, summarized in Table 2.2. The pharmacokinetic profile of classical liposomes occurs as a consequence of a dose-dependent saturation of the uptake of the classical liposomes by the mononuclear phagocyte system (MPS), in particular, saturation of the liver Kupffer cells. SL are long circulating, having a reduction in both rate and extent of uptake into the MPS, and exhibit no saturation of the MPS in the clinical dose range, which explains their very useful dose-independent pharmacokinetics. Dose-independent, i.e., log-linear, pharmacokinetics in a drug carrier has the advantage of making it much easier to predict plasma drug levels as a function of dose than does nonlinear pharmacokinetics. CL may exhibit log-linear pharmacokinetics in several circumstances: at high, MPS-saturating doses or in a situation where the encapsulated drug has substantial toxicity to the MPS, as for example, in the case of DXR in CL.[17] In this latter case the CL, by delivering DXR to the Kupffer cells, may result in sufficient toxicity to this cell population to convert the liposomes to long circulating ones with pharmacokinetics which approach log-linear.[18] In neither of these cases should the resulting clearance profiles be confused with the dose-independent pharmacokinetics exhibited by SL.

Association of oncologic agents with liposomes has other effects on their pharmacokinetics, including a sometimes dramatic decrease in their volume of distribution and a decrease in their elimination rate constant. The reduction in the

Table 2.2. Total body clearance versus dose in mice for CL compared to SL

Liposome dose (mg/kg)	CL (ml/hr)	SL (ml/hr)
4	1.08	0.08
20	0.54	0.10
40	0.39	0.09
80	0.21	0.07
200	0.15	0.08
400	0.11	0.06

Liposomes were 100 nm in diameter and were composed of PC:CHOL, 2:1 molar ratio (CL) or SM:PC:CHOL:PEG-DSPE, 1:1:1:0.2 molar ratio (SL).

volume of distribution can be as much as several hundred-fold for drugs like the anthracyclines (Table 2.3A)[19] and vincristine (Table 2.3B).[20,21] This occurs because particulate carriers like liposomes are not easily able to exit the circulation through intact capillary endothelium. This restricts the volume of distribution of the drug in large part to the central compartment as long as it remains associated with liposomes. Drug encapsulation in long circulating SL, in particular, can result in significant reductions in the plasma clearance rate of drugs by confining them to the central compartment where they can remain in circulation with half-lives of as long as 2 d in humans (Table 2.3A).[19] The consequence of all of these alterations in drug pharmacokinetics as a result of liposome association is a highly significant increase in the plasma area under the time-concentration curve (AUC), which provides a measure of increased drug bioavailability (Table 2.3A,B). It should be realized, however, that such increased bioavailability does not necessarily mean increased therapeutic availability.

Applications of Liposome-Entrapped Oncologic Agents

The practical consequences of the ability of liposomes to alter the pharmacokinetics of a drug are numerous, as illustrated by the list of liposomal oncologic agents in clinical development (Table 2.4). The restricted biodistribution of liposome-associated drugs can result in the distribution of an oncologic agent away from some sensitive tissues, e.g., the heart in the case of anthracyclines.[22] It can also result in the distribution of the drug towards solid tumor nodules when they exhibit increased vascular permeability, e.g., during the process of angiogenesis.[19,23,24] Both of these effects will tend to increase the therapeutic index of the drug—the first by reducing toxicity, the second by increasing tumor localization and efficacy. However, the altered biodistribution can also lead to the appearance of new toxicities. Toxicity to the mononuclear phagocyte[25,26] and hand and foot syndrome appear to be among potential new toxicities reported to date for SL liposomal formulations of anticancer drugs such as DXR.[19,27]

It is sometimes difficult to predict what liposome encapsulation, with the consequential alterations in the pharmacokinetics, will do to the maximum tolerated dose (MTD) of antineoplastic drugs. Increased drug half-life, and delayed clearance of liposome-associated drugs from the plasma can lead to a substantial decrease in the

Table 2.3. A comparison of pharmacokinetics for free drug and drug entrapped in CL or SL

A. Pharmacokinetics of doxorubicin in man at a dose of 50 mg/m^2.

	free DXR	DXR-CL	DXR-SL
$T_{1/2\alpha}$ (hr)	0.05	0.29	2.3
$T_{1/2\beta}$ (hr)	10.4	6.7	46.2
C_O (L/hr)	25.3	23.0	0.09
AUC (mg•hr/L)	3.5	58.5	893
V_D (L)	365	18.8	6.4

B. Pharmacokinetics of vincristine in rats at a dose of 0.25 mg/kg

	free VCR	VCR-CL	VCR-SL
$T_{1/2\alpha}$ (hr)	0.22	0.3	10.5
C_O (L/hr)	45.5	1.3	0.082
AUC (mg•hr/L)	2.2	75.8	1216

$T_{1/2}$, plasma half-life; AUC, area under the time concentration curve; V_D, volume of distribution; C_O, clearance rate

MTD, as found for the rapidly degraded drug ara-C.[1] Note that in this case, the efficacy increased even more, and the end result is an increase in therapeutic index.[1] On the other hand, the anthracycline drugs and vincristine showed modest increases in MTD, likely as a consequence of distribution of the drug away from sensitive tissues. This, in combination with an increase in efficacy as a consequence of the passive localization of the drug-liposome package in solid tumor nodules, leads to an increased therapeutic index.[21,28]

Future Developments: Targeted Delivery of Liposomal Oncologic Agents

Passive targeting of oncological agents entrapped in long circulating liposomes to solid tumors is attributed to a combination of prolonged circulation and increased capillary permeability of newly forming blood vessels. This passive targeting effect can, as seen above, lead to significant improvements in the therapeutic efficacy of the entrapped drug, as described in chapters 5-11 for DXR or daunorubicin in the treatment of Kaposi's sarcoma and other cancers.[25] However, there are circumstances in which the passive targeting approach will likely be ineffective, e.g., in the treatment of hematologic malignancies or in attempts to eradicate metastatic cells present in the blood or lymph prior to their development into solid tumors. Further improvements in the localization of liposomal oncologic agents specifically in the treatment of these conditions might be achieved through ligand (e.g., antibody)-mediated targeting of the liposomal carriers to cancer-associated molecules at the cell surface. Previous attempts to target CL in vivo by means of antibodies (Ab) coupled to the liposome surface (Ab-CL) proved largely ineffective due to the rapid removal of the Ab-CL from circulation by phagocytic cells of the liver, enhanced or triggered by the presence of the Ab on the liposome surface. Development of coupling methodologies for attaching Ab at the terminus of the PEG molecule in SL,[29,30] has

Table 2.4. Liposomal oncologic agents in clinical development

Drug	Indication	Investigational Status	Company
Long-circulating liposomal	Refractory Kaposi's sarcoma	Approved (US) Approvable (Europe)	SEQUUS Pharmaceuticals Inc. (formerly Liposome Technology, Inc)
	Advanced breast, ovarian and lung cancer	Phase II (US and Israel)	
Liposomal daunorubicin	Refractory Kaposi's sarcoma	Approved (US)	NeXstar Inc. (formerly Vestar)
Liposomal MTP-PE	Relapsed osteosarcoma	Phase III (US)	NCI
Liposomal doxorubicin	Advanced breast cancer	Phase II-III (US)	The Liposome Co.
	Soft tissue sarcomas (in combination with G-CSF)	Phase II (US)	
Liposomal vincristine	Non-Hodgkin's lymphoma, pancreatic cancer	Phase II (Canada)	Inex Inc.
Liposomal	Refractory Kaposi's sarcoma	Phase II (US)	Argus Pharm. Inc.
Liposomal	Advanced breast cancer	Phase I (US)	Argus Pharm. Inc.

resulted in long circulating Ab-SL that retain Ab-mediated binding. This has provided impetus to the development of targeted formulations of oncologic agents.

The utility of targeted delivery has been demonstrated for Ab-SL containing entrapped DXR (Ab-SL-DXR) in a pseudometastatic murine squamous lung carcinoma model[31,32] and also in the treatment of human B lymphoma growing in immune-deficient (SCID) mice.[32] In the latter model of a hematological malignancy, access of the Ab-SL to the cancer cells is relatively unimpeded. Targeting anti-CD19-SL with entrapped DXR to an internalizing epitope (CD19) on the B lymphoma cells led to a significant increase in the mean survival times of the mice compared to free DXR or nontargeted SL-DXR given at the same total dose.[32,33] In these experiments, targeting to an internalizing epitope is thought to be critical as a maximum amount of drug can be delivered to the cancer cell upon internalization of the liposome along with its contents (Fig. 2.1). In the case of therapy targeted to a noninternalizing epitope, delivery of drug into the cell would depend upon leakage of the drug from the liposome, followed by either passive diffusion or active transport of the drug into the cancer cell.[38] This approach is less desirable due to loss of drug by diffusion away from the cell surface, particularly in the turbulent environment of the blood, resulting in exposure of nontarget tissues (Fig. 2.1).

In treating solid tumors with Ab-SL the results appear to depend on whether the tumor is treated at the micrometastatic stage or at a later stage of disease. In the murine squamous lung carcinoma model, cancer cells injected iv colonize the lung beginning 3 d postinjection where they develop into solid tumor nodules.[34,35] Treatment of the mice on day 3 with a single injection of Ab-SL-DXR resulted in cure in 60% of the animals, while treatment with free drug was almost completely ineffective, and treatment with SL-DXR (no Ab) was intermediate with only modest increases in survival but no long-term survivors.[31,32] However, delaying the treatment until after the development of solid tumor nodules to the multi-cellular stage resulted in no advantage of the Ab-SL-DXR over the other treatments.[32] The failure of targeted therapy to show significant advantage in the treatment of more advanced solid tumors may be due to the existence of a "binding-site barrier."[36] It has been hypothesized that targeted carriers bind to the first target receptors they encounter, usually those on the exterior of the tumor nodule, which prevents their penetration into the tumor interior, negating any advantage relative to nontargeted carriers. Indeed, in the treatment of human ovarian cancer grown subcutaneously as solid tumors in nude mice, we have found that SL-DXR (no Ab) was more effective in decreasing tumor growth than Ab-SL-DXR,[32] likely because passive localization of the liposomes to the solid tumor allows for greater penetration into the tumor interior.

Our ability to treat more advanced solid tumors with Ab-targeted therapy may be critically dependent on the particular Ab, epitope and specific tumor. Recently, for example, Dr. Papahadjopoulos and colleagues have reported increased efficacy of Ab-SL-DXR in the treatment of a model of human breast cancer.[37] They used an antibody fragment (Fab) of the monoclonal antibody anti-HER2 coupled to the PEG terminus of the SL and demonstrated increased efficacy of this preparation compared to nontargeted liposomes or free drug.[37] In this case, not only does the antibody fragment have high affinity for its epitope, but it appears to trigger rapid internalization of the liposome package into the cell interior, both of which properties may be necessary to the success of this therapy.

Conclusions

Association of oncologic agents with carriers such as liposomes has dramatic effects on the biodistribution and pharmacokinetics of the drug. An understanding of the underlying principles leading to these changes not only will allow better liposomal carriers to be developed, but most importantly, will allow the newly approved liposomal oncologic agents to be used more rationally in the clinic without bias from the traditional use of the original formulations. Also critical to the rational use of liposomal oncologic agents is an understanding of the underlying cell biology of the disease and how that will affect the ability of the carrier to localize at the site of disease. Liposomes can be useful carriers for oncologic agents from such simple applications as serving as a 'solubilizing agent' for difficult to administer drugs or protecting rapidly degraded drugs from premature breakdown to more sophisticated applications such as mediating the passive targeting of liposome-entrapped

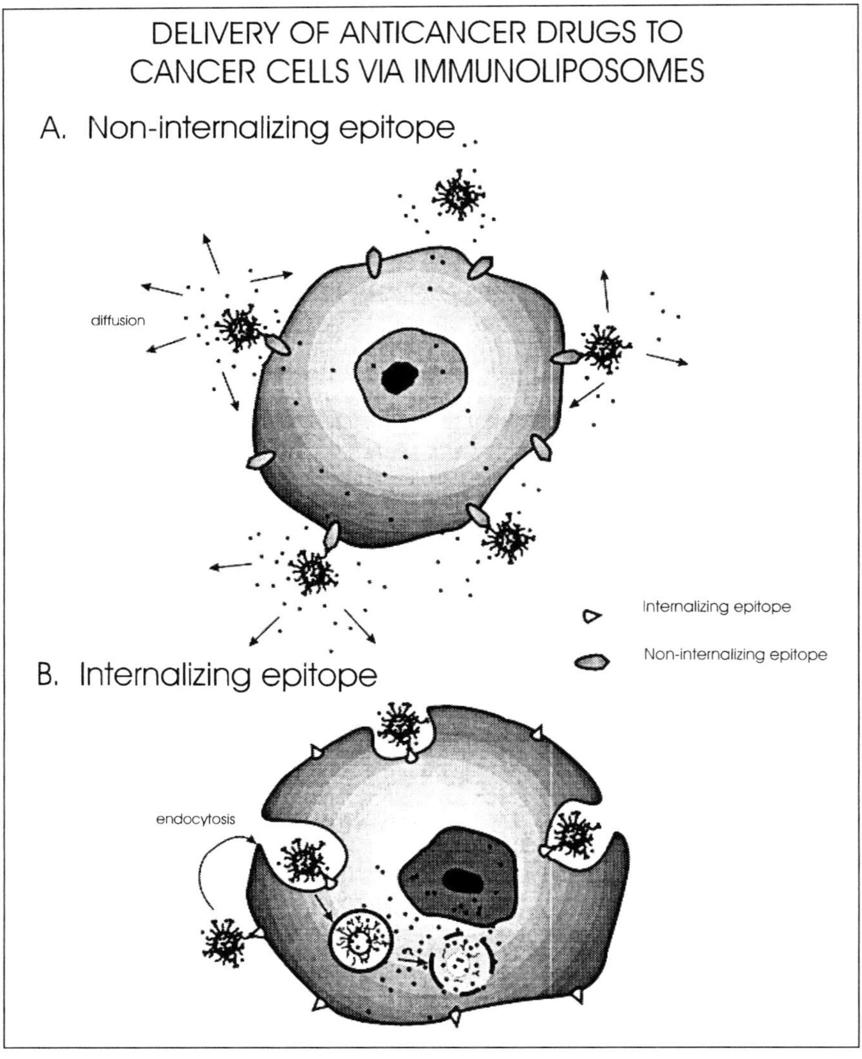

Fig. 2.1. A schematic representation of uptake of liposomal anticancer drugs targeted to noninternalizing versus internalizing epitopes on the cancer cell surface.

drugs to solid tumors or allowing the selective targeting of drugs specifically to diseased cells. Overall, these new formulations may behave in ways not predicted by the behavior of the free drug and should be viewed as new therapeutic agents.

Acknowledgments

Our work is supported by the Medical Research Council of Canada (MA-9127, UI-12411) and by Sequus Pharmaceuticals Inc., Menlo Park, CA. I thank Christian Hansen for editorial assistance in the preparation of the manuscript.

References

1. Allen TM, Mehra T, Hansen C et al. Stealth liposomes: an improved sustained release system for 1-beta-D-arabinofuranosylcytosine. Cancer Res 1992; 52(9):2431-2439.
2. Ellens H, Rustum Y, Mayhew E et al. Distribution and metabolism of liposome-encapsulated and free 1-beta-D-arabinofuranosylcytosine (Ara-C) in dog and mouse tissues. J Pharm Exp Therap 1982; 222(2):324-330.
3. Perez-Soler R, Francis K, al-Baker S et al. Preparation and characterization of liposomes containing a lipophilic cisplatin derivative for clinical use. J Microencapsulation 1994; 11[1]:41-54.
4. Sharma A, Mayhew E, Straubinger RM. Antitumor effect of taxol-containing liposomes in a taxol-resistant murine tumor model. Cancer Res 1993; 53(24):5877-5881.
5. Baguley BC, Falkenhaug EM. Plasma half-life of cytosine arabinoside (NSC-63878) in patients treated for acute myeloblastic leukemia. Cancer Chemotherapy Rep 1971; 55(3):291-298.
6. Borsa J, Whitmore GF, Valeriote FA et al. Studies on the persistence of methotrexate, cytosine arabinoside, and leucovorin in serum of mice. J Natl Cancer Inst 1969; 42(2):235-242.
7. Burke TG, Gao X. Stabilization of topotecan in low pH liposomes composed of distearoylphosphatidylcholine. J Pharm Sci 1994; 83(7):967-969.
8. Defrise-quertain F, Chatelain P, Delmelle M et al. Model studies for drug entrapment and liposome stability. In: Gregoriadis G, ed. Liposome Technology. Vol. 2. Boca Raton: CRC Press, 1984:1-17.
9. Mayer LD, Tai LC, Ko DS et al. Influence of vesicle size, lipid composition, and drug-to-lipid ratio on the biological activity of liposomal doxorubicin in mice. Cancer Res 1989; 49(21):5922-5930.
10. Allen TM. Stealth liposomes as a drug sustained release system for 1-β-D-arabinofuranosylcytosine (cytosine arabinoside). In: Lasic D, Martin F, eds. Stealth Liposomes. Boca Raton: CRC Press, 1995:193-202.
11. Huang SK, Stauffer PR, Hong K et al. Liposomes and hyperthermia in mice: increased tumor uptake and therapeutic efficacy of doxorubicin in sterically stabilized liposomes. Cancer Res 1994; 54(8):2186-2191.
12. Allen TM, Hansen CB, Lopes de Menezes DE. Pharmacokinetics of long circulating liposomes. Adv Drug Del Rev 1995; 16(2-3):267-284.
13. Lasic DD, Frederik PM, Stuart MC et al. Gelation of liposome interior. A novel method for drug encapsulation. FEBS Lett 1992; 312(2-3):255-258.
14. Mayer LD, Bally MB, Cullis PR. Uptake of adriamycin into large unilamellar vesicles in response to a pH gradient. Biochim Biophys Acta 1986; 857(1):123-126.
15. Zou YY, Priebe W, Stephens LC et al. Preclinical toxicity of liposome incorporated annamycin: selective bone marrow toxicity with lack of cardiotoxicity. Clin Cancer Res 1995; 1(11):1369-1374.
16. Hwang KJ. Liposome pharmacokinetics. In: Ostro MJ, ed. Liposomes: From Biophysics to Therapeutics. New York: Marcel Dekker, 1987:109-156.
17. Gabizon A, Meshorer A, Barenholz Y. Comparative long-term study of the toxicities of free and liposome-associated doxorubicin in mice after intravenous administration. J Natl Cancer Inst 1986; 77(2):459-469.
18. Allen TM, Hansen C. Pharmacokinetics of stealth versus conventional liposomes: effect of dose. Biochim Biophys Acta 1991; 1068(2):133-141.
19. Gabizon A, Catane R, Uziely B et al. Prolonged circulation time and enhanced accumulation in malignant exudates of doxorubicin encapsulated in polyethylene-glycol coated liposomes. Cancer Res 1994; 54(4):987-992.
20. Embree L, Gelmon KA, Tolcher T et al. Clinical pharmacokinetics of vincristine sulphate liposome injection (VSLI). Proc Am Assoc Cancer Res 1996; 37:179.

21. Allen TM, Newman MS, Woodle MC et al. Pharmacokinetics and anti tumor activity of vincristine encapsulated in sterically stabilized liposomes. Int J Cancer. Jul 1995; 62(2):199-204.
22. Gabizon A, Isacson R, Libson E et al. Clinical studies of liposome-encapsulated doxorubicin. Acta Oncologica 1994; 33(7):779-786.
23. Wu NZ, Da D, Rudoll TL et al. Increased microvascular permeability contributes to preferential accumulation of Stealth liposomes in tumor tissue. Cancer Res 1993; 53(16):3765-3770.
24. Huang SK, Lee KD, Hong K et al. Microscopic localization of sterically stabilized liposomes in colon carcinoma-bearing mice. Cancer Res 1992; 52(19):5135-5143.
25. Bogner JR, Kronawitter U, Rolinski B et al. Liposomal doxorubicin in the treatment of advanced AIDS-related Kaposi sarcoma. J Acquired Immune Deficiency Syndromes 1994; 7(5):463-468.
26. Bally MB, Nayar R, Masin D et al. Studies on the myelosuppressive activity of doxorubicin entrapped in liposomes. Cancer Chemo Pharm 1990; 27(1):13-19.
27. Uziely B, Jeffers S, Isacson R et al. Liposomal doxorubicin: antitumor activity and unique toxicities during two complementary phase I studies. J Clin Oncology 1995; 13(7):1777-1785.
28. Gabizon AA. Selective tumor localization and improved therapeutic index of anthracyclines encapsulated in long circulating liposomes. Cancer Res 1992; 52(4):891-896.
29. Blume G, Cevc G, Crommelin MD et al. Specific targeting with poly(ethylene glycol)-modified liposomes: coupling of homing devices to the ends of the polymeric chains combines effective target binding with long circulation times. Biochim Biophys Acta 1993; 1149(1):180-184.
30. Hansen CB, Kao GY, Moase EH et al. Attachment of antibodies to sterically stabilized liposomes: evaluation, comparison and optimization of coupling procedures. Biochim Biophys Acta 1995; 1239(2):133-144.
31. Ahmad I, Longenecker M, Samuel J et al. Antibody-targeted delivery of doxorubicin entrapped in sterically stabilized liposomes can eradicate lung cancer in mice. Cancer Res 1993; 53(7):1484-1488.
32. Allen TM, Ahmad I, Lopes de Menezes DE et al. Immunoliposome-mediated targeting of anti-cancer drugs in vivo. Biochem Soc Trans 1995; 23:1073-1079.
33. Lopes de Menezes DE, Pilarski LM, Allen TM. Selective cytotoxicity of immunoliposomal to B lymphocytes. Proceedings of the American Association for Cancer Res 1995; 35:307.
34. Ding L, Samuel J, MacLean GD et al. Effective drug-antibody targeting using a novel monoclonal antibody against the proliferative compartment of mammalian squamous carcinomas. Cancer Immunol Immunother 1990; 32(2):105-109.
35. Samuel J, Noujaim AA, Willans DJ et al. A novel marker for basal (stem) cells of mammalian stratified squamous epithelia and squamous cell carcinomas. Cancer Res 1989; 49(9):2465-2470.
36. Weinstein JN, van Osdol W. Early intervention in cancer using monoclonal antibodies and other biological ligands: micropharmacology and the "binding site barrier". Cancer Res 1992, 52:2747-2751.
37. Kirpotin D, Park JW, Hong K et al. Binding and endocytosis of sterically stabilized anti-HER2 immunoliposomes by human breast cancer cells. Proc Am Assoc Cancer Res 1996; 37:467.
38. Storm G, Näsander UK, Vingerhoeds MH et al. Atibody targeted liposomes to deliver doxorubicin to ovarian cancer cells. J Liposome Res 1997; 4(1):641-666.

CHAPTER 3

Liposomal Vincristine: The Central Role of Drug Retention in Defining Therapeutically Optimized Anticancer Formulations

Nancy L. Boman, Pieter R. Cullis, Lawrence D. Mayer, Marcel B. Bally, Murray S. Webb

Introduction

Vincristine is an alkaloid derived from *Vinca rosea* that is effective against a wide variety of human carcinomas.[1,2] When used in conjunction with corticosteroids, vincristine is the treatment of choice to induce remissions in childhood leukemia. Vincristine is also part of a complex protocol used in the treatment of adult patients with Hodgkin's disease or non-Hodgkin's lymphomas. In addition, vincristine has some effectiveness against Wilms' tumor, neuroblastoma, brain tumors, rhabdomyosarcoma, and carcinomas of the breast and the bladder, as well as the male and female reproductive systems.

Vincristine is a cell-cycle specific antiproliferative agent, arresting cell growth exclusively during metaphase by attaching to the growing end of microtubules and terminating further assembly.[3,4] As a consequence of this cell-cycle specificity, methods which extend exposure of vincristine to neoplastic cells should increase therapeutic activity. This prediction is supported by preclinical experiments by Jackson and Bender[5] and later by our research group.[6] We demonstrated that the concentration of vincristine required to achieve a 50% inhibition in cell proliferation (IC_{50}) in vitro decreased 10^5-fold when the drug exposure time was increased from 1-72 h. In contrast, the anthracycline doxorubicin, another anticancer drug, shows only a 40-fold decrease in the IC_{50} for the same increase in exposure time (see Table 3.1).

Systemic chemotherapy treatment is required to treat systemic disease and a central premise of cancer chemotherapy is to achieve maximum dose intensity under conditions where drug related toxicities are manageable. In the absence of a defined drug delivery technology, attempts to optimize both plasma concentration and the time of exposure to the diseased tissue to chemotherapy agents have relied on the manipulation of variables such as drug dose and dosing schedules. However, such strategies are of limited benefit when the drugs used are rapidly eliminated from the

Long Circulating Liposomes: Old Drugs, New Therapeutics, edited by Martin C. Woodle and Gerrit Storm.

Table 3.1. Effect of exposure time on the cytotoxicity of doxorubicin and vincristine on L1210 cells in vitro.

Exposure Time (hours)	IC_{50} (nM)	
	Doxorubicin	Vincristine
1	370	12,000
6	55	2,400
24	18	7.3
72	9.2	0.12

Data obtained from Mayer et al[6] and Masin and Nayer (unpublished).

plasma compartment, actively metabolized into inactive species, rapidly released from cells exhibiting low drug binding affinities and/or are toxic due to poor disease tissue specificity. Since most drugs used to treat cancer are afflicted with one or more of these problems, methods to maximize drug exposure following systemic administration must fundamentally improve both drug specificity and the bioavailability of the active drug.

For vincristine, a drug that exhibits a dose limiting neurotoxicity, two separate approaches have been pursued clinically to enhance drug exposure:

1) Use of intravenous infusions.
2) Applications of drug carrier technology.

Long-term infusions attempt to achieve a balance between toxicity and maintenance of therapeutic drug levels in the plasma compartment over extended infusion times (typically in excess of 96 h). Drug carriers have the potential to increase plasma concentration and AUC while also facilitating increased specificity of drug delivery to the disease site. The latter attribute distinguishes drug carrier technology from intravenous drug infusions, since drug delivery to the disease site is a consequence of the characteristics of the drug carrier as well as the vasculature within the disease tissue.

We strongly believe that increases in drug exposure in regions of disease can best be achieved using long circulating liposomal carriers, a systemic delivery system that can be designed to give exquisite control over drug levels in the plasma, the site of disease progression and, most importantly, drug bioavailability within the diseased tissue. In this context, we wish to emphasize that systems described as long circulating include those with minimal interactions with serum proteins and RES cells which also must have good drug retention properties. They may refer to liposomes with PEG surface coatings but also include other forms such as with highly cohesive bilayers, specific glycolipids, or drugs that effect RES uptake (see chapter 1). In this chapter we review our studies leading to a liposomal vincristine formulation that is now undergoing clinical testing. In addition, we will use vincristine as a model drug to illustrate how regulated blood circulation and drug release can be achieved by careful control of liposomal lipid composition and trapping characteristics. In turn, optimized release characteristics are correlated to significant improvements in therapeutic activity. It is important to recognize that the membrane permeability characteristics of vincristine are such that it has been a challenge to develop a carrier system that effectively retains the drug for extended time periods. This research has been developed on the basis of two assumptions:

1) Rapid release of the drug from the carrier will be of limited benefit, resulting in formulations that are therapeutically no better than free vincristine.
2) Complete retention of the drug will result in a therapeutically inactive liposomal formulation.

Therefore the goal of this research is to identify long circulating liposomal formulations that release vincristine at an optimal rate for maximum therapeutic activity. The focus of the following sections is:

1) Attributes that effect drug release.
2) Liposome circulation and disease exposure.
3) Clinical utility of the optimized formulation.

Vincristine Encapsulation and In Vitro Drug Release

Vincristine can be loaded into liposomes by several methods. So-called passive techniques rely on entrapping the drug during liposome formation, followed by removal of drug that was not sequestered in the liposome. Alternatively, vincristine can be actively encapsulated in liposomes that exhibit ion gradients.[7-9] As noted by Mayer et al,[7] ion gradient based loading procedures significantly improve encapsulation efficiency and decrease the rate of drug release.

We have focused our research on the use of transmembrane pH gradients to actively load vincristine into liposomes. It is well established that many drugs which are weak bases can be encapsulated within liposomes in response to a transmembrane pH gradient.[8] Vincristine is one of these compounds that can be rapidly loaded into liposomes and the physicochemical basis for encapsulation of weak bases such as vincristine using a transmembrane pH gradient is well characterized.[10,11] Encapsulation efficiencies approaching 100% in response to a transmembrane pH gradient have been achieved for formulations with vincristine/lipid (wt/wt) ratios in the range between 0.05 and 0.20. Typically, we prepare liposomes in a citric acid buffer and increase the external pH of the liposomes by titration with sodium phosphate. Alternatively, the liposomes can be eluted through a desalting column to exchange unencapsulated citrate with a defined buffer adjusted to a pH that is higher than that of the encapsulated citrate solution.

Release of vincristine from liposomes has been measured by the change of the vincristine/lipid ratio during dialysis. This formulation property is influenced by a number of factors dictated by the loading procedure, including the initial internal pH, the magnitude of the pH gradient, the internal buffering capacity and the residual pH gradient after drug loading. In general, the rate of vincristine release is a consequence of the ability of the formulations to maintain the transmembrane pH gradient during loading and under conditions used to evaluate drug release. As shown in Figure 3.1, increasing the magnitude of the transmembrane pH gradient by decreasing the internal pH, significantly decreased vincristine release from DSPC/Chol liposomes.[12] This observation is consistent with a proposed mechanism of leakage where the rate constant for vincristine movement across a lipid bilayer is proportional to the inverse square of the proton concentration. This mechanism predicts a 100-fold reduction in leakage rate for an increase in the transmembrane pH gradient by one unit.[12]

The most important factor controlling vincristine release in vitro and in vivo, however, is the lipid composition of the vesicles. For in vitro studies, liposomal lipid

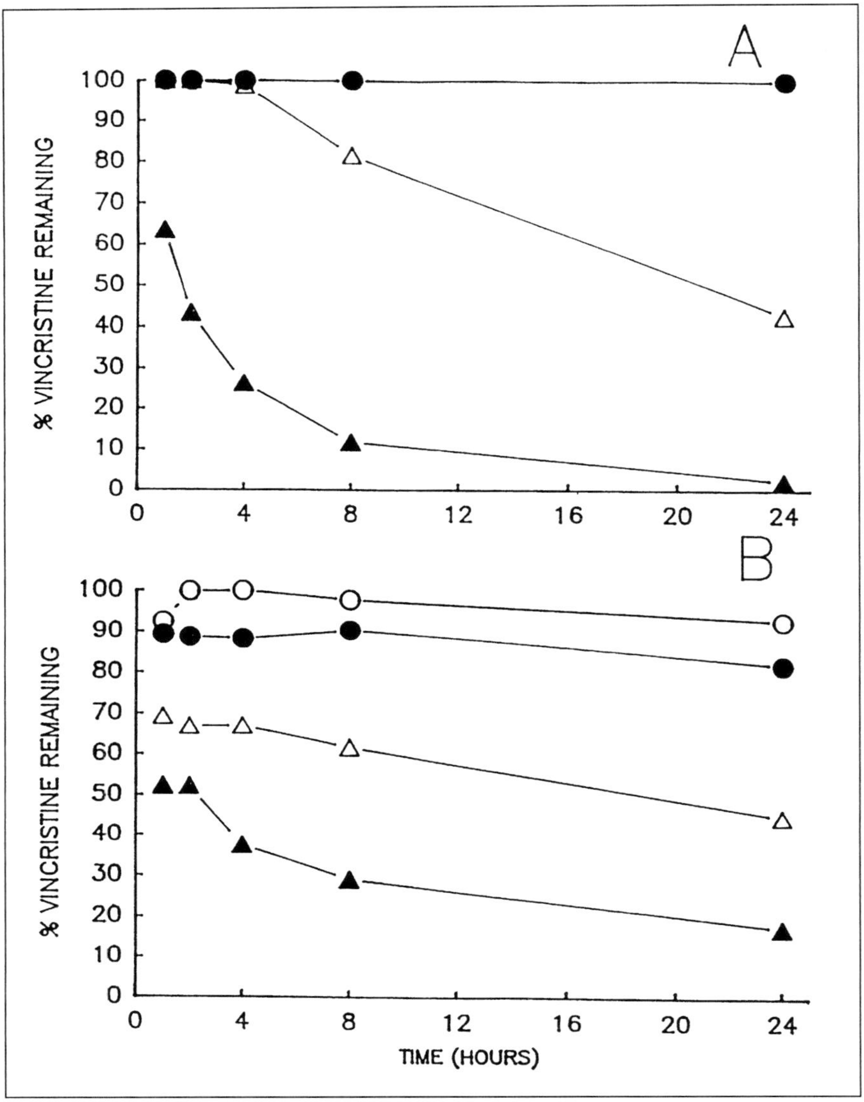

Fig. 3.1. Vincristine release from 100 nm DSPC/Chol vesicles incubated in buffer (A) and mouse serum (B) at 37°C for internal pH of 2.0 (O), 3.0 (●), 4.0 (Δ), and 5.0 (▲). Internal buffering capacity was 300 mM citrate for all systems. Initial drug/lipid ratios were 0.1/1 (wt/wt). Graph from Boman et al[12] with permission.

composition influences drug permeability rates as well as stability of the pH gradient which together determine drug accumulation. The intrinsic permeability of the drug through the liposomal membrane controls both drug loading rate and its subsequent release rate while the pH gradient stability affects both the liposome loading capacity as well as the rate constant for drug release. In vitro, the slower release

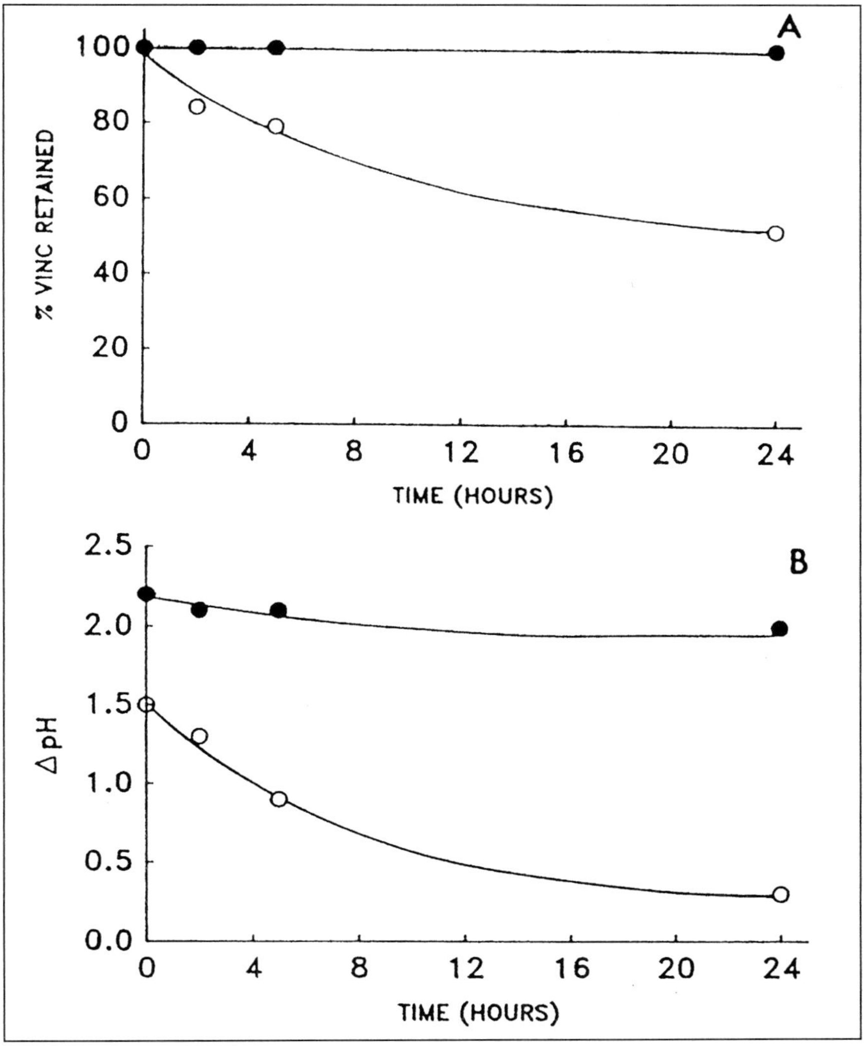

Fig. 3.2. A; vincristine release from DSPC/cholesterol (●) and EPC/cholesterol (O) vesicles subsequent to ΔpH-dependent drug accumulation. Vincristine was encapsulated at a drug/lipid ratio (w/w) of 0.19 (± 0.01). Liposomes were incubated at 21°C and at the indicated times liposome-associated drug was determined. B; transmembrane pH gradient dissipation at 21°C in DSPC/cholesterol (●) and EPC/cholesterol (O) vesicles subsequent to vincristine encapsulation at a drug/lipid ratio (w/w) of 0.19 (± 0.01). Incubation conditions were identical to those described in A. Graph from Mayer et al[13] with permission.

of vincristine from DSPC/Chol liposomes compared to that from EPC/Chol liposomes (Fig. 3.2A)[13] is due, in part, to enhanced stability of the transmembrane pH gradient (Fig. 3.2B) with the saturated acyl chain. Generally, in vitro release of vincristine is decreased in liposomes containing longer saturated acyl chain phosphatidylcholines, such as DAPC/Chol and DBPC/Chol, in comparison to liposomes containing short chain phosphatidylcholine, such as DMPC/Chol.[12] These in vitro studies have been extremely useful in defining mechanisms which govern transmembrane transport of vincristine in response to pH gradients, and thus provide methods to control its loading and release.

As described in the following section, the factors which decrease the rate of vincristine release in vitro have proven to have similar effects on the release of vincristine from liposomes in vivo. However, drug leakage from liposomes in vivo is usually much more rapid than observed in vitro. Altogether, these studies, and the subsequent findings in vivo, have defined the parameters which influence vincristine release from long circulating liposome formulations permitting optimization of the therapeutic value of these liposomal formulations.

Drug and Liposome Plasma Levels and Disease Exposure

It is well established that encapsulation of drugs in long circulating liposomes is an effective means of increasing the circulation lifetime of a drug, the concentration of drug in the blood compartment, and the amount of therapeutic agents delivered to tumors (see chapter 1). Increased blood levels and circulation longevity of liposomal drugs are controlled primarily by the characteristics of the carrier which, when administered iv, persist in the blood compartment for extended time periods. Elimination of a liposome encapsulated drug from the circulation is dependent on both accumulation in the RES through recognition of the liposomes as foreign and on extravasation of the liposomes at sites where the blood vessels exhibit pores or gaps in the vascular endothelium large enough to allow the liposome to cross.

It is important, however, to distinguish plasma elimination of the drug carrier from release of drug from the carrier and its elimination. In this section we illustrate the importance of drug release rates, showing that increased liposome circulation longevity is of little value when the entrapped drug is released too quickly. It is our opinion that the permeability characteristics of encapsulated compounds from liposomes residing within the blood compartment and within sites of disease are the most important attributes defining an efficacious liposomal anticancer agent. As summarized below, we have demonstrated this for vincristine, a drug for which it has been difficult to achieve adequate in vivo retention within liposomes. Importantly, progressive increases in antitumor efficacy are observed as the rate of vincristine release from the liposomal formulation decreases.

Liposome Elimination and Vincristine Blood Levels

When using long circulating liposomes as drug carriers, as outlined in the Introduction, increased drug concentrations within the blood compartment are affected by both the physical and chemical attributes of the carrier. In practice, decreased liposome elimination rates are obtained when using liposomes that are small (~100 nm diameter), neutral (presence of selected anionic or cationic lipids significantly enhances liposome elimination), and contain polyethylene glycol (PEG)-containing lipids.[14-16] Decreases in the rate of vincristine release have been achieved by careful selection of the encapsulation parameters (as summarized above) and the choice of lipids used in the preparation of the liposomes. The influence of liposome encapsulation on the blood levels of vincristine achieved following a single intravenous dose of 2 mg/kg is shown in Figure 3.3. These results illustrate two important points. First, free drug is rapidly eliminated from the plasma compartment in comparison to the liposomal drug formulations, with the exception of the EPC/Chol liposomal lipid composition. Second, retention of vincristine in the blood is strongly dependent on the choice of liposomal lipid composition. Lower drug levels are observed when the drug is encapsulated in EPC/Chol liposomes in comparison to DSPC/Chol liposomes. Furthermore, lower levels are observed at 24 h with DSPC/Chol compared to DSPC/Chol/PEG-PE or SM/Chol. At least some of these results were predicted on the basis of the data shown in Figure 3.2A. Thus, differences in plasma drug levels are due in a large part to differences in drug release rates. This is supported by parallel measurement of liposome levels, permitting calculation of the vincristine/lipid ratios within the plasma compartment, shown in Figure 3.4. These results indicate that vincristine is released rapidly from the EPC/Chol liposomes, slowly from DSPC/Chol liposomes, and slowest from SM/Chol liposomes (Fig. 3.4). Therefore the observed differences in plasma clearance of vincristine (Fig. 3.3) are due primarily to differences in leakage during liposome circulation rather than differences in liposome circulation.

Since the liposome elimination rates are comparable for the various long circulating liposomal vincristine formulations studied, our emphasis here is on how changes in the vincristine/lipid ratio and vincristine release influence both drug delivery to sites of disease progression and antitumor efficacy. It is important to note, however, that all the liposomes used in these studies exhibit extended circulation lifetimes in comparison to large liposomes (>200 nm) or liposomes that contain anionic lipids such as phosphatidylserine, phosphatidylglycerol and cardiolipin. Further, the pres-

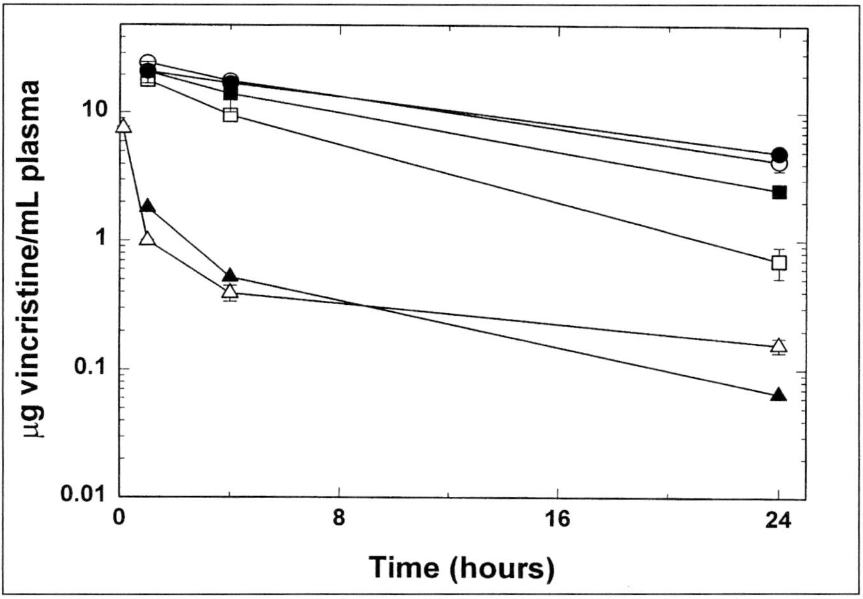

Fig. 3.3. In vivo plasma concentrations of vincristine after the intravenous administration of free vincristine (△) or vincristine encapsulated in liposomes composed of EPC/Chol (▲), DSPC/Chol (❐), DSPC/Chol/PEG_{2000}-DSPE (■), SM/Chol (○) or SM/Chol/PEG_{2000}-DSPE (●). Vincristine dose was 2 mg/kg in BDF1 (△, ❐, ■, ○, ●) or DBA/2J mice (▲). Data summarized from Webb et al[24] (△, ❐, ○), Webb, Bally and Mayer (unpublished data) (■, ●) and from Mayer et al[6] (▲).

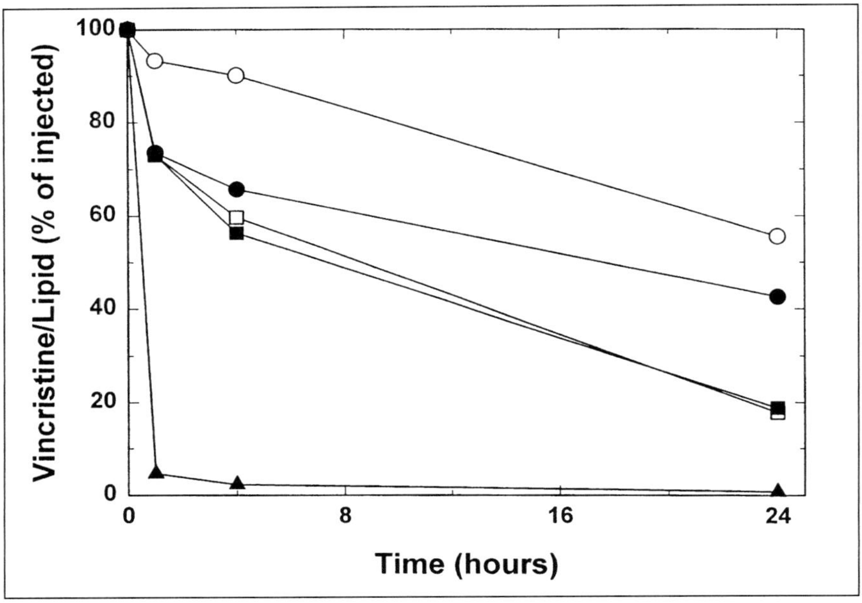

Fig. 3.4. Retention of vincristine in liposomal carriers after intravenous administration in mice. Vincristine retention is expressed as vincristine/lipid ratio in the plasma at various times after administration as a percentage of the administered vincristine/lipid ratio. Vincristine was encapsulated in liposomes composed of EPC/Chol (▲), DSPC/Chol (□), DSPC/Chol/PEG_{2000}-DSPE (■), SM/Chol (○) or SM/Chol/PEG_{2000}-DSPE (●). Vincristine dose was 2 mg/kg in BDF1 (□, ■, ○, ●) or DBA/2J mice (▲). Data summarized from Webb et al[24] (□, ○), Webb, Bally and Mayer (unpublished data) (■, ●) and from Mayer et al[6] (▲).

ence of entrapped vincristine has the effect of decreasing carrier elimination rate. This effect has been attributed to the now well-established RES blockade phenomenon that occurs with selected drugs encapsulated in liposomes.[17,18] RES blockade is observed for liposomal carriers of doxorubicin and vincristine with and without surface associated PEG. Longer times are, however, required following iv administration to observe the maximum level of RES blockade when drug is administered in PEG liposomes.

Factors Influencing Drug-to-Lipid Ratio Following iv Administration

The factors which increase vincristine retention in vitro also improve the retention of vincristine in vivo. For example, release of vincristine from DSPC/Chol liposomes after iv administration in DBA/2J mice was significantly slower than that from EPC/Chol liposomes (see Figs. 3.2 and 3.3). Similarly, vincristine release from DSPC/Chol liposomes after iv administration in mice was significantly slower when the transmembrane pH gradient was increased by two orders of magnitude. A comparison of the data in Figure 3.1 with that shown in Figure 3.5 indicates that decreasing the internal pH (pH_i) to 2.0 from 4.0 leads to reductions in drug release in vitro (Fig. 3.1) and in vivo (Fig. 3.5).

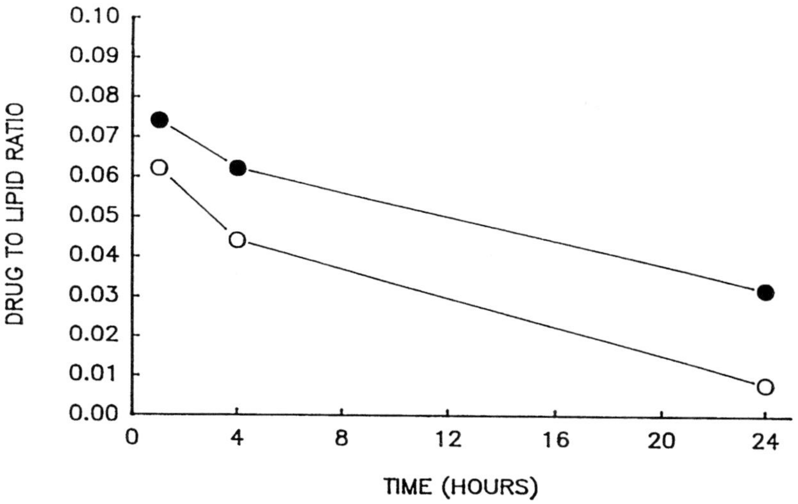

Fig. 3.5. Vincristine/lipid ratios for DSPC/Chol vesicles in vivo with internal pH of 2.0 (●) and 4.0 (O). Both systems were loaded at an initial drug/lipid ratio of 0.1/1 (wt/wt). Each point represents the average value obtained from four BDF1 mice. Graph from Boman et al[12] with permission.

Importantly, lipid composition and loading conditions can act together to achieve significant improvements in drug retention and a slower drug elimination rate. This important point is supported by recent data obtained with liposomal vincristine formulations prepared using G_{M1} containing and SM/Chol liposomes.[19-24] The SM/Chol formulation was initially developed for pharmaceutical reasons. Specifically, phospholipids such as DSPC are susceptible to acyl hydrolysis when stored at low pH. We therefore chose SM, which is known to be less sensitive to extended storage at pH 4, for reasons summarized elsewhere.[24] Fortuitously, however, the SM/Chol formulation also exhibited a drug release rate that was significantly slower than that observed for a DSPC/Chol formulation (Figs. 3.3, 3.4, 3.6). That is, the vincristine concentration in the plasma was significantly greater after administration of the drug in SM/Chol liposomes (Figs. 3.3, 3.6).

The balance between liposome circulation and drug permeability is well illustrated by studies attempting to further increase the vincristine concentration in plasma by addition of sterically-stabilized lipids (PEG_{2000}-DSPE) to the SM/Chol formulation. The results, shown in Figure 3.7, demonstrate that incorporation of PEG_{2000}-DSPE into SM/Chol liposomes engendered a significant decrease in the liposome elimination rate (Fig. 3.7A) but facilitated increased drug release from the liposomes (Fig. 3.7B). These two opposing effects resulted in no net change in plasma vincristine concentrations (Fig. 3.7C) and we concluded from these data that there would be no pharmacokinetic benefit achieved through incorporation of PEG-modified lipids in this formulation.[24]

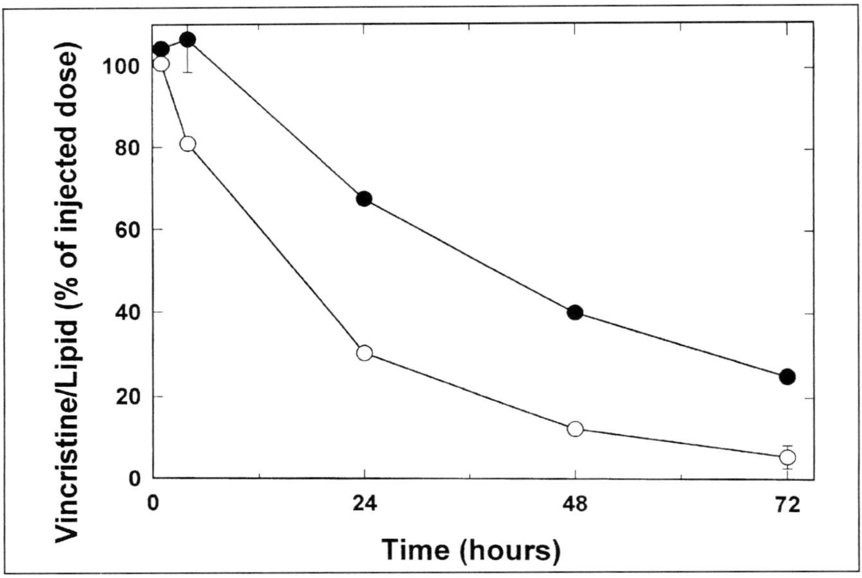

Fig. 3.6. Vincristine/lipid ratio, expressed as a percentage of the injected ratio, in the plasma of BDF1 mice at various times after the injection of large unilamellar liposomes of DSPC/Chol (O) or SM/Chol (●). Mice were injected with liposomes at a vincristine/lipid ratio of approximately 0.1/1, corresponding to a lipid dose of 20 mg/kg and a vincristine dose of 2.0 mg/kg. Total amounts injected were approximately 430 μg of lipid and 43 μg of vincristine. Data represent means (± s.e.) of three mice; where standard error bars are not visible, they are smaller than the size of the symbol. Graph from Webb et al[24] with permission.

Influence of Drug Retention on the Therapeutic Activity of Vincristine

An initial antitumor study comparing EPC/Chol to DSPC/Chol liposomal vincristine demonstrated the importance of increased drug plasma levels in achieving increased anticancer activity. This study clearly indicated that the DSPC/Chol formulation was significantly more active than either free vincristine or vincristine encapsulated in EPC/Chol liposomes when tested against the murine L1210 lymphocytic leukemia model (Fig. 3.8).[6] The EPC/Chol formulation exhibited no significant activity, consistent with data showing that drug encapsulation in this liposome formulation effected negligible increases in the level of vincristine in the plasma (Fig. 3.3). We interpreted these results as indicating the inability of the vincristine-permeable EPC/Chol formulation (Fig. 3.4) to maintain vincristine levels in the plasma above a therapeutic threshold (Fig. 3.8). Of greater interest, however, was the remarkable improvement in therapeutic activity observed for the DSPC/Chol liposomal formulation of vincristine. We have subsequently demonstrated that the DSPC/Chol liposomal formulation is significantly more active than free drug in treating a variety of murine and human xenograft tumor models.[6,23-25]

In the preceding example, the antitumor efficacy of vincristine in DSPC/Chol liposomes was compared to that of a vincristine-permeable formulation, EPC/Chol. Two studies comparing the therapeutic activity of DSPC/Chol liposomal vincristine to formulations with even slower drug release characteristics are, however, impor-

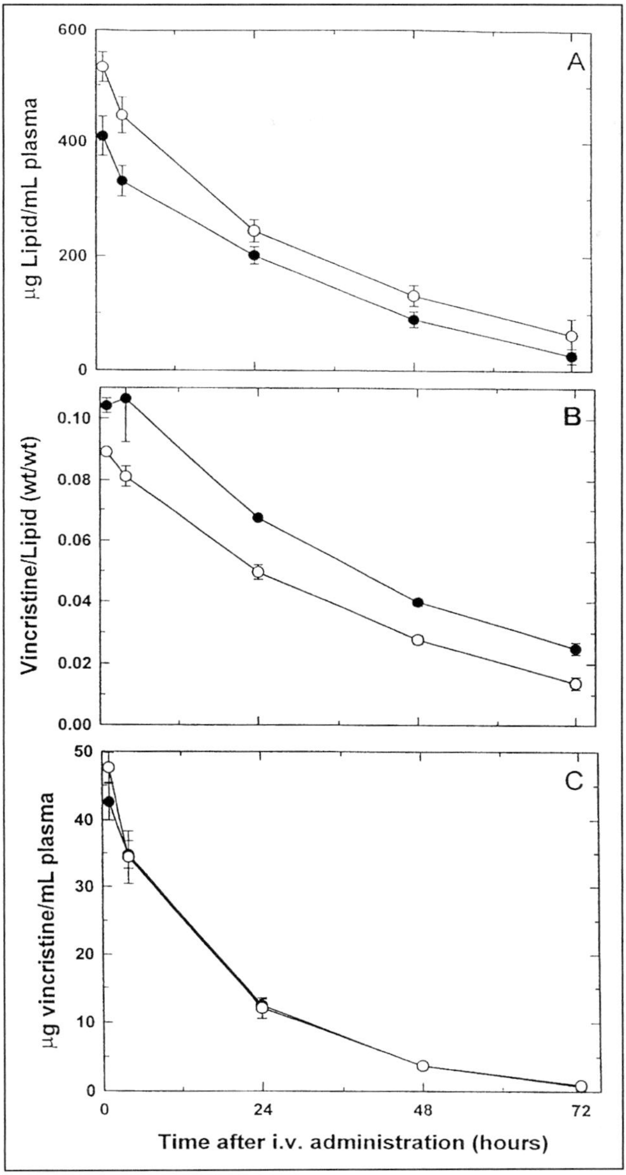

Fig. 3.7. Lipid clearance (A), vincristine/lipid ratio (B) and vincristine clearance (C) in the circulation of mice after the iv administration of vincristine encapsulated at a drug/lipid ratio of 0.1/1 (wt/wt) in liposomes composed of either SM/Chol (●) or SM/Chol/PEG_{2000}-DSPE (O). Vincristine was injected at a dose of 2.0 mg/kg, representing a lipid dose of 20 mg/kg. Data represent means (± s.e.) of three mice; where standard error bars are not visible, they are smaller than the size of the symbol. Data from Webb et al[24] and Webb, Masin, Bally and Mayer (unpublished).

tant to note. The incorporation of the ganglioside G_{M1} and use of an initial pH_i of 2.0 to enhance circulation and decrease the rate of vincristine release from DSPC/Chol liposomes also significantly increased the antitumor activity of vincristine when tested against the murine P388 leukemia model (Fig. 3.9). These studies indicated that 100% of tumor-bearing animals could be effectively cured of the disease when using the formulation which exhibited the slowest rate of drug release. The DSPC/

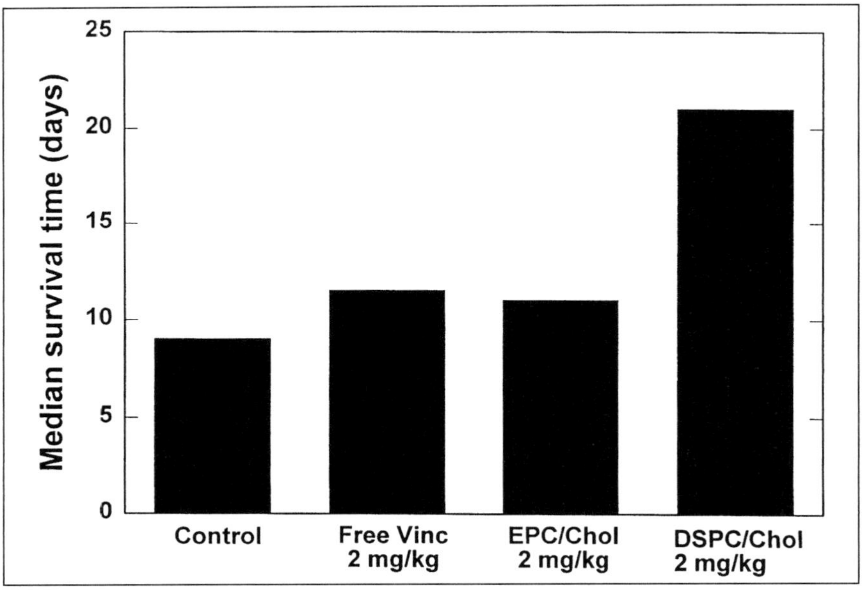

Fig. 3.8. Antitumor efficacy of free and liposomal vincristine in DBA/2J mice bearing the L1210 ip leukemia and treated iv with saline, free vincristine or vincristine encapsulated in either EPC/Chol or DSPC/Chol liposomes. Data summarized from Mayer et al.[6]

Chol/G_{M1} (pH_i 2.0) formulations administered at vincristine doses of 2, 3, and 4 mg/kg all produced long-term survivors with median survival times of > 70 d. It must be stressed that this increase in antitumor activity was due to decreased drug release from the liposomes and was not simply a consequence of increased liposome circulation longevity. This conclusion was also supported by studies evaluating the more pharmaceutically viable SM/Chol liposomal vincristine formulation; another formulation with drug release that is significantly slower than that observed for DSPC/Chol liposomes (Figs. 3.4, 3.6). This result predicted that the therapeutic activity of SM/Chol liposomal vincristine would be improved in comparison to DSPC/Chol. Studies using the murine P388 tumor model (data not shown) confirmed this prediction.[24] In addition, we demonstrated that the SM/Chol liposomal formulation exhibited significant therapeutic activity when tested against the human A431 xenograft tumor model (Fig. 3.10). Treating these animals with free vincristine resulted in a 2-3 d delay in initiation of tumor growth while treatment with DSPC/Chol liposomal vincristine effected a 15 d delay. The SM/Chol liposomal vincristine treated animals, however, exhibited no tumor growth for at least 40 d after drug administration (Fig. 3.10).

In an attempt to directly correlate vincristine release from liposomes and plasma levels of vincristine with antitumor activity we have summarized data obtained from P388 tumor bearing mice treated with the nine different liposome vincristine formulations tested to date. The results are presented in Figure 3.11. The efficacy of free or liposomal vincristine against the P388 tumor, as determined from the median days of survival after a single administration at 2 mg vincristine/kg, was very well

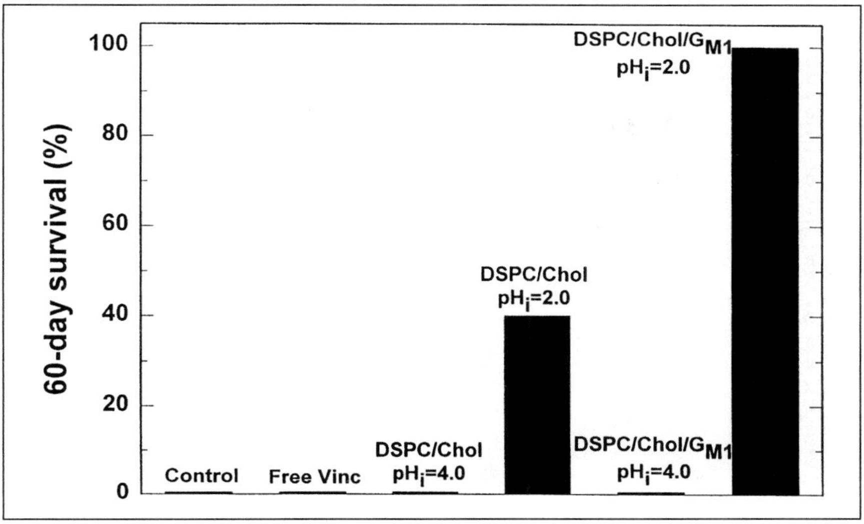

Fig. 3.9. Antitumor efficacy of free and liposomal vincristine in BDF1 mice bearing the P388 ip leukemia and treated iv with saline, free vincristine or vincristine encapsulated in either DSPC/Chol or DSPC/Chol/G_{M1} liposomes with pH_i values of either 2.0 or 4.0. Vincristine dose was 3 mg/kg and the vincristine/lipid ratio was 0.1/1 for the liposomal formulations. For saline, free vincristine, DSPC/Chol pH_i = 4.0 and DSPC/Chol/G_{M1} pH_i = 4.0 treatments, the 60-day survival was 0%; small bars are plotted solely for visibility. Data summarized from Boman et al.[23]

correlated ($r^2 = 0.95$) with increases in the circulation half-life of the drug (Fig. 3.11A). It is worth noting in this analysis the similarity of SM/Chol and SM/Chol/PEG_{2000}-DSPE liposomal vincristine (points 6 and 9) in both vincristine half-life and efficacy, despite the differences between these formulations in both lipid circulation lifetime and vincristine retention (Fig. 3.7A,B). In our experience, the primary factor influencing the vincristine half-life in the circulation is the retention of vincristine by the liposomal carrier. Consequently, for a variety of different liposomal formulations of vincristine, the antitumor efficacy is also strongly correlated with the half-life for the release of vincristine from the liposomes in the circulation ($r^2 = 0.84$) (Fig. 3.13B). In contrast, antitumor efficacy was poorly correlated with the circulation lifetime of the lipid carriers ($r^2 = 0.49$) (Fig. 3.11C). Taken in sum, these data clearly indicate that the primary pharmacokinetic parameter associated with increased antitumor efficacy is the circulation half-life of vincristine.

Drug Exposure at the Disease Site

Any analysis of antitumor efficacy and circulation half-life of vincristine must also take into consideration the propensity for liposomes to move from the plasma compartment into extravascular sites in regions of disease progression. This phenomena has been well documented for liposomal vincristine formulations[6,23-25] and is best illustrated by results obtained using the SM/Chol liposomal vincristine formulation. As indicated in the previous section, the therapeutic activity of vincristine in SM/Chol liposomes is better than that observed for a DSPC/Chol formulation. This has been demonstrated for ascitic tumors (Fig. 3.11A) as well as solid tumors

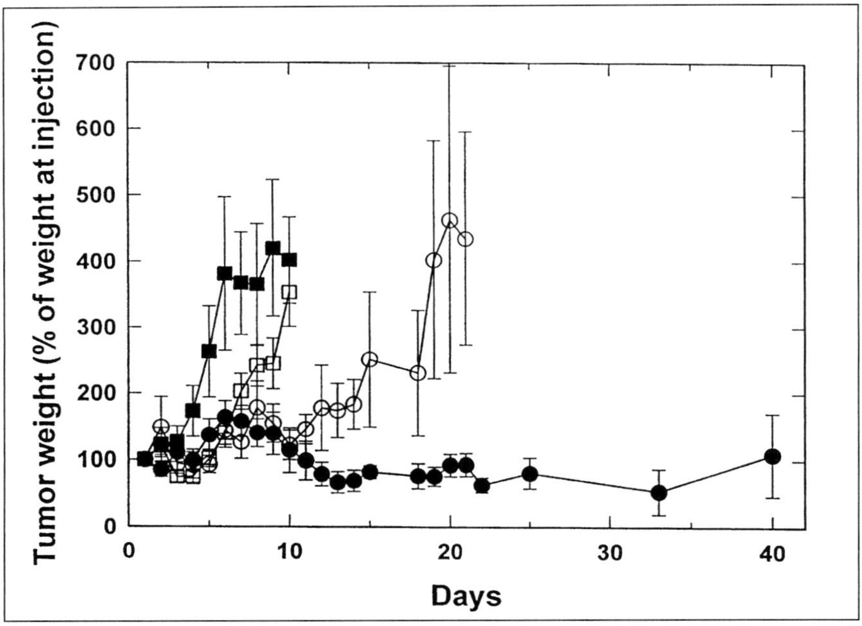

Fig. 3.10. Antitumor efficacy of free and liposomal vincristine in SCID mice bearing A431 tumors. SCID mice bearing two A431 tumors received no treatment (■) or were injected iv with free vincristine (❑) or with large unilamellar liposomes of DSPC/Chol (O) or SM/Chol (●) containing vincristine at a drug/lipid ratio of 0.1/1 (wt/wt). Vincristine was injected at a dose of 2.0 mg/kg, representing a lipid dose of 20 mg/kg. Data represent the weight of A431 tumors (expressed as a percentage of the tumor weight immediately before treatment) and are the means (± s.e.) of 8-10 tumors in 4-5 mice. Graph from Webb et al[24] with permission.

Fig. 3.11. (Opposite page) Summary of the pharmacokinetic parameters of free and liposomal formulations of vincristine that are associated with alterations in antitumor efficacy. Pharmacokinetic parameters are the half-life ($T_{1/2}$) for vincristine in the circulation (A); half-life ($T_{1/2}$) for the retention of vincristine during liposome circulation (B), and; half-life ($T_{1/2}$) for the circulation of the liposomal carriers in the circulation (C). Antitumor efficacy is reported for activity against an ip P388 tumor in BDF1 mice and is expressed as median days of survival after treatment. All animals were treated with an iv administration of either free or liposomal vincristine at a vincristine dose of 2.0 mg/kg, all liposomes were 0.1 µm. Data are presented for 1) free vincristine from Boman et al,[23] Mayer et al,[13] and Webb et al;[24] 2) DSPC/Chol ($pH_i = 4.0$) from Webb et al;[24] 3) DSPC/Chol ($pH_i = 4.0$) from Boman et al;[23] 4) DSPC/Chol ($pH_i = 2.0$) from Boman et al;[23] 5) DSPC/Chol/G_{M1} ($pH_i = 4.0$) from Boman et al;[23] 6) SM/Chol ($pH_i = 4.0$) from Webb et al;[24] 7) DSPC/Chol/G_{M1} ($pH_i = 2.0$) from Boman et al;[23] 8) DSPC/Chol/sphingosine ($pH_i = 2.0$) from Boman (unpublished); and 9) SM/Chol/PEG_{2000}-DSPE ($pH_i = 4.0$) from Webb, Masin, Bally and Mayer (unpublished).

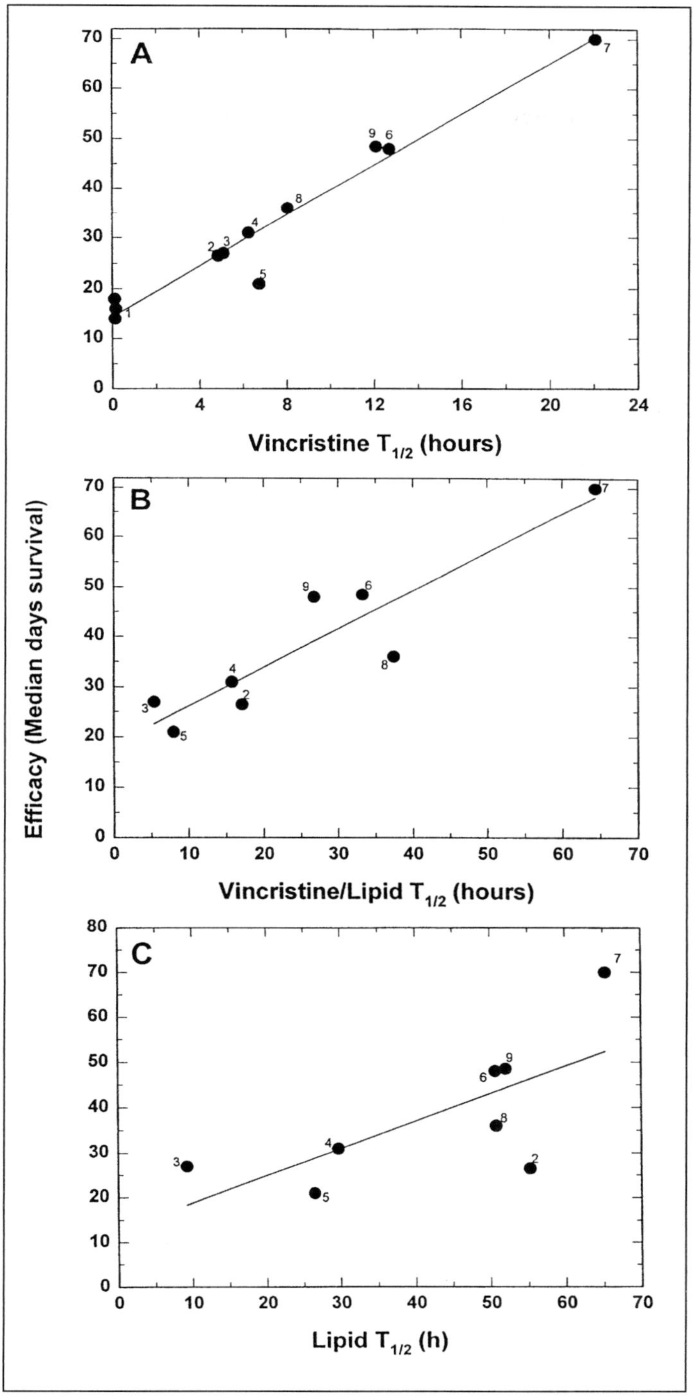
A
Vincristine $T_{1/2}$ (hours)
B
Vincristine/Lipid $T_{1/2}$ (hours)
C
Lipid $T_{1/2}$ (h)
Efficacy (Median days survival)

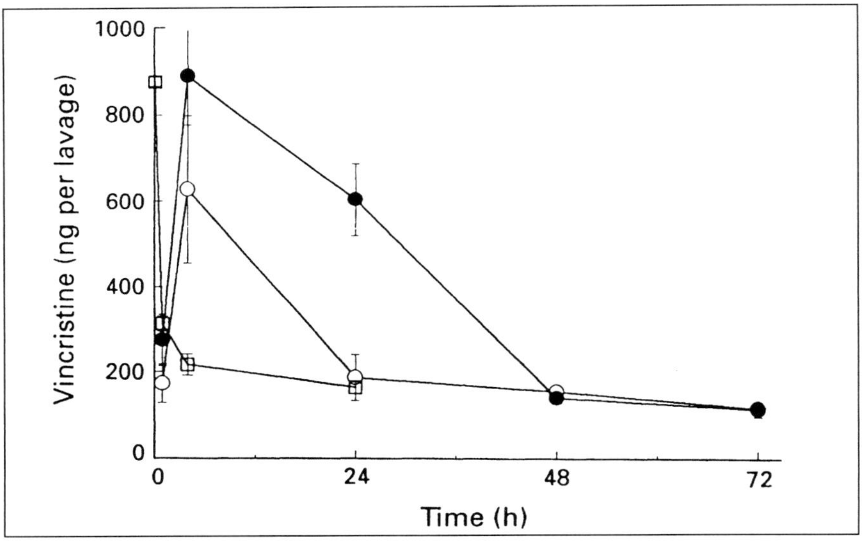

Fig. 3.12. Accumulation of vincristine in the peritoneal cavity of BDF1 mice bearing peritoneal P388 tumor cells after iv administration of free vincristine (❑) or of large unilamellar liposomes of DSPC/Chol (O) or SM/Chol (●) containing vincristine at a drug/lipid ratio of 0.1/1 and a vincristine dosage of 2.0 mg/kg. Data represent means (± s.e.) of four mice; where standard error bars are not visible, they are smaller than the size of the symbol. Graph from Webb et al[24] with permission.

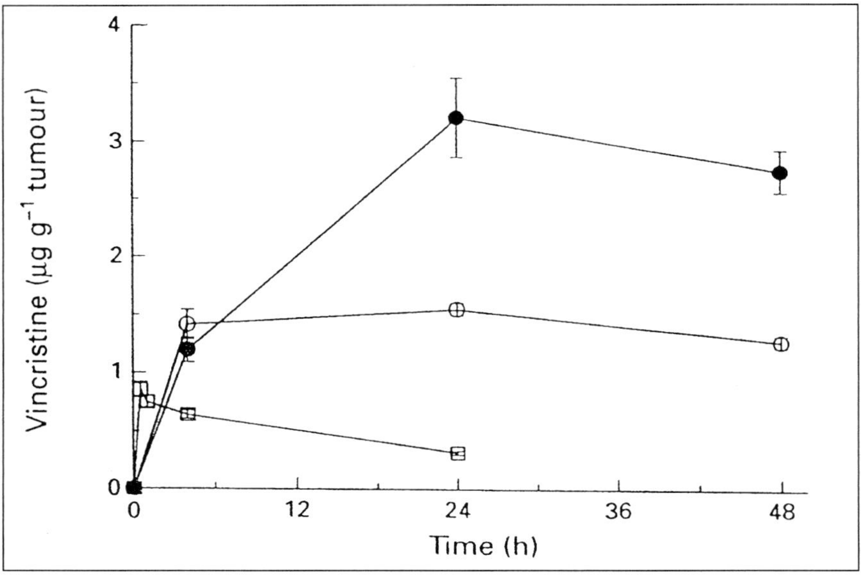

(Fig. 3.10). Biodistribution studies that have measured vincristine levels in the site of ascitic tumor development (Fig. 3.12) and within the solid tumor (Fig. 3.13) have clearly demonstrated that increased drug levels in the plasma are associated with increased drug accumulation at the disease site.

It was anticipated that liposomal vincristine formulations that deliver more vincristine to the tumor for an extended time would also be associated with increased antitumor efficacy. However, it must be emphasized that increased vincristine accumulation in sites of tumor growth occurs as a consequence of liposome accumulation.[26] Further, an analysis of the vincristine/lipid ratio in the tumors of treated animals indicated that liposomes which have extravasated to the tumor site release drug at a rate comparable to that observed within the plasma compartment.[24]

We conclude from these data that successful optimization of the therapeutic activity of liposomal anticancer drugs, through changes in drug release rates, must balance two conflicting attributes. Since the process of extravasation of the liposomes out of the blood compartment to the tumor site is slow, drug release rates must also be sufficiently slow to ensure that maximal quantities of drug are contained within the extravasated liposomes. In contrast, once the liposomes have extravasated the drug must be released at rates sufficient to achieve therapeutic activity.

Alternative Liposomal Formulations of Vincristine

The pharmacokinetics and therapeutic activity of a liposomal vincristine formulation prepared using sterically stabilized PEG-containing liposomes has recently been described.[27] Encapsulation of vincristine into these sterically stabilized liposomes increased the vincristine half-life in the circulation from the initial half-life of 20 min seen for a conventional liposome formulation to approximately 10.5 h (Fig. 3.14).[27] The cause of the extremely rapid clearance of vincristine in the conventional liposome formulation used in this report is unclear and not representative of the clearance seen for the PC/Chol and SM/Chol formulations described by our group.[6,13,24,25] We, however, would suggest that such differences are due to either faster liposome elimination rates or increases in the rate of drug release. Allen's studies, for example, administered liposomes prepared with 5 mol% EPG to rats. It has been shown that PG liposomes bind the complement protein C3bi and induce platelet aggregation mediated through the presence of the C3bi receptor on rat platelets.[28,29] Alternatively, we demonstrated that the presence of anionic lipids facilitates drug release from liposomes that have been prepared using the pH gradient drug loading procedure.[7] It would be of interest to determine whether the differences in plasma drug levels reported by Allen et al were due to liposome clearance or faster drug release rates. Although this study did not compare the efficacy of sterically-stabilized liposomes to conventional liposomes, it did show that the efficacy of vincristine

Fig. 3.13. (Opposite page, bottom) Tumor levels of vincristine after administration of free and liposomal vincristine in SCID mice bearing A431 tumors. SCID mice bearing two A431 tumors were injected iv with free vincristine (❑) or with large unilamellar liposomes of DSPC/Chol (O) or SM/Chol (●) containing vincristine at a drug/lipid ratio of 0.1/1 (w/w). Vincristine was injected at a dose of 2.0 mg/kg, representing a lipid dose of 20 mg/kg. Data represent means (± s.e.) of three mice (six tumors); where standard error bars are not visible, they are smaller than the size of the symbol. Graph from Webb et al[24] with permission.

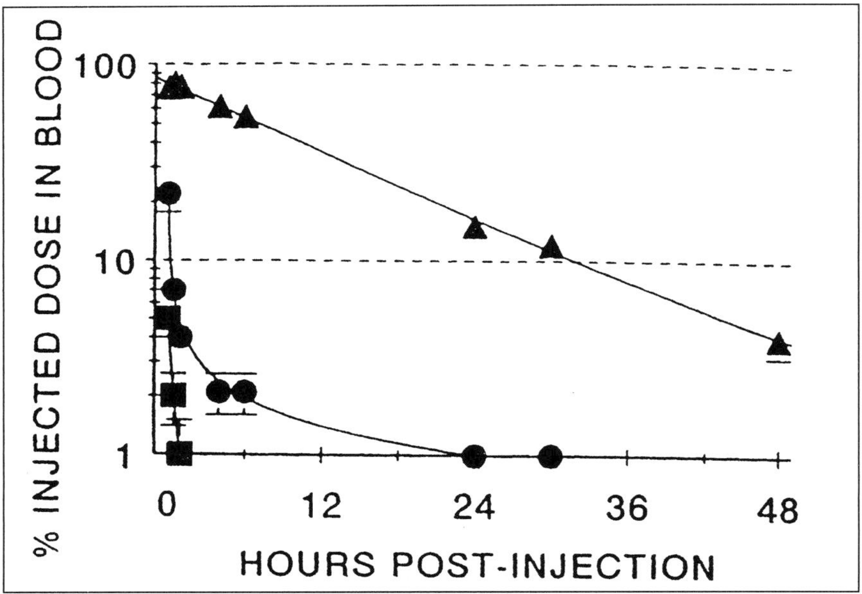

Fig. 3.14. Pharmacokinetics of long circulating liposomal vincristine (▲), conventional liposomal vincristine (●), and free vincristine (■); 0.25 mg vincristine/kg body weight was injected iv into Sprague-Dawley rats. Graph from Allen et al[27] with permission.

encapsulated in PEG-containing liposomes against ip and sc tumors was greater than that of free vincristine.[27] The median survival times for the sterically stabilized liposomes[27] were comparable to the median survival times seen previously with the DSPC/Chol pH_i 4.0 formulation against P388 leukemia and shorter than for the nonsterically stabilized SM/Chol formulation.[24] The PEG-containing formulation showed lower efficacy, in terms of median survival times against L1210 leukemia, compared to that of a DSPC/Chol formulation.[6] The PEG-containing formulation was also shown to decrease growth of the C26 colon carcinoma solid tumor, compared to free drug, and increase the survival time of the mice.[27] Similar results have also been reported by Vaage et al,[9] for sc mouse mammary carcinomas (MC2) treated with free vincristine or vincristine encapsulated in sterically stabilized liposomes.

Another modification which has been shown to produce long circulation times for liposomes is the incorporation of the uronic acid derivative palmitoyl-D-glucuronide (PGlcUA).[30,31] Liposomes composed of DPPC/Chol/PGlcUA (4:4:1 mol%) and containing vincristine were observed to inhibit the growth of the Meth A sarcoma in Balb/c mice to a greater extent than observed for vincristine encapsulated in "control" liposomes.[32] However, in this study PG-containing liposomes were used as a control. As mentioned above, negatively charged liposomes containing PG activate complement in the circulation, rendering them less stable in terms of both drug retention and circulation longevity.[33] In order to clearly determine the factors responsible for tumor efficacy, we suggest that it is necessary to determine the serum clearance rates for the liposomes and the drug/lipid ratios. Our summary results, shown in Figure 3.11, would suggest that regardless of the liposome clearance rate, the plasma concentration must be elevated for extended time periods in order for improved therapeutic activity to be observed. We would argue that this can be achieved using relatively simple liposomal formulations that provide adequate plasma circulation and stability of encapsulation.

Clinical Utility of Liposomal Vincristine

The preclinical toxicity and efficacy data obtained by our research group were sufficient to warrant clinical development of a liposomal vincristine formulation. Comprehensive summaries of these preclinical studies of liposomal vincristine have been published elsewhere.[11,26] It is important to recognize that the beneficial effects observed preclinically which occur as a consequence of vincristine encapsulation within liposomes are not due to a decrease in toxicity but rather to improved antitumor activity. As emphasized in this chapter, the latter is associated with increased vincristine circulation longevity and accumulation of the liposomal drug at the tumor site. The rationale advancing this research effort continues to be based on the fact that vincristine is a cell-cycle specific agent whose activity can be enhanced by prolonged cell exposure. This rationale, supported by the exceptional pharmacokinetic, toxicologic and antitumor properties of liposomal vincristine, led to development of a Phase I Clinical Trial at the British Columbia Cancer Agency. The results from the Phase I toxicity study, briefly summarized elsewhere,[26] were encouraging and prompted initiation of a Phase II efficacy study that is presently ongoing.

Summary

The antitumor activity of vincristine in a variety of preclinical murine tumor models is strongly correlated with increased vincristine circulation longevity and increased vincristine tumor accumulation. Dramatic improvements in vincristine circulation lifetime and vincristine accumulation at tumor sites have been effected by vincristine encapsulation in liposomal carriers. However, vincristine is distinct from antineoplastic agents such as doxorubicin in that it leaks much more readily from liposomal carriers. Consequently, further increases in vincristine circulation longevity, drug accumulation at tumors and antitumor efficacy have been achieved primarily through improvements in vincristine retention within the liposomal carrier. The relationship between circulation lifetime and antitumor efficacy is the direct result of extravasation of the intact liposome, with the encapsulated vincristine, from the circulation to the tumor site. Liposomal carriers with enhanced drug retention characteristics will carry a greater therapeutic payload per extravasated liposome than those with poor drug retention characteristics. We have observed no additional pharmacokinetic or therapeutic benefit in liposomal vincristine formulations that have been rendered "long circulating" by the presence of PEG_{2000}-DSPE. Rather, these preclinical studies with well-characterized "conventional" but still relatively long circulating formulations which exhibit long drug circulation lifetimes (i.e. not necessarily equivalent to long circulation lifetimes) are the foundation supporting an encouraging Phase I/II Clinical Trial of liposomal vincristine.

References

1. Carter SK, Livingston RB. Plant products in cancer chemotherapy. Cancer Treat Rep 1976; 60:1141-1156.
2. Sieber SM, Mead JAR, Adamson RH. Pharmacology of antitumor agents from higher plants. Cancer Treat Rep 1976; 60:1127-1139.
3. Owellen RJ, Owens AH Jr, Donigian DW. The binding of vincristine, vinblastine, and colchicine to tubulin. Biochem Biophys Res Commun 1972; 47:685-691.

4. Owellen RJ, Hartke CA, Dickerson RM, Hains, FO. Inhibition of tubulin-microtubule polymerization by drugs of the *Vinca* alkaloid class. Cancer Res 1976; 36:1499-1502.
5. Jackson DV Jr, Bender RA. Cytotoxic thresholds of vincristine in a murine and a human leukemia cell line in vitro. Cancer Res 1979; 39:4346-4349.
6. Mayer LD, Nayar R, Thies RL, Boman NL, Cullis PR, Bally MB. Identification of vesicle properties which enhance the antitumour activity of liposomal vincristine against murine L1210 leukemia. Cancer Chemother Pharmacol 1993; 33:17-24.
7. Mayer LD, Bally MB, Hope MJ, Cullis PR. Uptake of antineoplastic agents into large unilamellar vesicles in response to a membrane potential. Biochim Biophys Acta 1985; 816:294-302.
8. Madden TD, Harrigan PR, Tai LCL, Bally MB, Mayer LD, Redelmeier TE, Loughrey HC, Tilcock CPS, Reinish LW, Cullis PR. The accumulation of drugs within large unilamellar vesicles exhibiting a proton gradient: a survey. Chem Phys Lipids 1990; 53:37-46.
9. Vaage J, Donovan D, Mayhew E, Uster P, Woodle, M. Therapy of mouse mammary carcinomas with vincristine and doxorubin encapsulated in sterically stabilized liposomes. Int J Cancer 1993; 54:959-964.
10. Harrigan PR, Wong KF, Redelmeier TE, Wheeler JJ, Cullis PR. Accumulation of doxorubicin and other lipophilic amines into large unilamellar vesicles in response to transmembrane pH gradients. Biochim Biophys Acta 1993; 1149:329-338.
11. Boman NL, Bally MB, Cullis PR, Mayer LD, Webb MS. Encapsulation of vincristine in liposomes reduces its toxicity and improves its anti-tumor efficacy. J Liposome Res 1995; 5:523-541.
12. Boman NL, Mayer LD, Cullis PR. Optimization of the retention properties of vincristine in liposomal systems. Biochim Biophys Acta 1993; 1152:253-258.
13. Mayer LD, Bally MB, Loughrey H, Masin D, Cullis PR. Liposomal vincristine preparations which exhibit decreased drug toxicity and increased activity against murine L1210 and P388 tumors. Cancer Res 1990; 50:575-579.
14. Papahadjopoulos D, Allen TM, Gabizon A, Mayhew E, Matthay K, Huang SL, Lee K-D, Woodle MC, Lasic DD, Redemann C, Martin FJ. Sterically stabilized liposomes: improvements in pharmacokinetics and antitumor therapeutic efficacy. Proc Natl Acad Sci USA 1991; 88:11460-11464.
15. Mayhew EG, Lasic D, Babbar S, Martin FJ. Pharmacokinetics and antitumor activity of epirubicin encapsulated in long circulating liposomes incorporating a polyethylene glycol-derivatized phospholipid. Int J Cancer 1992; 51:302-309.
16. Allen TM, Mehra T, Hansen C, Chin Y-C. Stealth liposomes: an improved sustained release system for 1-β-D-arabinofuranosylcytosine. Cancer Res 1992; 52:2431-2439.
17. Bally MB, Nayar R, Masin D, Hope MJ, Cullis PR, Mayer LD. Liposomes with entrapped doxorubicin exhibit extended blood residence times. Biochim Biophys Acta 1990; 1023:133-139.
18. Parr MJ, Bally MB, Cullis PR. The presence of G_{M_1} in liposomes with entrapped doxorubicin does not prevent RES blockade. Biochim Biophys Acta 1993; 1168:249-252.
19. Allen TM, Chonn A. Large unilamellar liposomes with low uptake into the reticuloendothelial system. FEBS Lett 1987; 223:42-46.
20. Allen TM, Hansen C, Rutledge J. Liposomes with prolonged circulation times: factors affecting uptake by reticuloendothelial and other tissues. Biochim Biophys Acta 1989; 981:27-35.
21. Gabizon A, Papahadjopoulos D. Liposome formulations with prolonged circulation time in blood and enhanced uptake by tumors. Proc Natl Acad Sci USA 1988; 85:6949-6953.

22. Liu D, Huang L. pH-sensitive, plasma-stable liposomes with relatively prolonged residence in circulation. Biochim Biophys Acta 1990; 1022:348-354.
23. Boman NL, Masin D, Mayer LD, Cullis PR, Bally MB. Liposomal vincristine which exhibits increased drug retention and increased circulation longevity cures mice bearing P388 tumors. Cancer Res 1994; 54:2830-2833.
24. Webb MS, Harasym TO, Masin D, Bally MB, Mayer LD. Sphingomyelin-cholesterol liposomes significantly enhance the pharmacokinetic and therapeutic properties of vincristine in murine and human tumour models. Br J Cancer 1995; 72:896-904.
25. Mayer LD, Masin D, Nayar R, Boman NL, Bally MB. Pharmacology of liposomal vincristine in mice bearing L1210 ascitic and B16/BL6 solid tumors. Br J Cancer 1995; 71:482-488.
26. Mayer LD, Gelmon K, Cullis PR, Boman N, Webb MS, Embree L, Tolcher T, Bally MB. Preclinical and clinical studies with liposomal vincristine. In: Progress in Drug Delivery Systems IV, Biomedical Research Foundation, Tokyo, 1995;151-161.
27. Allen TM, Newman MS, Woodle MC, Mayhew E, Uster PS. Pharmacokinetics and anti-tumor activity of vincristine encapsulated in sterically stabilized liposomes. Int J Cancer 1995; 62:199-204.
28. Loughrey HC, Bally MB, Reinish LW, Cullis PR. The binding of phosphatidylglycerol liposomes to rat platelets is mediated by complement. Thrombosis & Haemostasis 1990; 64:172-176.
29. Reinish LW, Bally MB, Loughrey HC, Cullis PR. Interactions of liposomes and platelets. Thrombosis & Haemostasis 1988; 60:518-523.
30. Namba Y, Sakakibara T, Masada M, Ito F, Oku N. Glucuronate-modified liposomes with prolonged circulation time. Chem Pharm Bull 1990; 38:1663-1666.
31. Oku N, Namba Y, Okada S. Tumor accumulation of novel RES-avoiding liposomes. Biochim Biophys Acta 1992; 1126:255-260.
32. Tokudome Y, Oku N, Doi K, Namba, Y, Okada S. Antitumor activity of vincristine encapsulated in mice bearing Meth A sarcoma. Biochim Biophys Acta 1996; 1279:70-74.
33. Chonn A, Cullis PR, Devine DV. The role of surface charge in the activation of the classical and alternative pathways of complement by liposomes. J Immunol 1991; 146:4234-4241.

CHAPTER 4

Long Circulating Liposome-Encapsulated Tumor Necrosis Factor in the Treatment of Solid Tumors

Timo L. M. ten Hagen, Alexander H. van der Veen, Alexander M. M. Eggermont

Introduction

The first observations which led to discovery of tumor necrosis factor alpha (TNF-α) was the spontaneous regression of tumors in patients with a bacterial infection. A serum factor, induced by bacterial products, was found to be responsible for the antitumor activity. After isolation and purification it was found that TNF-α also exerts regulatory effects on various cell types, e.g., stimulation of cell growth and hemopoiesis. However, TNF-α has also cytostatic and cytotoxic effects specifically on tumor cell lines in vitro, indicating a possible usefulness for antitumor therapy. However, in vivo administration of TNF-α is accompanied by severe toxicity. At relatively low doses a dramatic drop in blood pressure follows administration of TNF-α, and a septic shock-like syndrome may develop which can be fatal. Upon administration of nonlethal dosages weight loss was noticed, which eventually demonstrated that TNF-α and cachectin are the same molecule.

The severe toxicity of TNF-α in vivo resulted in diminished enthusiasm concerning its usefulness. However, local application of TNF-α, like intratumoral injections, or circumventing systemic release, as is achieved in an isolated limb perfusion (ILP), showed promising results. Efficacy of TNF-α application in combination with cytostatic agents in an ILP setting for treatment of melanomas or sarcomas has now been well established. These trials demonstrated synergy between TNF-α and melphalan, indicating that TNF-α is very useful if systemic exposure can be minimized, local concentrations in tumor increased and the toxicity problem solved. In this chapter we will discuss possible ways of reducing (systemic) toxicity of TNF-α and increasing concentration of TNF-α and cytotoxic agents in tumors, resulting in effective treatment. One way of doing this is by setting of an ILP, which will be discussed in more detail below. Another way, which might have broader application than the isolated perfusion setting, is the use of liposomes, which will be discussed subsequently. Liposomes might be useful in reducing toxicity of TNF-α through encapsulation of the agent, while preserving its antitumor activity. Conceivably,

Long Circulating Liposomes: Old Drugs, New Therapeutics, edited by Martin C. Woodle and Gerrit Storm.

liposomal application of TNF-α will also be possible for tumors located at sites at which isolated perfusion is either impossible or may render severe morbidity.

Due to positive results obtained with ILP in humans with solid tumors of the extremities, rat models which closely resemble the clinical setting were developed in our laboratory. A soft tissue sarcoma model, using a rapidly growing, metastasizing, nonimmunogenic, grade III sarcoma (BN 175),[1] and a tumor model with an osteosarcoma which originated spontaneously in the tibia of a rat (ROS-1) were used.[2] Important advantages of these sarcomas are evident from the nature of the tumors. The tumors are nonimmunogenic, as is the case in human sarcomas and most human solid tumors. This means that the host defense system in tumor-bearing animals is intact and interactions between therapy and immune system can be studied. In contrast, immunogenic tumors can only be implanted in immunosuppressed rats, as is the case with xenografted tumors, making studies on interactions between host immune system and tumor treatment impossible. Moreover, although in our laboratory treatment of nonimmunogenic tumor in rats resulted in complete regression, regrowth of tumors has been observed in some cases in a later stage. This has also been seen in some humans after ILP. One might have serious doubts whether this is true for immunogenic tumors, as the weakened tumor cells after therapy are an easy target for the remaining host defense of the experimental animal. The rat tumor models are used to study antitumor effects of TNF-α in combination with cytotoxic agents when applied locally or systemically after liposomal encapsulation.

Experimental and Clinical Findings with TNF-α in Regional Treatment of Solid Tumors

Characteristics of Solid Tumors

Effective treatment of solid tumors is complicated by factors due to the nature of the tumor and by side-effects which accompany treatment. For antitumor agents to be effective the agents must be effective in the in vivo micro-environment of solid tumors, and the agents must be able to reach tumor cells in vivo in concentrations high enough to have an effect. This must all take place with minimal toxicity on healthy tissues. The penetration of solid tumors and the heterogeneity with which this takes place is very much influenced by the vascularization of the tumor. It is likely that high drug levels can be reached especially in well vascularized regions, whereas drug levels will be minimal in poorly perfused regions.[3] Tumors also present an abnormally high interstitial pressure, and especially inhibit passage of large molecules and particles such as liposomes.[3] Poor delivery of therapeutic agents to tumors is also influenced by an abundant matrix in tumors. One might speculate therefore that the resistance of solid tumors to the penetration of drugs is responsible for treatment failure.

However, solid tumors also exhibit characteristics which benefit targeting and efficacy of antitumor therapy. The typical characteristics of tumor associated vasculature (TAV) allows treatment which is targeted towards the endothelium of solid tumors. TNF-α appears to strongly influence TAV, explaining why TNF-α has its major effect on larger tumors with well developed vasculature.[4,5] Another important factor which might benefit tumor treatment is the increased vascular permeability of tumors as compared with various normal tissues.[6] Tumor growth is accompanied by a process called angiogenesis, resulting in new vessels which are leaky and allow large particles to extravasate. On the other hand, tumors are unable to form their own lymphatic system, resulting in poor drainage of the tumor.[7] High permeability of the TAV allows particles such as liposomes to extravasate. This increased possibility of particles and agents to leave the bloodstream, and the poorly developed drainage of tumors results in preferential localization of liposomes in tumors; this is further discussed below.

Tumor Necrosis Factor α (TNF-α)

Already in 1891 Coley observed spontaneous regression of tumors in patients undergoing a bacterial infection.[8] It was found that a serum factor which was named tumor necrosis factor alpha caused hemorrhagic necrosis (reviewed in ref. 9). This cytokine is synthesized and secreted mainly by activated monocytes and macrophages and exists in soluble form as a homotrimer of 52 kDa (human). It interacts with two receptors, of which type I is held responsible for cytotoxic effects of TNF-α on tumor cell lines, whereas type II is believed to be involved in cell proliferation.

TNF-α is a mediator of inflammatory responses and was shown to regulate growth and differentiation of a wide variety of cell types. On the other hand, TNF-α is selectively cytotoxic for many transformed cells, especially in combination with IFN-γ. It is believed that effects on endothelium especially (i.e., upregulation of adhesion molecules, increasing microvasculature permeability, changing endothelial morphology and conversion of endothelial cells into a procoagulant state) play a role in antitumor activity of TNF-α in vivo, as discussed below.

Although TNF-α has profound cytotoxic activity to some tumor cells in vitro, and is shown to have antitumor activity in vivo, the activity of TNF-α is by no means tumor specific. Some tumor cells lack sensitivity for TNF-α whereas non-neoplastic cells can be sensitive for TNF-α. Systemic administration of TNF-α is associated with severe toxicity such as rigors, fever, fatigue, blood pressure changes, and WBC count decrease, which is dose limiting. These side effects are responsible for the failure of systemic treatment with TNF-α at a maximal tolerated dose, which is only $^1/_{50}$ of doses that cause tumor regression in animal models.[10] This illustrates a need for modes of administration which circumvent the toxic side effects. Possible ways of doing this are by applying the setting of an ILP, or encapsulation of the agent in liposomes.

Clinical and Experimental Findings

As discussed above, TNF-α has been found to be very toxic when administered systemically.[11] In phase I/II studies the maximal tolerated dose was reported to vary between 200 and 400 μg/m2, depending on treatment schedule. Due to limitation of TNF-α dose phase II trials revealed disappointing results with complete remissions observed in only about 1%.[12-15] Therefore, clinical use of TNF-α was thought to be restricted to locoregional applications such as intratumoral or intracavitary administration, which allows high local concentrations.

The first consistent antitumor results with TNF-α were obtained after ILP in man when used in combination with IFN-γ and melphalan.[16-21] A major advantage of ILP is that it permits high regional concentrations: 15-20 times higher than those reached after systemic administration.[22] Lejeune and Lienard at the Jules Bordet Institute in Brussels pioneered the use of TNF-α in ILP.[16] Because of the new antitumor mechanism of TNF-α, which was shown to be synergistic with melphalan, interest in ILP with this cytokine grew. ILP with TNF-α in combination with IFN-γ and melphalan in 122 patients with multiple in transit metastases (melanoma stage IIIA/AB) resulted in a tumor response rate of 96%.[17,23] TNF-α also demonstrated impressive antitumor activity in patients when given at high dose in combination with melphalan with IFN-γ[17] or without IFN-γ[20] in ILP in 125 soft tissue sarcoma patients with large extremity tumors. The overall response rate of >80% resulted in a limb salvage percentage of >80%. Similar results have been obtained in 16 types of sarcomas as well as melanomas and a range of rare tumors in extremities (unpublished observations).[19] These trials clearly demonstrate the usefulness of TNF-α in antitumor treatment.

The outcome of the clinical trials led to development of experimental rat models to evaluate ILP and systemic solid tumor treatment in greater detail. ILP with TNF-α and melphalan resulted in 100% response both in the rat osteosarcoma and soft tissue sarcoma model.[24,25] The combination of TNF-α and melphalan showed synergistic antitumor activity in rats, as similarly observed in humans. However, ILP with TNF-α alone or melphalan alone resulted in progressive disease or no change in all rats, confirming the necessity of combination. However, in vitro no synergy in cytotoxicity on the fibrosarcoma or the osteosarcoma cell line was found between melphalan and TNF-α.[24,25] TNF-α induced destruction of TAV, as well as induction of hemorrhagic necrosis and coagulative ischemic necrosis seemed to be most important. Electron microscopy revealed platelet aggregation and hemorrhagic necrosis, resulting from TNF-α induced TAV damage[26] as was also observed in melanoma patients treated by ILP with TNF-α, IFN-γ and melphalan.[27]

As described above, isolated perfusion of the limb has many advantages. A likely development therefore is the application of isolated perfusion for other parts of the body or organs (e.g., lung,[28,29] kidney or liver[30-32]). These developments are still in their infancy, however, and are much more complicated than previously expected. Therefore, other methods have to be applied to treat solid tumors which are located on less accessible sites. One such method is the application of liposomes for encapsulation of agents for systemic treatment.

Application of Liposomes in Solid Tumor Treatment

As stated previously, systemic administration of TNF-α has failed due to severe toxicity. However, tumors located in extremities are quite successfully treated with TNF-α in the setting of an ILP, when used in combination with melphalan. Treatment of systemic metastatic disease on the other hand is more complicated and requires systemic application of agents. ILP-like concentrations of drugs may be achieved after systemic administration by targeting these agents to tumors using long circulating liposomes (e.g., Stealth® or sterically stabilized liposomes) as drug carriers.

The first liposomes (so-called conventional or classical liposomes) had a natural fate to mainly end up in cells of the mononuclear phagocyte system (MPS), depending on size, lipid composition (charge etc.), and route of injection. Decreasing liposome size (e.g., to 100 nm) and use of neutral lipids diminishes uptake by MPS cells. However, with the introduction of so-called long circulating (or Stealth®) liposomes a dramatic reduction in MPS uptake has been reached, as well as an improved localization of these liposomes in tumor-tissue.[33,34] Stealth® liposomes are coated with polyethylene glycol (PEG) which bind water which decreases binding of opsonins, resulting in diminished recognition by MPS cells. Localization of Stealth® liposomes in tumors is related to two factors:

1) An increased permeability of the TAV arising during the process of angiogenesis.
2) An enhanced capability of Stealth® liposomes to travel unhindered through and across the vessels as compared with conventional liposomes, enabling easier transit across the microvasculature and interstitial space.

Exposure of tumors to TNF-α and cytotoxic agents is only short during ILP (30 min in rat and 90 min in human). This might explain why especially high dose TNF-α is effective. Liposomal targeting might not always result in high levels seen in ILP. However, a major advantage of liposomes is accumulation of liposomal TNF-α forming a depot in the tumor from which TNF-α is slowly released, resulting in prolonged significant levels of TNF-α in the tumor. The proposed mechanism of delivery of antitumor agents by liposomes to tumors might therefore be an increased extravasation of liposomes at the tumor site due to leakiness of vessels, resulting in accumulation of liposomal drug, followed by a gradual release from liposomes.

Several different anthracyclines and other cytotoxic agents have been examined in tumor models in liposome-encapsulated form (e.g., doxorubicin, daunorubicin, epirubicin, annamycin vincristine taxanes and cisplatin).[35] For instance, cisplatin has a broad spectrum of activities on tumors which is accompanied by many toxicities (mainly nephrotoxicity and neurotoxicity). Liposome-encapsulation results in improved efficacy and reduced toxicity in vivo in treatment of liver and spleen metastases. Phase I and II trials consistently demonstrate their safety. The most extensively studied cytotoxic agents encapsulated in liposomes are doxorubicin and cisplatin. The encapsulation of doxorubicin has been shown to reduce cardiotoxicity and myelo-suppression in vivo, and reduces damage to other (nonmalignant) tissues,[36,37] and to reverse multi-drug resistance (MDR).[38]

We demonstrated that TNF-α can be encapsulated in Stealth® liposomes with good encapsulation efficiency.[39] Other cytokines such as IFN-γ, IL-1 and IL-2 may also be encapsulated in liposomes and when applied in vivo have increased half-life and antitumor effect.[40,41] Studies in rats showed reduced toxicity of liposome-encapsulated TNF-α as compared with TNF-α in the free form, which was confirmed by us.[42,43] Whereas multiple administrations of free TNF-α alone already resulted in a weight loss of 40%, multiple administrations of liposome-encapsulated TNF-α (with or without DOXIL®) had no significant effect on rat weights (Table 4.1). Data on liposomal TNF-α in vivo in cancer therapy are limiting. Antitumor effect of coencapsulated IL-1 and TNF-α was studied in melanoma bearing mice.[41] In these studies, however, the cytokines were encapsulated in large liposomes which are primarily taken up by the macrophages.

As was seen with melphalan, ILP studies in the rat using doxorubicin in combination with TNF-α also resulted in complete regression of tumor although the combination of TNF-α with doxorubicin seemed less potent. Due to clinical availability of doxorubicin encapsulated in long circulating liposomes (Stealth® liposomal doxorubicin, registered as DOXIL® in the US, SEQUUS Pharmaceuticals, Menlo Park, CA), we started with systemic treatment of solid tumors with combination of DOXIL® and liposome-encapsulated TNF-α (TNF-SL) in the rat extremity soft tissue sarcoma model. This model is very useful, as a positive control is incorporated in the form of ILP. Drug levels which can be reached with ILP are very high. Therefore ILP is used for the screening of antitumor activity of agents, as one may expect little chance of effective systemic treatment when already, ILP with a certain agent or combination of agents is without effect on the tumor.

The TNF-SL liposomes showed good long circulating characteristics with a blood residence half-time of approximately 22 h. In our rat soft tissue sarcoma model a localization of Stealth® liposomes in tumors was observed, reaching around 8% of the injected dose at 24 h after injection 45 (Table 4.2). Seventy-two hours after injection of liposomes up to 15% of the injected dose could still be detected in circulation, and a tumor localization of 4% at that time point. With the combination of Stealth® liposome-encapsulated TNF-α and DOXIL® for the first time we observed a significant antitumor effect in our model after systemic treatment.[44,45] Preliminary results in rat show that tumor growth can be inhibited for several weeks when TNF-SL and DOXIL® are combined, while each treatment alone failed to induce any tumor control (Table 4.3). These results clearly show that effective systemic treatment of solid tumors with TNF-α is possible, when the agent is encapsulated in liposomes. Secondly, as was shown in the ILP with melphalan and with doxorubicin, the combination of TNF-α with a cytotoxic agent is mandatory for systemic treatment.

Table 4.1. Weight loss of rats treated with multiple intravenous injections of free or liposome-encapsulated TNF-α in combination with doxorubicin or DOXIL

Treatment	Percentage of Weight at End of Treatment[a]
TNF-SL[b]	104
TNF-SL + DOXIL	90
TNF-α free + DOXIL	61
TNF-SL + doxorubicin free	95

a Percentage was calculated by comparing rat weight before treatment with rat weight at end of treatment.
b TNF-α encapsulated in long circulating (Stealth®) liposomes.

Table 4.2. Biodistribution of long circulating (Stealth®) liposomes in soft tissue sarcoma-bearing rats, calculated as percentage of injected dose

	Time (h) After Intravenous Injection					
Organ/tissue	0	4	12	24	48	72
Blood[a]	~100	82	66	43	28	13
Liver[b]	5	6	8	8	9	14
Spleen	1	4	6	8	9	12
Tumor	1	5	9	7	5	3

a Percentage of injected dose per calculated blood volume
b Percentage of injected dose per whole organ or tissue
Each group had a minimum of three rats.

Table 4.3. Tumor response in rats to systemic treatment with free or liposome-encapsulated TNF-α and doxorubicin

Treatment	Response Rate[a]	Response Type[b]
TNF-α free	0	PD
TNF-SL[c]	0	PD
Doxorubicin free	0	PD
DOXIL	0	PD
TNF free + DOXIL	0	PD
TNF-SL + DOXIL	100%	SD

a Percentage of rats responding to treatment
b PD progressive disease, SD stable disease (> 14 days)
c TNF-α encapsulated in long circulating (Stealth®) liposomes
A minimum of three rats were treated per group.

Concluding Remarks

Studies with TNF-α in locoregional treatment, particularly in isolated perfusion setting, clearly demonstrate the usefulness of this cytokine for treatment of solid tumors. However, reduction of systemic toxicity and an increase in local concentration—essential to effective treatment—could not be achieved following systemic administration of the cytokine in the free form. However, liposomal encapsulation of TNF-α reduces toxicity and results in increased concentrations at the tumor site. Moreover, we have shown that inhibition of tumor growth can be reached after systemic administration of liposome-encapsulated TNF-α in combination with DOX-SL®. These agents applied in the free form were without effect on the tumor. We conclude therefore that TNF-α is very much alive as a drug in regional perfusion and may be administered systemically in the future in a liposomal setting for the treatment of solid tumors.

References

1. Kort WJ, Zondervan PE, Hulsman LOM, van Schalkwijk WP, Wijma IM, Westbroek DL. Spontaneous tumor incidence in female Brown Norway rats after lifelong diets high and low in linoleic acid. J Natl Cancer Inst 1985; 74:529-436.
2. Barendsen GW, Janse HC. Differences in effectiveness of combined treatments with ionizing radiation and vinblastine, evaluated for experimental sarcomas and squamous cell carcinomas in rats. Int J Radiat Oncol Biol Phys 1978; 4:95-102.
3. Jain RK. Barriers to drug delivery in solid tumors. Scientific American 1994; 271:58-65.
4. Mule JJ, Asher A, McIntosh J et al. Antitumor effect of recombinant tumor necrosis factor-a against murine sarcomas at visceral sites: tumor size influences the response to therapy. Cancer Immunol Immunother 1988; 26:202-208.
5. Weidner N, Folkman J. Tumor vascularity as a prognostic factor in cancer In: VT DeVita, S Hellman, SA Rosenberg, eds. Important Advances in Oncology. Philadelphia: Lippincott-Raven Publishers, 1996:167-190.
6. Dvorak HF, Brown LF, Detmat M, Dvorak AM. Vascular permeability factor/vascular endothelial growth factor, microvascular hyperpermeability, and angiogenesis. Am J Pathol 1995; 146:1029-1039.

7. Gullino PM. Techniques in tumor pathophysiology In: H Busch, ed. Methods in Cancer Research. New York: Academic Press, 1970:45-92.
8. Coley WB. Contribution to the knowledge of sarcoma. Ann Surg 1891; 14:199-200.
9. Sidhu RS, Bollon AP. Tumor necrosis factor activities and cancer therapy—A perspective. Pharmac Ther 1993; 57:79-128.
10. Asher A, Mule JJ, Reichert CM, Shiloni E, Rosenberg SA. Studies on the anti-tumor effciacy of systemically administered recombinant tumor necrosis factor against several murine tumors in vivo. J Immunol 1987; 138:963-974.
11. Spriggs DR, Sherman ML, Michie H et al. Recombinant human tumor necrosis factor administered as a 24 h intravenous infusion. A phase I and pharmacologic study. J Natl Cancer Inst 1988; 80:1039-1044.
12. Feldman ER, Creagan ET, Schaid DJ, Ahmann DL. Phase II trial of recombinant tumor necrosis factor in disseminated malignant melanoma Am J Clin Oncol 1992; 15:256-259.
13. Rinehart J, Balcerzak SP, Hersh E. Phase II trial of tumor necrosis factor in human sarcoma: A Southwest Oncology Group Study. Proc Annu Meet Am Soc Clin Oncol 1990; 9:A1229.
14. Bartsch HH, Pfizenmaier K, Schroeder M, Nagel GA. Intralesional application of recombinant tumor necrosis factor alpha induces local tumor regression in patients with advanced malignancies. Eur J Cancer Clin Oncol 1989; 25:287-291.
15. Kahn JO, Kaplan LD, Volberding PA et al. Intralesional recombinant tumor necrosis factor-alpha for AIDS-associated Kaposi's sarcoma: a randomized, double-blind trial. J Acquir Immune Defic Syndr 1989; 2:217-223.
16. Liénard D, Ewalenko, Delmotte JJ, Renard N, Lejeune FJ. High-dose recombinant tumor necrosis factor alpha in combination with interferon gamma and melphalan in isolation perfusion of the limbs for melanoma and sarcoma. J Clin Oncol 1992; 10:50-62.
17. Eggermont AMM, Schraffordt Koops H, Lienard D, Kroon BBR, van Geel AN, Hoekstra HJ, Lejeune FJ. Isolated limb perfusion with high dose Tumor Necrosis Factor-a in combination with interferon-g and Melphalan for irresectalble extremity soft tissue sarcomas: a multicenter trial. J Clin Oncol 1996; 14:2553-2565.
18. Lejeune FJ, Liénard D, Leyvraz S, Mirimanoff RO. Regional therapy of melanoma. European Journal of Cancer 1993; 29A:606-612.
19. Eggermont AMM. Treatment of melanoma intransit metastases confined to the limb. Cancer Surveys 1996; 26:335-349.
20. Eggermont AMM, Schraffordt Koops H, Klausner JM, Schlag PM, Kroon BBR, Ben-Ari G, Lejeune FJ. High dose TNFα in isolation perfusion of the limb: highly effective treatment for melanoma in transit metastases and unresectable sarcomas. Ann Surgery 1996; 224:756-765.
21. Creech OJ, Krementz ET, Ryan RF, Winblad JN. Chemotherapy of cancer: regional perfusion utilizing an extracorporeal circuit Ann Surg 1958; 148:612-632.
22. Benckhuijsen C, Kroon BBR, Van Geel AN, Wieberdink J. Regional perfusion treatment with melphalan for melanoma in a limb: an evaluation of drug kinetics. Eur J Surg Oncol 1988; 14:157-163.
23. Liénard D, Eggermont AMM, Schraffordt Koops H, Kroon BBR, Lejeune FJ. Results of a randomised phase II trial on the use of TNF + Melphalan vs TNF + Melphalan + IFNγ in patients with melanoma in transit metastases confined to the limb. Eur Cytokine Netw 1996; 7:299.
24. Manusama ER, Stavast J, Durante NMC, Marquet RL, Eggermont AMM. Isolated limb perfusion with TNFα and melphalan in a rat osteosarcoma model: A new antitumour approach. Eur J Surg Oncol 1996; 22:152-157.

25. Manusama ER, Nooijen PTGA, Stavast J, Durante NMC, Marquet RL, Eggermont AMM. Synergistic antimour effect of recombinant human tumour necrosis factor α with melphalan in isolated limb perfusion in the rat. Br J Surg 1996; 83:551-555.
26. Nooijen PTGA, Manusama ER, Eggermont AMM, Schalkwijk L, Stavast J, Marquet RL, De Waal RMW, Ruiter DJ. Synergistic effects of TNF-α and melphalan in an isolated limb perfusion model of rat sarcoma A histopathological, immunohistochemical and electronmicroscopical study. Bri J Cancer 1996; 74:1908-1915.
27. Renard N, Nooijen PTGA, Schalkwijk L et al. VWF release and platelet aggregation in human melanoma after perfusion with TNFα. J Pathol 1995; 176:279-287.
28. Weksler B, Blumberg D, Lenert JT, Ng B, Fong Y, Burt ME. Isolated single-lung perfusion with TNF-α in a rat sarcoma lung metastases model Ann Thorac Surg 1994; 58:328-332.
29. Prograbniak HW, Witt CJ, Terrill R et al. Isolated lung perfusion with tumor necrosis factor: a swine model in preparation of human trials. Ann Thorac Surg 1994; 57:1477-83.
30. van der Veen AH, Manusama ER, van Kampen CA, Durante NMC, Marquet RL, Eggermont AMM. Tumor Necrosis Factor-α (TNFα) in isolated kidney perfusions in rats: toxicity and antitumor effects. Eur J Surg Oncol 1994; 20:404-405.
31. de Vries MR, Borel Rinkes IHM, Wiggers T, Jonker AM, Weijers-Hille MJ, Eggermont AMM. Hyperthermic isolated hepatic perfusion with TNFα with and without melphalan in pigs. Eur J Surg Res 1995; 27(S1):81.
32. Fraker DL, Alexander HR, Thom AK. Use of tumor necrosis factor in isolated hepatic perfusion. Circulatory Shock 1994; 44:45-50.
33. Huang SK, Lee K-D, Hong K, Friend DS, Papahadjopoulos D. Microscopic localization of sterically stabilized liposomes in colon carcinoma-bearing mice. Cancer Res 1992; 52:5135-5143.
34. Klibanov AL, Murayama K, Torchilin VP, Huang L. Amphipathic polyethyleneglycols effectively prolong the circulation time of liposomes. FEBS 1990; 268:235-237.
35. Kim S. Liposomes as carriers of cancer chemotherapy: Current status future prospects. Drug 1993; 46:618-638.
36. Gabizon A, Peretz T, Sulkes A, Ben-Yosef R, Be-Baruch N, Carane R, Biran S, Barenholtz Y. Systemic administration of doxorubicin-containing liposomes in cancer patients: A phase I study. Eur J Cancer Clin Oncol 1989; 25:1795-1803.
37. Hengge UR, Brockmeyer NH, Baumann M, Reimann G, Goos M. Liposomal doxorubicin in AIDS-related Kaposi's sarcoma. Lancet 1993:342:497.
38. Oudard S, Theirry A, Jorgensen TJ, Rahman A. Sensitization of multidrug-resistant colon cancer cells to doxorubicin encaspulated in liposomes. Cancer Chemother Pharmacol 1991; 28:259-265.
39. ten Hagen TLM, van der Veen AH, Marquet RL, Eggermont AMM. Biodistributionof TNF-α encapsulated in long circulating liposomes in the BN 175 extremity soft tissue sarcoma model in Brown Norway rats Symposium. TNF and Related Cytokines: Clinical Utility and Biology of Action and Advances in Congestive Heart Failure, Research and Therapeutics March 10-16, 1996, Hilton Head, South Carolina.
40. Kedar E, Rutkowski Y, Braun E, Emmanuel N, Barenholz Y. Delivery of cytokines by liposomes I. Preparation and characterization of interleukin-2 encapsulated in long circulating sterically stabilized liposomes. J Immunotherapy 1994; 16:47-59.
41. Saito M, Fan D, Lachman LB. Antitumor effects of liposomal IL1α and TNFα against the pulmonary metastases of the B16F10 murine melanoma in syngeneic mice. Clin Exp Metastasis 1995; 13:249-259.
42. Debs RJ, Fuchs HJ, Philip R, Brunette EN, Duzgunes N, Shellito JE, Liggitt D, Patton JR. Immunomodulatory and toxic effects of free and liposome-encapsulated tumor necrosis factor α in rats. Cancer Res 1990; 50:375-380.

43. ten Hagen TLM, van der Veen AH, van Tiel ST, Seynhaeve ALB, Eggermont AMM. Treatment of soft tissue sarcoma of the limb with liposome-encapsulated tumor necrosis factor a in the rat. Biological Therapy of Cancer XI, October 23-26, 1996, Washington DC.
44. van der Veen AH, ten Hagen TLM, Marquet RL, Eggermont AMM. Treatment of solid limb tumors with liposome-encapsulated Tumor Necrosis Factor α in the rat. Proc Am Assoc Cancer Res 1996; 37:483.
45. Eggermont AMM, Manusama ER, ten Hagen TLM. Regional application of TNFα in the treatment of cancer: A preclinical-clinical interactive program. J Inflammation 1996; 47:104-113.

CHAPTER 5

Therapy of Human Carcinoma Xenografts with Doxorubicin Encapsulated in Sterically Stabilized Liposomes (DOXIL®): Efficacy and Safety Studies

Jan Vaage, Paul S. Uster, Peter K. Working

Studies in animal tumor models have found that the therapeutic effects of anticancer drugs can be enhanced and prolonged, and their toxic side effects reduced, when the drugs are encapsulated in liposomes.[1,2] The effectiveness of drugs in conventional liposomes is limited, however, by their rapid uptake by the cells of the mononuclear phagocytic system (MPS), reducing the amount of drug that reaches the tumor.[3] By the covalent attachment of polyethylene glycol (PEG) to the lipid bilayers of the liposomes, the uptake of liposomes by the MPS is reduced, and the circulation time is increased.[4,5] Such liposomes are often referred to as sterically stabilized in analogy to stabilization of organic colloid particles, in which steric modifications of the particle surface result in reductions in particle-to-particle interactions that could lead to aggregation or fusion. In colloids, steric stabilization is accomplished by addition of surface charge or via coating of the surface with various molecules such as starch or PEG.[6] In the case of liposomes in biological environments, steric stabilization not only reduces particle-to-particle interactions, but also decreases adsorption of various macromolecules onto the liposome surface, loss of liposomal components to other particles, and interactions of the liposomes with cells, all of which provide greater liposome stability compared to conventional, nonpegylated liposomes.[7,8] This liposome property has been given the registered trade name "STEALTH".

DOXIL® (Doxorubicin HCl liposome injection) is a formulation of doxorubicin HCl encapsulated in long circulating liposomes containing methoxypolyethylene glycol (MPEG). Sterically stabilized liposomes exhibit prolonged circulation and altered tissue distribution relative to conventional liposomes after intravenous injection.[4,5,9] Owing to stable encapsulation, the plasma pharmacokinetics and tissue distribution of encapsulated material is characteristic of the liposome, not the internalized drug, i.e., the behavior of sterically stabilized liposomes is similar regardless of their encapsulated contents.[8] As a result, the plasma pharmacokinetics

Long Circulating Liposomes: Old Drugs, New Therapeutics, edited by Martin C. Woodle and Gerrit Storm.

and tissue distribution of encapsulated drugs are frequently quite different than those of the free, nonliposomal drug, with attendant changes in their safety and therapeutic effectiveness profiles.[10] When encapsulated in sterically stabilized liposomes, the therapeutic effect of doxorubicin against several mouse mammary carcinomas[11] and xenografted human ovarian,[12] and prostatic,[13] and panceatic[14] carcinomas was increased and its toxicity reduced in comparison with doxorubicin either encapsulated in conventional, nonpegylated liposomes[11] or free in saline.[11-13] Several xenograft studies are described below, together with data on the increased localization of doxorubicin in tumors, quantified with fluorescent confocal laser microscopy. The toxicity of DOXIL is also summarized, and a detailed description of a comparative cardiotoxicity study in rabbits is provided to illustrate the effect of encapsulation of doxorubicin in sterically stabilized liposomes on its safety profile.

Studies in Human Xenograft Models

The therapeutic effectiveness of free doxorubicin hydrochloride in saline (Adriamycin RDF™) and encapsulated in sterically stabilized STEALTH liposomes (DOXIL) was compared in three human xenograft carcinomas: The human ovarian carcinoma HEY (London, Ontario, Regional Cancer Center collection of human tumors), the pancreatic adenocarcinoma AsPC-1 (cell repository line CRL 1682 from the ATCC collection of human tumors), and the prostatic carcinoma PC-3 (cell repository line CRL 1435 from the ATCC). In all three studies, the tumors were implanted subcutaneously (sc) in mature female Swiss nude mice; the ovarian carcinoma was also implanted intraperitoneally (ip). For sc implants, tumor tissue from donor mice was cut into 1 mm^3 pieces and rinsed in cold culture medium prior to the sc implantation of 2 randomly selected pieces in the right and left posterior flanks. For ip implantation, a single-cell suspension was prepared by mechanical disassociation of tumor tissue in cold culture medium containing 10% fetal calf serum. The cells were washed once, and approximately 10^5 cells were injected.

In each study tumor-bearing mice were randomly assigned to therapy and control groups. Body weights, incidence of tumor growth, and tumor size were recorded weekly. Mice were euthanized when one or both sc tumors had grown to a size beyond possible regression (mean 0.35 cm^3), before showing signs of discomfort. Mice with ip tumor implants were euthanized and necropsied when palpation determined progressive tumor growth. Tumor-free mice were necropsied at the termination of each experiment, and the absence of live tumor cells at the implantation site was confirmed by histological examination. Differences in incidence of tumors were evaluated with the 2 x 2 contingency test (Fisher's exact test). Differences in mean tumor volume and mean survival time were evaluated by Student's t test. Tumor volumes were calculated by the formula $0.4(ab^2)$ where "a" is the larger and "b" the smaller diameter.

DOXIL was provided by SEQUUS Pharmaceuticals, Inc., Menlo Park, California. Adriamycin was from Farmitalia Carlo Erba, Milan, Italy.

Ovarian Carcinoma

This study compared the therapeutic effects of Adriamycin and DOXIL against the human ovarian carcinoma HEY.[12] The tumor has an average in vivo doubling time of 12 d (from 50 mm^3 to 100 mm^3 in 12 d) and had in this study an average 90% probability of growth in untreated nude mice. Treatments at a dose level of 6.0 mg/kg doxorubicin were given 1, 8 and 15 d after tumor implantation; injections were iv or ip to treat sc or ip implanted tumors, respectively.

Subcutaneous Implants

DOXIL inhibited tumor growth more effectively than Adriamycin, Day 42, $p < 0.001$; Fig. 5.1), and a significantly greater number of DOXIL-treated mice were tumor-free at necropsy (15/18 versus 5/18; $p = 0.002$). Over the course of the treatments the mice receiving DOXIL had an average 5% weight loss, and the mice receiving Adriamycin in saline had an average 7% weight loss; body weights in both groups returned to normal by 5 weeks post-treatment.

Intraperitoneal Implants

Palpable ip tumor growth developed in 8 of 10 placebo control mice from 18-25 d after implantation (Table 5.1). All of the DOXIL treated mice were tumor-free at necropsy, 70 d after tumor implantation ($p < 0.005$). Of the mice treated with free doxorubicin, 6 of 10 died with bloody peritoneal exudate from 7-25 d after the first therapeutic ip injection; deaths were considered due to Adriamycin toxicity. The 4 mice that survived were all tumor-free when the test was terminated on day 70. Mice receiving DOXIL experienced an average 3% weight loss, which was recovered 2 weeks after the last injection.

Pancreatic Carcinoma

This study compared the therapeutic effects of Adriamycin and DOXIL against the pancreatic adenocarcinoma AsPC-1.[14] The tumor has an average in vivo doubling time of 18 d, and an average 95% probability of growth in untreated nude mice. Treatments, at a dose level of 3.0 mg/kg doxorubicin, were given iv 1, 8, 15, 22, and 29 d after sc tumor implantation. Intraperitoneal tumor implantation and ip therapy was not attempted in this model because in the clinical disease ip drug therapy could interfere with the healing of intestinal anastomoses and is therefore not a recommended procedure.[15] Over the course of the treatments, the mice receiving DOXIL had an average 3% weight loss, and the mice receiving Adriamycin in saline had an average 2% weight loss. Body weights recovered within 5 weeks.

Table 5.1. Ovarian carcinoma HEY in nude mice. Incidence of intraperitoneal growth and survival with treatment on days 1, 8, 15

Treatment	Mean Survival (days)	Final Incidence[a]
Saline	50 ± 4.5	8/10
Adriamycin,9 mg/kg	25 ± 6.2	0/4[b]
DOXIL,9 mg/kg	70 ± 0.0[c]	0/10[d]

Incidence of ip tumor growth per 10 mice.
Each mouse received an ip injection of 10^5 cells on day 0.
[a] Cumulative incidence of ip tumor growth on day 70 necropsy.
[b] 6/10 mice died before day 26.
[c] Significantly longer than saline ($p < 0.001$).
[d] Significantly less than saline ($p < 0.005$).

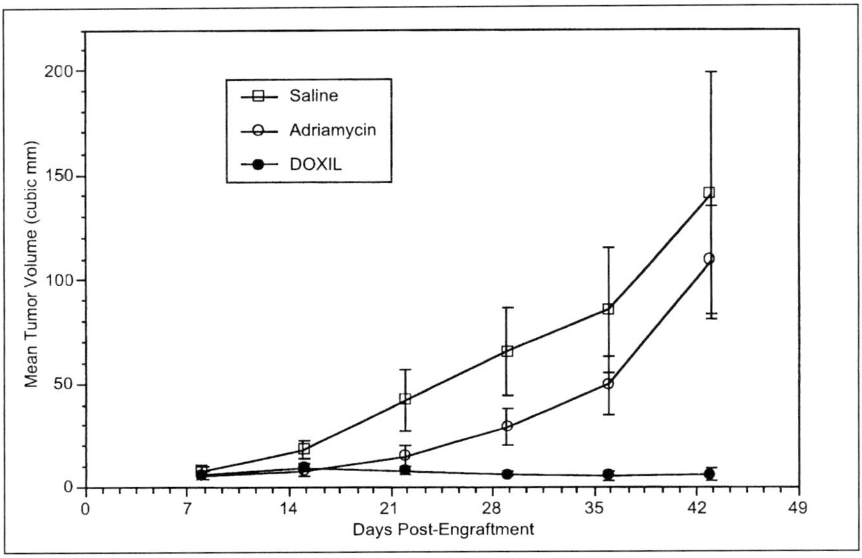

Fig. 5.1. Effect of treatment with saline, free doxorubicin in saline (Adriamycin), and doxorubicin in Stealth liposomes (DOXIL) on the growth of sc implants of the ovarian carcinoma HEY. Values are the mean volumes of 16-20 tumors per group of 8-10 mice.

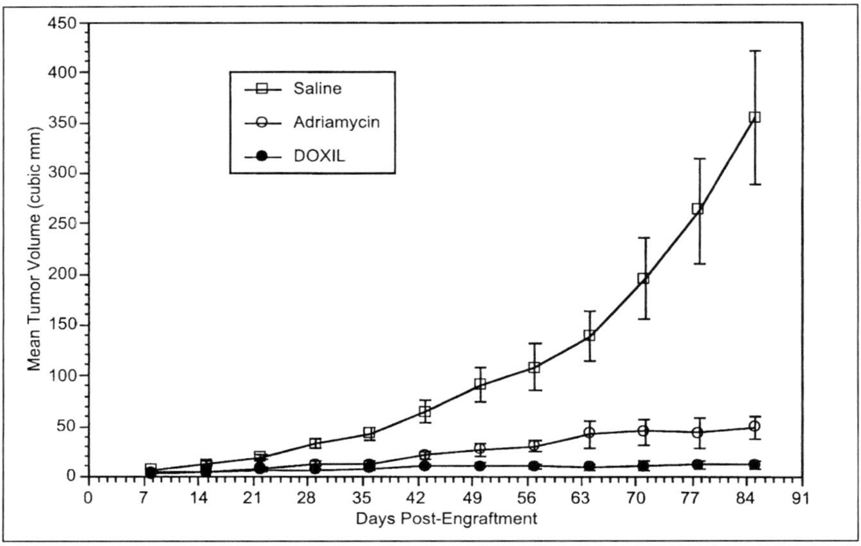

Fig. 5.2. Effect of treatment with saline, free doxorubicin in saline (Adriamycin), and doxorubicin in Stealth liposomes (DOXIL) on the growth of sc implants of the pancreatic carcinoma AsPC-1. Values are the mean volumes of 20 tumors per group of 10 mice.

DOXIL inhibited the growth of AsPC-1 more effectively than Adriamycin (day 85, $p < 0.0001$; Fig. 5.2), and a significantly greater number of DOXIL-treated mice were tumor-free at necropsy (7/20 versus 1/20, $p = 0.02$). In view of the current opinion that "*In the absence of a clear cut advantage of any therapy for pancreatic cancer, one must consider chemotherapy for this disease still to be experimental,*"[15] the present observations on the high therapeutic efficacy and low systemic toxicity of DOXIL against xenografts of a human pancreatic carcinoma are encouraging.

Doxorubicin, and consequently Adriamycin and DOXIL, are fluorescent, and this property can be used to track the disposition of liposomal and nonliposomal doxorubicin in tissues. To a limited extent the fluorescence of doxorubicin can also be used to quantify tumor and tissue levels of drug, although the measurement may be confounded by the auto-quenching of fluorescence that occurs when the drug is encapsulated in liposomes. Confocal laser scanning fluorescence microscopy and microfluorimetry were utilized to quantify drug uptake by sc tumor implants after iv injection of Adriamycin or DOXIL. The tumors had been implanted 30 d previously and had reached a volume of 0.03-0.04 cm^3. Measurement of the uptake of DOXIL and Adriamycin was conducted on cryostat sections of tumor removed from mice euthanized periodically for up to 168 h after iv injection of 3.0 mg/kg of either formulation.

Adriamycin was detectable in the tumor for just 24 h after treatment. In contrast, doxorubicin was measured for 168 h after treatment with the same dose of DOXIL. The relative values for the areas-under-the-curves (AUC) for free doxorubicin and DOXIL were 26 and 462, respectively, when the DOXIL valves are corrected with the auto-quenching factor 2.8 ± 0.15[12] (which assumes that all of the drug was contained within the liposomes). This represents an 18-fold increase in the drug at the disease site after DOXIL treatment (Fig. 5.3). When it is assumed that all liposomal drug was released from encapsulation and there was no quenching effect, the AUC values for Adriamycin and DOXIL were 26 and 165, representing a 6-fold increase in tumors after DOXIL treatment. The actual increase is likely intermediate to these two values, since the total fluorescence intensity will most likely include a component from both unquenched free doxorubicin and quenched liposomal doxorubicin. It seems likely that the actual increase may be closer to 18-fold initially, while most doxorubicin remains encapsulated, and nearer to 6-fold later, as the drug leaves the liposomes.

Prostatic Carcinoma

This study compared the therapeutic effects of Adriamycin and DOXIL against the prostatic carcinoma PC-3.[13] The tumor has an average in vivo doubling time of 6d, and an average 85% probability of growth in untreated nude mice. Treatments with DOXIL or Adriamycin, at a dose level of 6.0 mg/kg doxorubicin each, were given iv 1, 8, 15 and 22 d after tumor implantation. Over the course of therapy, the mice receiving DOXIL had an average 4% weight loss, and the mice receiving Adriamycin in saline had an average 9% weight loss. The weight losses were recovered within 5 weeks.

DOXIL inhibited the growth of PC-3 more effectively than Adriamycin (day 43, $p < 0.006$; Fig. 5.4). A significantly greater number of DOXIL-treated mice were tumor-free at necropsy (5/20 versus 0/18, $p = 0.03$). In view of the summary opinion expressed in the National Cancer Institute's (NCI) Meeting Report on Prostate Cancer, that "*prostatic cancer shows little if any responsiveness to chemotherapy*",[16] the present observations on the high therapeutic efficacy and low systemic toxicity of

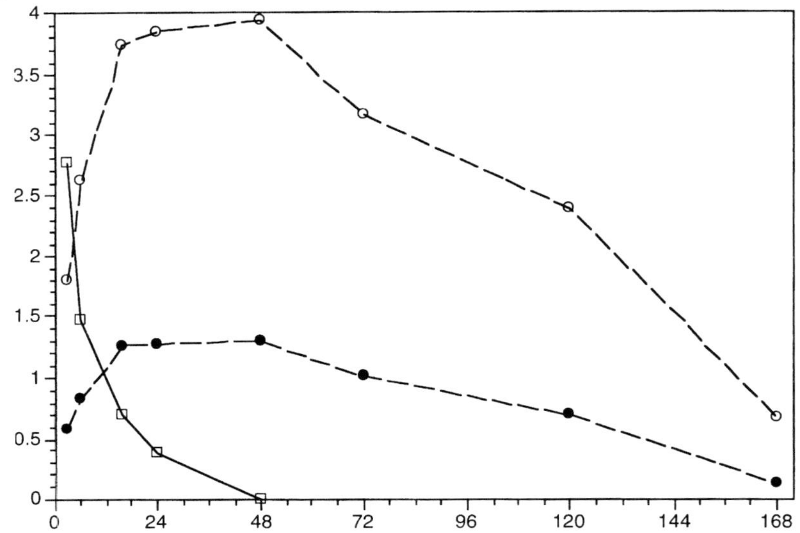

Fig. 5.3. Quantitation by microfluorimetry of doxorubicin in AsPC-1 tumor implants in nude mice. Open circles represent the quantification of doxorubicin in tumors from DOXIL-treated mice when corrected for autoquenching as described in the text; closed circles represent the amount when not adjusted for autoquenching. Open squares represent amount quantified in tumors from Adriamycin-treated mice.

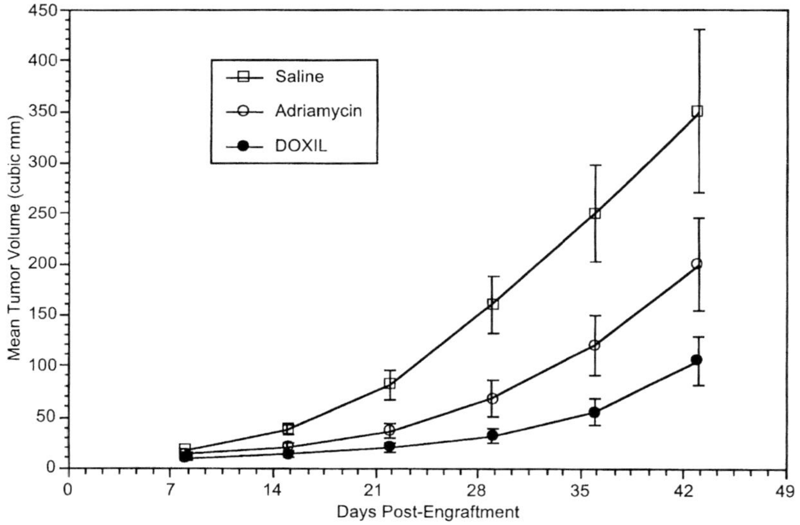

Fig. 5.4. Effect of treatments with saline, free doxorubicin in saline (Adriamycin), and doxorubicin in Stealth liposomes (DOXIL), on the growth of sc implants of the prostatic carcinoma PC-3. Values are the mean volumes of 20 tumors per group of 10 mice.

DOXIL against xenografts of a human prostatic carcinoma, are encouraging. Microfluorimetry studies with this xenograft also showed a significant increase in tumor AUC after DOXIL treatment, with the increase approaching 25-fold when corrected for auto-quenching.[13]

Therapeutic Activity Summary

DOXIL has been consistently shown to be more active than the same dose of Adriamycin in the treatment of murine and human xenograft tumors in mice. DOXIL is generally more effective in inhibiting or halting tumor growth, in effecting cures and/or in prolonging the survival of tumor-bearing animals. Most often, all three endpoints were improved in DOXIL-treated animals, and in no case has DOXIL been seen to be less effective than Adriamycin.

One can speculate on the mechanism that accounts for the increased effectiveness of DOXIL. From its intratumor location, doxorubicin would be slowly released from the liposomes, with the drug concentration consequently maintained at effective intracellular and extracellular cytotoxic levels for prolonged periods.[17] The stable encapsulation of doxorubicin within the sterically stabilized liposomes and the prolonged extravasation of "fresh" drug-loaded liposomes from the circulation, which result in the increased tumor levels, as well as the slow drug release from the liposomes inside the tumor, are the formulation characteristics responsible for the therapeutic efficacy of DOXIL. It is also possible that doxorubicin encapsulated in sterically stabilized liposomes may be less affected by the multidrug resistance P-glycoprotein efflux pump than is free doxorubicin.[18,19]

Earlier studies using human tumors implanted into nude mice showed that doxorubicin encapsulated in conventional liposomes was no more potent as a therapeutic agent than doxorubicin suspended in saline.[20,21] This was also observed in a mouse mammary tumor model.[11] In the present study, the encapsulation of doxorubicin in sterically stabilized liposomes significantly increased the therapeutic efficacy of the drug against three different human carcinomas growing in nude mice. The liposome formulation also had reduced systemic toxicity, in comparison with free doxorubicin administered in saline suspension. Because 3-5 weekly injections of doxorubicin in sterically stabilized liposomes produced significant therapeutic benefit with only minor and transient toxic side-effects, the observed therapeutic advantage of sterically stabilized long circulating liposomes as a vehicle for drug delivery has clinical relevance as a potential new method in cancer drug therapy.

Safety Studies in Animals with DOXIL

Formal safety studies supported the observation that DOXIL was significantly less toxic than Adriamycin. Toxicity observed following iv administration of single doses of DOXIL to mice, rats, rabbits and dogs was qualitatively similar in the nature of the response in each species.[9,10,22] Dogs were the most sensitive species, exhibiting severe gastrointestinal toxicity in response to high doses or rapid infusions of DOXIL; slowing the infusion rate almost completely alleviated the gastrointestinal toxicity. Treatment-related toxicity included hematologic changes, myelotoxicity, marginal clinical symptoms, gastrointestinal toxicity, reversible cutaneous lesions and alopecia, which, with the exception of the dermal lesions, were generally less severe in DOXIL-treated animals.

The toxicity profile of DOXIL following repeated administration was similar in rats and dogs and was an extension of the findings seen in the acute studies.

Treatment-related effects included skin ulcers, particularly on the legs and feet, body weight and food consumption changes, mild alopecia, myelotoxicity (bone marrow cellularity changes), testicular atrophy, hematologic effects (leukopenia and anemia) and, in rats and rabbits, limited cardiomyopathy. Dogs, again the more sensitive species, also exhibited gastrointestinal toxicity, but no pathologic signs of cardiotoxicity. With the exception of cardiotoxicity, testicular atrophy and persistent alopecia at the dermal lesion sites, adverse effects were fully resolved during the recovery periods that followed each repeat dose study.

There was no evidence of nephrotoxicity in DOXIL-treated animals, although nephrotoxicity is a known adverse effect of Adriamycin in animals, particularly rodents and dogs. Myelotoxicity was observed in rats and dogs following DOXIL treatment, but at reduced level of severity compared to animals that received the same cumulative dose of Adriamycin.

Doxorubicin-related cardiotoxicity seen in DOXIL-treated animals was both less severe and less frequent, and was not simply due to a longer latency period prior to its development. As much as 50% more DOXIL could be administered than Adriamycin without incurring an equivalent risk of cardiomyopathy. This was clearly demonstrated in a rabbit model of doxorubicin cardiotoxicity, discussed below. Doxorubicin-associated nephrotoxicity and cardiotoxicity are well correlated with high peak plasma levels. The decreased relative toxicity of DOXIL is believed to be related to the decreased peak concentration of free doxorubicin that has been measured in the plasma of treated animals compared to that measured after administration of the same dosage of Adriamycin.[22] Because anthracycline-related cardiotoxicity is the most significant toxicity associated with doxorubicin-based drugs, a more extensive study was conducted to evaluate the comparative cardiotoxicity of DOXIL and Adriamycin.

Cardiotoxicity of DOXIL in Rabbits

Rabbits are considered the most relevant species to assess the cardiotoxicity of anthracyclines because of the similarity of histologic changes in rabbits treated with anthracyclines to those that occur in humans after prolonged Adriamycin administration.[23,24] Rabbits also develop the delayed progressive cardiomyopathy that is often seen in the clinic.[24] In this phenomenon, cumulative Adriamycin doses that do not cause immediate significant myocardial disease may eventually lead to severe damage even in the absence of continued drug administration.

In this study, young adult male New Zealand White rabbits received a 1 mg/kg iv bolus dose of either DOXIL or Adriamycin at an approximate dose rate of 2 ml/minute. Control animals received sterile 0.9% NaCl. Rabbits were treated every five days for a total of 14 or 21 treatments. Animals that received a cumulative dose of 14 mg/kg (182 mg/m^2) were sacrificed 1, 5 or 13 weeks after the final dose. Animals that received a cumulative dose of 21 mg/kg (273 mg/m^2) were sacrificed 13 weeks after the 21st dose. The latter included the DOXIL and saline treatment groups only, since Adriamycin-treated animals were not expected to tolerate such a high cumulative dose. At necropsy, the heart was immersed in 4% buffered paraformaldehyde and grossly sectioned prior to additional fixation. Five regions (the septum near the auricles, each ventricle near the auricles, the left ventricular papillary muscle, and the apex) were imbedded in methylacrylate resin, sectioned, stained and examined. Scoring was quantitative using an adaptation of published criteria.[24] Lesions were assigned severity grades of 1, 2 and 3, and the overall cardiotoxicity score is the sum of the number of affected sites (of the five listed above) multiplied by the appropriate severity grade for each lesion.

Nine early deaths occurred in the DOXIL treatment group; 6 were considered secondary to the effects of dermal lesions (see below) and 3 due to cardiotoxicity. Five early deaths occurred in the Adriamycin treatment group, with strong evidence of cardiotoxicity and congestive heart failure in all 5 animals. Heart lesions (congestive heart failure and microscopic evidence of vacuolar degeneration of myocardial fibers and myocardial atrophy) were similar in the DOXIL and Adriamycin treatment groups, but were more common and more severe in the latter group (Table 5.2). DOXIL-treated animals with signs of cardiotoxicity (4/25; 16%) had a mean cardiotoxicity score of 3 (range 1-5), compared to a mean score of 4.3 (range of 1-11) in rabbits that received Adriamycin (10/15; 67%). The decreased cardiologic toxicity of DOXIL was not simply due to a longer latency period prior to its development, since cardiomyopathy was seen to increase in incidence and severity with time after treatment in the Adriamycin treatment group, but not in the DOXIL group. A higher cumulative dose of DOXIL (up to 50% more) could be given without incurring increased cardiotoxicity as compared to Adriamycin.

The incidence and severity of doxorubicin-related cardiomyopathy are well correlated with increasing cumulative dosage in humans. These studies showed that a several-fold higher dose of doxorubicin in liposomes (DOXIL) can be given than of Adriamycin without causing greater cardiotoxicity.

Treatment of rabbits with DOXIL was associated with the development of reversible dermal lesions, particularly on the feet and legs, as has also been reported in rats and dogs treated with DOXIL.[9] The formation of the lesions has been demonstrated to be related to dose intensity and can be readily controlled by lowering the DOXIL dosage or increasing the dosing interval.[25]

Safety Summary

The toxicity profile of DOXIL in animals appears similar to that reported for continuous or long-term infusion of Adriamycin in humans. Given as a standard intravenous bolus of repeated at 3-week intervals, cumulative doses of greater than 550 mg/m^2 of Adriamycin pose a significant risk of cardiotoxicity in humans.[26,27] Doxorubicin-related cardiotoxicity appears to be related to the peak plasma concentration achieved after administration of the drug. Repeated administration of high doses of Adriamycin as bolus injections on a weekly schedule, and the resultant high plasma concentrations of doxorubicin, are associated with increased risk of cardiotoxicity. Less rapid development of cardiotoxicity is seen when doxorubicin HCl is administered as fractionated doses or as long-term infusions, but without loss of antineoplastic activity.[28-30] Long-term or continuous infusions avoid the high peak plasma concentrations believed responsible for the development of cardiotoxicity, but maintain concentrations and AUCs in the plasma without loss of efficacy.[31,32] In continuously infused human patients, myelosuppression appears to be schedule-independent, and stomatitis is variably associated with dosing schedule. In fact, the most commonly reported dose-limiting side effect of Adriamycin administered as a long-term infusion has been stomatitis, often accompanied by hand-and-foot syndrome.[33,34]

In general, then, Adriamycin administered as a continuous or long-term infusion is characterized by reduced cardiotoxicity, increased stomatitis, the appearance of hand-and-foot syndrome and minimal effects on myelotoxicity.

Table 5.2. Cardiotoxicity of DOXIL in rabbits. Incidence (severity, range)

Weeks After Final Dose	DOXIL	Adriamycin
	Cumulative Dose: 12-14 mg/kg	
1 Week	1/5 (5)	6/7 (5, 1-11)
5 Weeks	0/5	1/3 (1)
13 Weeks	2/5 (3, 1-4)	3/5 (1, 1-4)
	Cumulative Dose: 21 mg/kg	
13 Weeks	1/10 (2)	nd
	Overall	
Incidence	4/25 (16%)	10/15 (67%)
Cardiotoxicity Score	3 (1-5)	4.5 (1-11)

nd = not done

Comparatively, administration of doxorubicin as long circulating DOXIL liposomes in animals is characterized by reduced cardiotoxicity, increased dermal ulcerations, particularly of the feet and legs, and marginally reduced myelosuppression. Thus, in its safety profile at least, DOXIL appears to mimic the effect of long-term infusions of Adriamycin.

Overall Summary

Studies in animals have shown that DOXIL is both safer and more effective than the same dose of Adriamycin. It is likely that both are related to the effect of the sterically stabilized liposomes on tissue distribution (lower levels in target tissues of toxicity and higher levels in tumors) as well as on the plasma levels of drug. The rate of uptake of DOXIL liposomes by the MPS is reduced by virtue of their decreased interaction with plasma components, resulting in long circulation time and, indirectly, increased extravasation at tumor sites. The stability of the Stealth liposome, and its associated low doxorubicin release (leakage) rate, results in low levels of free (nonliposomal) doxorubicin HCl in the bloodstream. In animal tumor models, both murine tumors and human xenografts, DOXIL was more effective than equivalent doses of Adriamycin or doxorubicin HCl encapsulated in conventional liposomes. These differences are also probably related to the pharmacokinetic differences between the formulations; both Adriamycin and doxorubicin in nonpegylated liposomes are quickly cleared from the bloodstream and do not accumulate in tumors, DOXIL does. Thus, DOXIL appears to be a safe and effective agent for the treatment of both solid and dispersed tumors.

References

1. Gabizon A, Goren D, Fuks Z et al. Superior therapeutic activity of liposome-associated Adriamycin in a murine metastatic tumor model. Br J Cancer 1985; 51:681-9.
2. Szoka F C. Liposome drug delivery. In: Wilschut J, Hoekston R editors. Membrane Fusion. New York: Marcel Dekker, 1991:845-90.
3. Gabizon A, Chisin R, Amselem S et al. Pharmacokinetic and imaging studies in patients receiving a formulation of liposome-associated adriamycin. Br J Cancer 1991; 64:1125-1132.
4. Gabizon A, Papahadjopoulos D. Liposome formulations with prolonged circulation time in blood and enhanced uptake by tumors. Proc Natl Acad Sci USA 1988; 85:6949-6953.
5. Allen TM, Hansen C, Martin FJ et al. Liposomes containing a synthetic lipid derivative of polyethylene glycol show prolonged circulation half-lives in vivo. Biochim Biophys Acta 1991; 1066:29-36.
6. Napper DH. Polymeric Stabilization of Colloidal Dispersions. New York: Academic Press, 1983.
7. Lasic DD, Martin FJ, Gabizon A et al. Sterically stabilized liposomes: a hypothesis on the molecular origin of the extended circulation times. Biochim Biophys Acta 1991; 1070:187-192.
8. Woodle MC, Lasic DD. Sterically stabilized liposomes. Biochim Biophys Acta 1992; 1113:171-199.
9. Woodle MC, Newman MS, Working PK. Biological properties of sterically stabilized liposomes. In: Lasic DD, Martin F, eds. Stealth Liposomes. Boca Raton: CRC Press, 1995:103-117.
10. Working PK, Dayan AD. Pharmacological-toxicological Expert Report. CAELYX™ (Stealth® liposomal doxorubicin HCl). Human Exp Toxicol 1996; 15:752-785.
11. Vaage J, Mayhew E, Lasic D et al. Therapy of primary and metastatic mouse mammary carcinomas with doxorubicin encapsulated in long circulating liposomes. Int J Cancer 1992; 51:942-8.
12. Vaage J, Donovan D, Mayhew E et al. Therapy of human ovarian carcinoma xenografts using doxorubicin encapsulated in sterically stabilized liposomes. Cancer 1993; 72:3671-5.
13. Vaage J, Barbera-Guillem E, Abra R et al. Tissue distribution and therapeutic effect of intravenous free or encapsulated liposomal doxorubicin on human prostate carcinoma xenografts. Cancer 1994; 73:1478-84.
14. Vaage J, Donovan D, Uster P et al. Tumor uptake of doxorubicin in polyethylene glyco-coated liposomes and therapeutic effect against a xenografted human pancreatic carcinoma. Br J Cancer 1997; 75:482-486.
15. Douglass HO Jr, Tepper J, Leichman L. Neoplasms of the exocrine pancreas. In: Holland JF, Frei E III, Bast RC et al, eds. Cancer Medicine. 3rd ed. Lea & Febiger, 1993:1466-84.
16. Chiarodo A. Meeting Report. National Cancer Institute Roundtable on Prostate Cancer: Future Cancer Research Directions. Cancer Res 1991; 51:2498-2505.
17. Vichi P, Tritton T R. Adriamycin: Protection from cell death by removal of extracellular drug. Cancer Res 1992; 52:4135-8.
18. Endicott J A, Ling V. The biochemistry of P-glycoprotein-mediated multidrug resistance. Ann Rev Biochem 1989; 58:137-171.
19. Gottesman MM. How cancer cells evade chemotherapy: Sixteenth Richard and Hilda Rosenthal Foundation Award Lecture. Cancer Res 1993; 53:747-754.
20. Kojima N, Ueno N, Takano M et al. Effect of adriamycin entrapped by sulfatide-containing liposomes on ovarian tumor-bearing nude mice. Biotechnol Appl Biochem 1986; 8:471-478.

21. Nagata J, Yamauchi M, Takagi H et al. Antitumor activity against human gastric cancers of sulfatide-inserted liposomes containing entrapped adriamycin. J Clin Biochem 1990; 8:111-119.
22. Working PK, Newman MS, Huang SK et al. Pharmacokinetics, biodistribution and therapeutic efficacy of doxorubicin encapsulated in Stealth® liposomes (Doxil®). J Liposome Res 1994; 4:667-687.
23. Iatropoulos MJ. Anthracycline cardiomyopathy: Predictive value of animal models. Cancer Treat Symp 1984; 3:3-17.
24. Jaenke RS. An anthracycline antibiotic-induced cardiomyopathy in rabbits. Lab Inves 1974; 30:292-304.
25. Newman MS, Amantea MA, Sullivan TM et al. Pharmacokinetics/pharmacodynamics of pegylated-liposomal doxorubicin and the appearance of dermal lesions in dogs. Clin Pharmacol Therap 1996; 59:96.
26. Benjamin RS, Wiernik PH, Bachur NR. Adriamycin chemotherapy—efficacy, safety, and pharmacologic basis of an intermittent singe high-dosage schedule. Cancer 1974; 33:19-27.
27. Minow RA, Benjamin RS, Lee ET et al. Adriamycin cardiomyopathy: Risk factors. Cancer 1977; 39:1397-402.
28. Gabizon A, Catane R, Uziely B et al. Prolonged circulation time and enhanced accumulation in malignant exudates of doxorubicin encapsulated in polyethylene-glycol coated liposomes. Cancer Res 1994; 54:987-92.
29. Chlebowski RT, Paroly WS, Pugh RP et al. Adriamycin given as a weekly schedule with a loading course: Clinically effective with reduced incidence cardiotoxicity. Cancer Treat Rep 1980; 64:47-51.
30. Lokich JJ, Bothe A, Zipoli T et al. Constant infusion schedule for Adriamycin: A phase I-II clinical trial of a 30-day schedule by ambulatory pump delivery system. J Clin Oncol 1983; 1:24-8.
31. Speth PAJ, van Hoesel QGCM, Haanen C. Clinical pharmacokinetics of doxorubicin. Clin Pharmacokin 1988; 15:15-31.
32. Speth PAJ, Linssen PCM, Holdrinet RSG et al. Plasma and cellular adriamycin concentrations in patients with myeloma treated with ninety-six-hour continuous infusion. Clin Pharm Therap 1987; 41:661-5.
33. Rudolph R, Stein RS, Pattillo RA. Skin ulcers due to Adriamycin. Cancer 1976; 38:1087-94.
34. Samuels BL, Vogelzang NJ, Ruane M et al. Continuous venous infusion of doxorubicin in advanced sarcomas. Cancer Treat Rep 1987; 71:971-2.

CHAPTER 6

Design and Development of Long Circulating Liposomal Daunorubicin for In Vivo Targeting of Solid Tumors: DaunoXome®

Eric A. Forssen, Richard T. Proffitt

Summary

This chapter reviews the formulation research, development, and characterization of a tumor-targeting daunorubicin liposome preparation with a long circulating half-life (DaunoXome®). The identification of liposome formulations capable of delivering their contents to solid tumors in vivo made use of an active loading technique for high efficiency entrapment of the gamma emitter, ^{111}In. With this approach, an array of lipid compositions and physical characteristics (size, phase transition temperature, surface charge, etc.) were evaluated in vivo for their capabilities to remain intact for prolonged periods while in the circulation but able to release their entrapped contents when delivered to solid tumors. During these investigations, it was found that liposomes composed of distearoylphosphatidylcholine (DSPC): cholesterol in a 2:1 mole ratio and with diameters less than 100 nm are particularly effective for maximizing tumor uptake. In clinical studies involving nearly 400 patients, these ^{111}In liposomes were used to image successfully a variety of primary cancers and their metastases. Tumors imaged in these studies include: breast, prostate, colon, kidney, cervix, thyroid, larynx, lung (small cell and non-small cell), lymphomas (malignant and Hodgkin's), sarcomas (soft tissue and Kaposi's). Of particular interest was the finding that secondary tumors (presumably metastases) imaged more intensely than did primary tumors, suggesting that secondary tumors may be more efficient at accumulating materials entrapped within these liposomes. This formulation served as a model for development of liposome drug-carrier systems designed to target antineoplastic agents to solid tumors in vivo. Anthracyclines became an early focus of this work since it had been demonstrated previously that the formulation of these compounds into liposomes resulted in lower toxicities for this drug class. As we will discuss, daunorubicin was selected over other anthracyclines for liposome development for several reasons, including a lower rate of cumulative cardiotoxicity and a level of cytotoxicity against tumor cells comparable

Long Circulating Liposomes: Old Drugs, New Therapeutics, edited by Martin C. Woodle and Gerrit Storm.

to or greater than that of doxorubicin. Preclinical investigations indicate that DaunoXome increases in vivo daunorubicin tumor delivery by about 10-fold over free drug, yielding a comparable increase in therapeutic efficacy. Investigations on the modes of delivery and of action indicate that DaunoXome arrives at and accumulates within tumor cells in an intact form. As with the tumor imaging preparation, it appears that DaunoXome extravasates selectively into solid tumors through discontinuities in the capillary beds in the tumor neovasculature. Once within tumor cells, the liposomes release the drug over a prolonged period (36 h or more), providing sustained, high levels of cytotoxic material. HIV-positive patients tolerate DaunoXome well and it compares favorably with the typical therapy of ABV (doxorubicin, bleomycin, vincristine), demonstrating reduced toxicity while retaining comparable antitumor activity. Several countries, including the United States, have approved DaunoXome for use in treating Kaposi's sarcoma in HIV-positive patients. In this chapter, we report on the formulation development, preclinical and clinical investigations of DaunoXome.

Background

Early investigators discovered the ability of liposomes to alter pharmacokinetics,[1-4] overcome barriers to cell entry,[5-7] and improve the physical properties of entrapped agents.[8-10] Researchers had prepared formulations with a wide range of lipid compositions and sizes, which contained a variety of compounds entrapped either within the inner aqueous compartment or within the lipid membrane bilayer. Until comparatively recently, however, they have had only limited successes in producing efficacious pharmaceutical products with clear clinical benefits on scales sufficient to meet the needs of a large number of patients. Crucial to developing formulations with acceptable therapeutic profiles has been the design of liposomes with appropriate lipid compositions and physical characteristics. For a number of years, investigators have focused on applications of liposomes in cancer treatment where there is encouraging potential to enhance tumor delivery and ameliorate drug toxicities.[11-15] Mauk and Gamble[16,17] were the first to develop active loading techniques for labeling specific liposome formulations that later were used in tumor imaging studies. These investigators prepared liposomes containing a weak chelator entrapped in the inner aqueous space and the cationic ionophore A23187 within the membrane bilayer. Under appropriate conditions, using an active loading method, these liposomes could be loaded efficiently (approaching 100%) with the gamma emitting radioisotope ^{111}In. These radiolabeled liposomes were evaluated to determine formulation parameters that would preserve their physical stability under physiological conditions in serum in vitro but which nevertheless would allow their breakdown and release of entrapped contents in tissues in vivo.[17-18] Monitoring ^{111}In radioactivity using Perturbed Angular Correlation (PAC) techniques,[19-22] these investigators identified formulation parameters critical for the tumor uptake. These parameters include use of pure phospholipids with phase-transition temperatures above physiological (>37°C), a net neutral to slightly negative surface charge, and the inclusion of cholesterol in the lipid membrane (to minimize lipid exchange and the leakage of entrapped substances).

Measurements using the PAC technique to monitor ^{111}In release showed that liposomes composed of cholesterol and phosphatidylcholine molecules containing fully saturated long-chain fatty acids (e.g., distearoylphosphatidylcholine, DSPC), in a 1:2 molar ratio, were stable to leakage of the ^{111}In under physiological conditions.[17] This result was confirmed by Wallingford and Williams[23] using PAC measurements to demonstrate that DSPC:cholesterol liposomes maintain their physical integrity with a half-life of approximately 130 h at 37°C in freshly drawn human plasma. In contrast, a liposome formulation prepared from egg lecithin rapidly leaked ^{111}In, and had an half-life for physical integrity of only 7 h under these conditions. Proffitt et al[24] demonstrated that small, stable liposomes were able to remain in circulation and deliver a greater percentage of the encapsulated ^{111}In to the tumor than did negatively or positively charged liposomes. Also, tumors could be imaged clearly by gamma camera scintigraphy on mice that had been injected with stable ^{111}In-labeled liposomes. It was hypothesized that such small stable liposomes could escape intact from the circulation due to increased capillary permeability at the tumor site. In a subsequent paper by these researchers,[25] it was shown that reticuloendothelial system (RES) blockade (with unlabeled liposomes that are cleared rapidly from the circulation) improved ^{111}In tumor accumulation by over 50% after subsequent administration of liposomes that were small, stable, and neutral. This result supports the hypothesis that intact circulating liposomes deliver their contents directly into the tumor. Furthermore, based on autoradiographic evidence, these authors proposed that the malignant cells themselves can accumulate and lyse these liposomes. Additional PAC studies revealed that, when these DSPC:cholesterol liposomes remain in circulation, essentially all of the encapsulated ^{111}In resides within the liposome for at least 48 h. In contrast, as the liposomes accumulate within the tumor, ^{111}In is released rapidly and retained by the tumor.[26] This result is consistent with the proposed lysis of intact liposomes within tumor cells.

Although a long circulating half-life is thought to be important for adequate accumulation of liposomes at the tumor site, there could be undesirable nonspecific uptake in normal tissues. For example, Hwang and co-workers[27] found that small liposomes composed of bovine brain sphingomyelin and cholesterol (2:1 mol ratio) were cleared very slowly from the bloodstream during the first three hours post injection. At 23 h post injection, nearly 16% of the initial dose still remained in the blood. In addition to RES clearance, normal tissues such as the skin and intestine had accumulated 5-10% of the total dose. This distribution pattern, in particular the long circulation times, could lead to a high background in the case of imaging studies. Recently, Ogihara-Umeda et al[28,29] showed that radiolabeled liposomes containing a small amount of biotin-conjugated phosphatidylethanolamine could be effectively cleared from the bloodstream with avidin. Thus, tumors could be visualized within 2 h after administration of the imaging liposomes. Also, increased uptake by normal tissues could lead to unwanted toxicities when anticancer drugs are incorporated into such liposomes. For example, "hand-and-foot syndrome" (a debilitating condition involving ulceration and desquamation of the palms and soles) has been reported in some patients receiving long circulating liposomal doxorubicin.[30]

Tumor Imaging Agent

An in vivo approach was developed in order to identify liposome compositions and their physical characteristics that would maximize delivery of entrapped substances to solid tumors in vivo. Mauk et al[17,18,31] made this approach readily usable by developing an active loading procedure for labeling liposomes with high levels of the gamma-emitting radionuclide ^{111}In,[16] Figure 6.1. These investigators evaluated the stability and release of entrapped contents from these labeled liposomes as a function of their composition. They were able to accomplish this by taking advantage of the radioactive decay pattern for ^{111}In in which the radionuclide sequentially emits two gamma rays as it decays to stable ^{111}Cd.[19,32] Each gamma ray is characterized by a specific energy level with a time delay between the first and second rays. There is an angular dependency between the two gamma-rays that is a function of the reorientation rate associated with the ^{111}In nuclear spin. This angular dependency can be perturbed by interactions of the nuclear moments in the intermediate state with fluctuating external fields.[19,32] The rate of reorientation is inversely proportional to the tumbling rate of the molecule to which the ^{111}In may be bound. A faster rate occurs when ^{111}In is free in solution or bound to a small molecule (MW << 1,000). This rate is reduced substantially, however, when ^{111}In becomes bound to large molecular weight species such as macromolecules or cell surfaces.

Key formulation parameters producing significant tumor accumulation were found to include:

1) Lipid phase transition temperatures above physiological range (> 37°C); (i.e., liposome bilayer membrane remains in the gel (solid) state under physiological conditions).
2) Inclusion of cholesterol in the membrane bilayer to minimize bilayer permeability and lipid exchange.
3) Net neutral surface charge.
4) Particle size less than 100 nm in diameter (by photon correlation spectroscopy, Nicomp).

Extensive screening studies have demonstrated that liposomes composed of DSPC and cholesterol in a 2:1 mole ratio and prepared with diameters between about 40 and 80 nm are particularly effective for delivery of entrapped ^{111}In to solid tumors in vivo.[24,26,33,34] These DSPC:cholesterol vesicles preferentially deliver entrapped ^{111}In to solid neoplasms in vivo as demonstrated in several murine tumor models including: mammary adenocarcinomas (EMT-6 and MA16C), B16 melanoma, Lewis lung carcinoma, P1798 lymphosarcoma, sarcoma 180, and colon carcinoma 51.[24-26,35] Delivery of ^{111}In to these tumors is found to range from 10-40% of the injected ^{111}In dose per gram tissue.

These formulation efforts led to development of an ^{111}In-based tumor imaging agent (VesCan®). This was the first product to enter into clinical studies that had been designed to deliver its contents selectively to solid tumors in vivo.[34,36] This imaging agent was investigated in clinical trials in about 375 patients and found to be capable of imaging a wide range of solid neoplasms (Table 6.1).[36,37] When tumors have been present, as confirmed by other means, the detection rate has been as high as 85% (*sensitivity* or rate of true positives). In conditions where a tumor has not been present, the rate of negative observations has been 96% (*specificity* or rate of true negatives).[38] An independent study with VesCan® further supported these findings with positive imaging in four of seven patients.[39] In this investigation, vesicles were loaded with 37 MBq ^{111}In and administered to patients with malignant tumors at a median lipid dose of 2.0 mg/kg. Sequential images showed early rapid blood clearance, with homogeneous uptake of ^{111}In in the liver and spleen. Total renal excretion of ^{111}In was less than 10% of the injected dose, occurring mainly as ^{111}In-EDTA in three patients. Gamma camera images at 24-48 h following administration revealed increased activity in the tumors of four patients.

During the course of clinical studies with this imaging agent, investigators observed that, based on the intensity of gamma-camera images, Kaposi's sarcoma (KS) demonstrates one of the highest levels of ^{111}In uptake of all tumors imaged.[37] This is

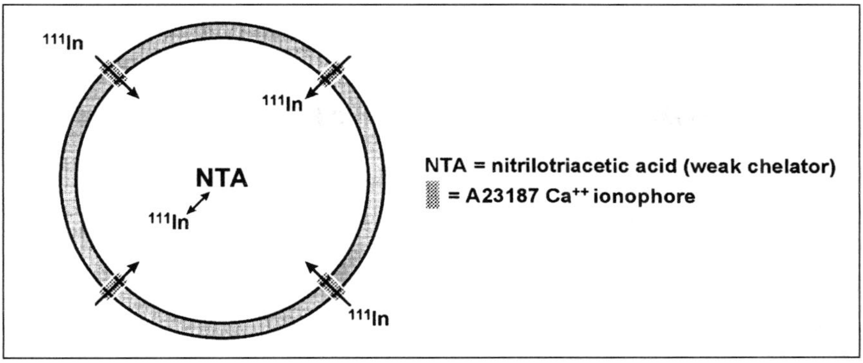

Fig. 6.1. Active loading of ^{111}In into preformed liposomes. Vesicles are first prepared containing the weak chelator nitrilotriacetic acid (NTA) entrapped within the inner aqueous space and the ionophore, A23187, included in the membrane bilayer. The ionophore allows the ^{111}In metal ions to pass through the bilayer by solubilizing the ions in the membrane's lipid environment. Just prior to administration, radioactive ^{111}In is added to the vesicle preparation, heated to 65°C and the active loading process takes place.

believed to be a function of the high degree of neoangiogenesis and vascular smooth muscle cell proliferation that is associated with this tumor.[40,41] Although its derivation is under current investigation, KS is thought to arise either from endothelium (vascular or lymphatic) or from vascular smooth muscle cells.[42-44] Not only are KS lesions highly vascular, they are also very permeable, allowing particulates such as liposomes to extravasate readily.[45,46] This observation of significant accumulation of liposome-entrapped ^{111}In within KS lesions led directly to the clinical evaluation of similarly formulated daunorubicin SUVs (DaunoXome) against this disease.[37,47] The DSPC:cholesterol vesicles of VesCan® have been well tolerated without significant adverse effects. Clinical parameters including hematologic values, respiratory functions, urinary output, and hepatic functions have all remained normal.

Tumor Physiology and Liposome Targeting

The growth of solid tumors to clinically relevant size is dependent upon the development and maintenance of a sufficient blood supply. Due to the inadequate capacity of oxygen and nutrients to diffuse more than about 60 µm through tissues, solid tumors would be significantly growth limited without their ability to induce neovascularization.[48-51] Upon growth of neoplasms beyond a certain size (about 100 µm or more in diameter), hypoxic regions occur, stimulating the production of angiogenic proteins, such as vascular endothelial growth factor, leading to extensive neovascularization.[52-54] Newly formed capillary networks are then able to provide a supply of nutrients and gas exchange that allows further tumor growth and development.[55,56] Progressive recruitment of blood vessels to the tumor site and reciprocal support of tumor expansion by the neovasculature are believed to result in a self-perpetuating cycle, helping to promote tumor growth.[57] Studies of experimental solid tumors indicate morphological abnormalities in tumor microcirculation relative to normal tissues. Typically, tumor vasculatures are poorly formed, displaying disarrayed capillary systems, with gaps between the endothelial cells, open-ended

Table 6.1. Tumors Imaged with ^{111}In-Labeled DSPC:Cholesterol Liposomes

Tumor	
Breast	Oropharyngeal
Cervix	Ovary
Kaposi's sarcoma	Prostate
Kidney	Small cell and non-small cell lung carcinomas
Larynx	Soft tissue sarcoma
Malignant lymphoma	Spindle cell sarcoma
Malignant melanoma	Testicular
Metastases to liver, lung, bone marrow, lymph nodes and soft tissues	Thyroid

and dead-end capillaries. Basement membranes are absent or have been damaged significantly by proteases elaborated by the tumors.[58] Vascular casts from benign and malignant tumors show extravasation from microvessels and an increased capillary density compared to normal tissue.[59] The presence of profusely anastomotic microvessels and frequent capillary sprouts is characteristic of active neovascularization.[58,60] For example, when Less et al[51] and Skinner et al[59] infused resins into the vasculature of colonic carcinomas, they found an increased capillary density and disorganized microvascular structure.

The mechanism of solid tumor selectivity for DSPC:cholesterol SUVs is not fully understood but is believed to be a function of liposome size and stability (resulting in prolonged circulation times), as well as the tumor vascular physiology described above. In tissues with continuous, non-fenestrated vascular endothelia, the escape of circulating particulates is greatly limited.[61,62] Large particles and liposomes with diameters greater than about 200 nm exhibit almost no transcapillary passage, even where endothelial junctions and/or fenestrae are quite large.[63] Tight intracellular junctions (2-6 nm) between the endothelial cells in most normal tissues prevent extravasation of even very small SUVs (down to 30 nm dia)[64] thus helping to protect these tissues from the toxic effects of the encapsulated drug. On the other hand, normal tissues with sinusoids and/or fenestrated capillaries (such as liver, spleen, and bone marrow) and also abnormal tissues with capillaries lined with damaged or poorly formed endothelia are relatively porous, allowing the escape of particulates with diameters below approximately 100 nm.[55,65] Many primary and metastatic tumors are known to have discontinuous and highly permeable vascular systems that allow the efflux of small particulates, including SUVs.[62,66] These tissue characteristics could, at least in part, account for the increased AUC levels observed in liver, spleen, and tumor tissues. The differences in the vascular systems of normal and neoplastic tissues are illustrated in Figure 6.2. Most normal tissues have capillary systems in which the lumen is surrounded by a continuous lining of endothelial cells. In some tissues such as the kidney, there are fenestrations that allow the passage of compounds with molecular weights of about 60,000 or less. Other tissues,

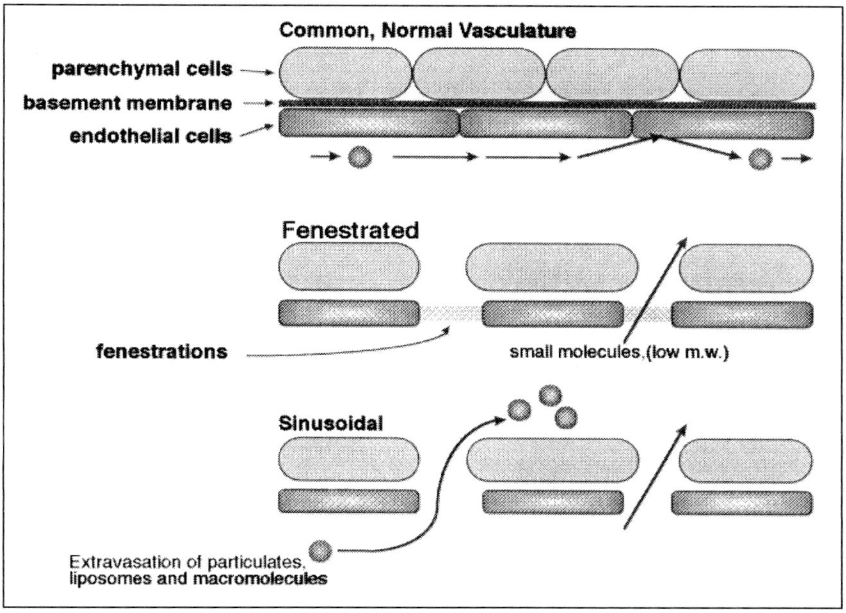

Fig. 6.2. Characteristics of normal and tumor vasculature. The vasculature in most normal tissues consists of a contiguous lining of endothelial cells without openings through or gaps between the cells. A basement membrane (basal lamina), which also acts to inhibit the extravasation of macromolecules and particulates, interfaces between the endothelium and surrounding parenchymal cells. Fenestrated endothelial systems (such as in the kidney) allow the efflux of smaller macromolecules (< 60,000 MW) but block the passage of particulates. Sinusoidal vasculatures (such as in the spleen, liver and many tumors) lack endothelial cell linings and basement membranes allowing the efflux of macromolecules and particulates.

such as liver, spleen, and many solid tumors, have sinusoidal structures lacking endothelial cell linings and basement membranes, which allow the circulating elements in the blood to come into direct contact with the tissue parenchyma.[62]

Liposome formulations based upon DSPC:cholesterol have unique advantages as carriers for delivery of antineoplastic drugs to solid tumors. First, vesicle membranes composed of these lipids are well below their phase transition temperature under physiological or ambient conditions. Compared to other formulations, such as those composed of egg phosphatidylcholines, these liposomes are more resistant to leakage of entrapped materials while stored under a wide range of conditions. Second, narrowing the lipid composition to two key constituents simplifies the determination of manufacturing conditions and formulation parameters that will be favorable for both the active species and the carrier. Third, the method by which these vesicles selectively deliver entrapped substance to solid tumors in vivo is believed to be dependent upon a physiological property, characteristic of the vasculature within many neoplasms (highly permeable or leaky capillaries) rather than specific antigenic epitopes or metabolic factors unique to specific tumor cell types. Since this delivery mode is not limited to a specific tumor, it is anticipated not only to

Table 6.2. Response rates for solid tumors to Daunorubicin

Tumor	Response Rate
Rabdomyosarcoma	21%
Soft tissue sarcomas	20%
Hodgkin's lymphoma	20%
Neuroblastoma	13%

from Von Hoff[82]

Table 6.3. Comparative IC_{50} values (ng/ml) for Daunorubicin and Doxorubicin

Cell Line	Daunorubicin	Doxorubicin	Ref.
Murine			
L1210	170	110	Iwasaki[86]
MST	10	70	Arcamone[87]
CC531 colon	50	210	Gheuens[88]
Human			
HL60	28	146	Nagasawa[89]
LOVO-109	10	15	Michieli[90]
RAJI	2.3	1.5	Michieli[90]

Table 6.4. Incidence of anthracycline-induced cardiotoxicity as a function of treatment and cumulative dose

Cumulative Dose (mg/m^2)	Incidence of Cardiotoxicity
Daunorubicin	
550	1-2%
800	5%
950	10%
1200	30%

From Dorr and Von Hoff[91]

target a wider range of solid tumors but also to be less susceptible to targeting failures such as occur with antibody targeting to antigenic sites, which may be subject to antigenic drift.

DaunoXome

Successful demonstrations of the use of ^{111}In SUVs for the imaging of a variety of clinical neoplasms lead to the formulation of a cytotoxic drug into liposomes of similar composition.[67,68] Selecting an agent with potential for a significantly improved therapeutic index, however, was crucial to achieving a meaningful expression of enhanced in vivo efficacy. Liposome entrapment can result in variable and sometimes unpredictable effects upon drug activity. Increased toxicities have been noted for some drugs such as methotrexate[69] and cytosine arabinoside,[70,71] while decreased toxicities have been observed for others including the anthracyclines in particular.[12,13,72,73] Previous investigators have demonstrated that when anthracyclines are entrapped in liposomes, decreased toxicities could be achieved without loss of antineoplastic activity.[11,74,75] Recently, however, there have been several reports of enhanced activity for encapsulated anthracyclines.[68,76-79]

Daunorubicin was the first of the anthracyclines in most common use to enter clinical trials, followed closely by doxorubicin. While not as widely used as doxorubicin, daunorubicin has demonstrated efficacy in the treatment of several adult solid tumors.[80-82] In spite of its demonstrated activity against solid tumors during the early clinical investigations, daunorubicin was superseded by doxorubicin, which had displayed greater potency in some preclinical solid tumor models.[83] Clinical investigations have continued to demonstrate significant activity for daunorubicin against various solid tumors (Table 6.2). In vitro tissue culture testing has demonstrated that the direct cytotoxicity of daunorubicin is comparable to or somewhat greater than that of doxorubicin (Table 6.3). In some cases, comparisons of the inhibitory concentration values for daunorubicin and doxorubicin have indicated that daunorubicin is 10-100 times more cytotoxic than doxorubicin. Being less polar, the cellular uptake of daunorubicin is more rapid than it is for doxorubicin,[84] potentially leading to better tumor penetration. An additional advantage of daunorubicin is that, on a cumulative basis, it is notably less cardiotoxic than doxorubicin (Table 6.4). A final consideration in our selection of daunorubicin is that it has an increased stability over doxorubicin in aqueous solutions.[85]

Formulation Approach

Early during the development of a daunorubicin liposome preparation, two formulation approaches were investigated closely. In one, daunorubicin was entrapped within the lipid bilayer as an association complex between the drug and the anionic phospholipid, distearoylphosphatidylglycerol. In the second, daunorubicin was entrapped within the inner aqueous core of the liposome as a citrate salt. Although both approaches yielded preparations with significant therapeutic improvements over free drug, entrapment within the aqueous space demonstrated superior biodistribution and antineoplastic activity.[68] The improved biodistribution was due, in part, to increased serum stability and to the decreased negative charge within the membrane bilayer (previous investigations had demonstrated that minimizing the negative charge on daunorubicin liposomes resulted in more drug delivered to tumor[67]). The lower serum stability of the membrane-associated formulation most likely is due to the increased exposure of drug to serum proteins and its exchange out of the lipid bilayer. In contrast, when entrapped within the aqueous interior, daunorubicin has less access to the vesicle surface and, consequently, less opportunity to exchange out through binding to serum proteins. Studies comparing the stabilities of these two formulations in fetal calf serum at 37°C, indicated a significantly greater retention of daunorubicin when entrapped in the interior.[68] These studies indicated a loss of over one-third of the daunorubicin in the membrane-associated formulation following an incubation of only three hours compared with a loss of only 2% after 50 hours when drug was entrapped in the aqueous interior. A net negative charge within the membrane bilayer had also been shown to be inversely related to tumor uptake and efficacy.[67] Entrapment within the inner aqueous space eliminated the need for anionic lipids as membrane-associated complexing agents, and resulted in enhanced tumor uptake and efficacy.

Once the superiority of the formulation containing daunorubicin within the inner aqueous compartment was established, efforts focused on its development and characterization. Preclinical studies were performed to demonstrate efficacy in two tumor models, ascertain toxicity, and determine the pharmacokinetic profile with particular reference to tumor targeting. These preclinical studies have been reported previously in detail[68,79,92] and are summarized below.

We used P1798 lymphosarcoma as our first tumor model to compare the single-dose efficacies of DaunoXome and free daunorubicin. This model is particularly useful as an initial formulation screen since it grows rapidly and the tumor take rate approaches 100%. For untreated controls the median survival time (MST) was 13 d, (Fig. 6.3). When compared at equivalent daunorubicin doses, DaunoXome produced significantly greater MST values than free drug at all dose levels tested. The maximum MST observed for free daunorubicin was 18.5 d at 30 mg/kg. At 40 mg/kg, free drug was toxic, reducing the MST to 12.5 d. No long-term survivors (LTS) were observed in any of the free drug treatment groups. In contrast, the maximum MST for DaunoXome was 21.5 d, at 30 mg/kg. Of particular note for DaunoXome treatment was the observation of three LTS, (>60 d).

We used a more slowly growing mammary adenocarcinoma, MA16C,[93,94] as our second tumor model. In this study, untreated control mice had an MST of 36 d, (Fig. 6.4). Two of twenty control mice survived through the end of the study (114 d) tumor-free while the remaining eighteen controls developed tumors that eventually were fatal. In the free drug treatment groups, tumor-free LTS were noted at only 10 and 20 mg/kg (one and four animals, respectively). All other free daunorubicin treated mice either developed detectable tumors or died of drug toxicity at higher doses. The incidence of tumor-free LTS for the DaunoXome treatment groups was dramatically higher. Of the 49 LTS treated with DaunoXome at doses ranging from 2-35 mg/kg, nearly all were tumor free (three mice had small tumors: two at 2 mg/kg and one at 20 mg/kg). Comparison of equivalent daunorubicin doses indicated that DaunoXome produced significantly greater tumor growth inhibition than did free drug. DaunoXome was associated with apparent complete cures at 25 mg/kg, with 100% of this group becoming tumor-free LTS. In contrast, only 40% of the mice treated with free drug at 20 mg/kg were tumor-free LTS.

The direct cytotoxic effect of DaunoXome on tumors in vivo is supported by the significant reduction of tumor volume, relative to free drug. Across all free drug dose levels, each established tumor eventually grew to considerable size (>500 mg). Of 40 mice treated with DaunoXome at therapeutic doses (15-30 mg/kg), 14 had confirmed tumors that regressed and disappeared completely. When compared to free drug at equivalent daunorubicin doses, DaunoXome significantly increased tumor growth suppression throughout the dose range of 2-25 mg/kg ($p < 0.05$, at 2 and 5 mg/kg for tumors measured on day 29 of the study; at 15, 20 and 25 mg/kg on day 50). Tumor growth inhibition for DaunoXome at 2 mg/kg was approximately equivalent to that observed for free drug at 15 mg/kg. DaunoXome was thus more effective than free daunorubicin in terms of MST and tumor growth inhibition, throughout the

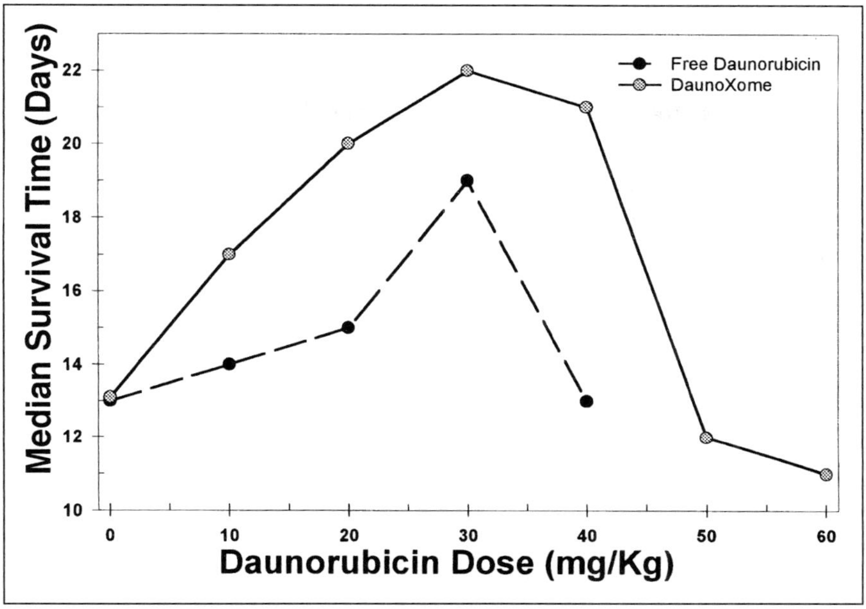

Fig. 6.3. Murine P1798 lymphosarcoma solid tumor, comparative treatment results. CD_2F1 mice received 10^6 P1798 lymphosarcoma tumor cells intradermally and were treated four days later with a single dose of free or liposomal daunorubicin at the indicated doses. Log-dose survival time studies (data not shown) have indicated that a three day difference in survival time corresponds to about a 10-fold difference in tumor cell burden.[68]

2-25 mg/kg dose range. Of particular interest was the observation that the MST and percent LTS observed for the lowest DaunoXome dose (2 mg/kg) were equivalent to those seen at the optimal free drug dose (20 mg/kg).

Significant differences between the pharmacokinetics of free daunorubicin and DaunoXome were observed in plasma and in P1798 solid tumor. DaunoXome produced one hour plasma levels of daunorubicin-eq (daunorubicin fluorescent equivalents; parent drug plus fluorescent metabolites) of 268 µg/ml, an increase of 185-fold over free drug levels of 1.4 µg/ml (Fig. 6.5). DaunoXome increased plasma AUC values by a similar amount, 227-fold over free drug. Tumor levels for daunorubicin-eq at one hour post injection of free drug were 9.6 µg/g and demonstrated no significant increase thereafter. In contrast, DaunoXome treatment produced an accumulation of daunorubicin-eq within the tumor that continued through eight hours, peaking at 100 µg/g (Fig. 6.6). The kinetics for tumor accumulation of daunorubicin-eq followed an apparent first order absorption process.[95] Comparison of tissue daunorubicin AUC ratios for DaunoXome to free drug indicated that DaunoXome produced the greatest increase (10-fold) in tumor tissue.[68] Hepatic and splenic tissues showed smaller AUC ratio increases (1.6-fold each). The increased brain AUC (2.1-fold) may not be significant since the absolute levels were low for both free and entrapped drug. Additionally, previous reports have indicated that liposomes do not cross the blood-brain barrier.[7] Minor AUC increases were noted for kidneys and small intestines. DaunoXome produced decreased AUC values for both heart and

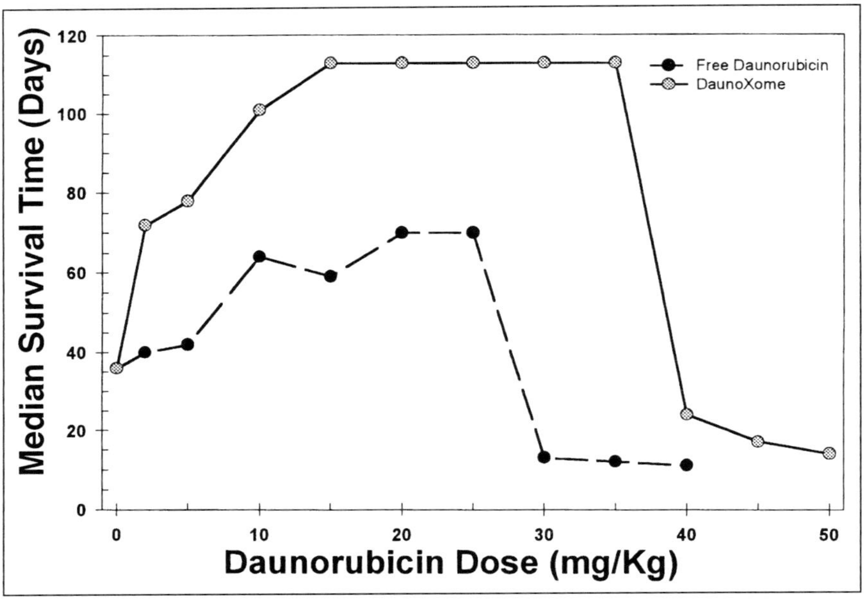

Fig. 6.4. MA16C mammary adenocarcinoma: Comparative median survival times as a function of treatment. In this study, BCF1 mice received subcutaneous 0.1 ml implants of a 1:10 brei of MA16C tumor cells followed by a single intravenous treatment at four days of either free or liposomal daunorubicin at the indicated doses.

lungs (0.4- and 0.7-fold, respectively). For all tissues examined in this study, P1798 tumor demonstrated the greatest increased accumulation of DaunoXome-delivered daunorubicin.

In vitro cytotoxicity testing of DaunoXome and free drug against P1798 tumor cells indicates that the relative potencies of these two forms of daunorubicin vary widely as a function of incubation times (Fig. 6.7).[92] For short incubation times (≤ 8 h), DaunoXome is *less* cytotoxic than free daunorubicin. This is not surprising since the free drug is highly permeable to cell membranes and is fully available when added to the tissue culture medium. For longer incubation times, on the other hand, DaunoXome is *more* cytotoxic than the free drug. Confocal microscopy of P1798 cells incubated with free drug or DaunoXome has demonstrated differences in intracellular availability and trafficking. While free daunorubicin is taken up rapidly by the cells entering the nucleus in about one to two hours, liposome-delivered daunorubicin is released slowly from vesicles over a prolonged time, not becoming visible within the nucleus at levels comparable to free drug until more than twenty-four hours later.

We have conducted mechanism of action studies in our laboratory, which suggest that DaunoXome liposomes may enter tumor cells by endocytosis. In one study, we encapsulated the indigo dye precursors 5-bromo-4-chloro-3-indolyl-β-D-galactopyranoside (X-gal) or 5-bromo-4-chloro-3-indolyl-β-D-phosphate (BCIP) into DSPC: cholesterol liposomes.[96] In their free state, these compounds are poorly taken up by cells either in vitro or in vivo. In the presence of the lysosomal enzymes

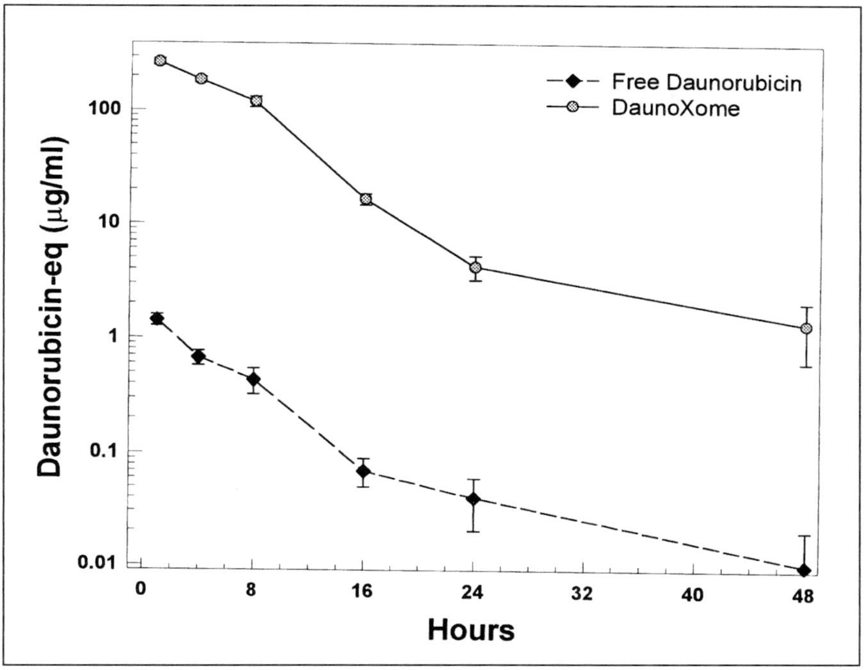

Fig. 6.5. Plasma pharmacokinetic profile of DaunoXome and free daunorubicin. $CD2F_1$ mice received intravenous free or liposomal daunorubicin at a daunorubicin dose of 20 mg/kg. DaunoXome was noted to increase the plasma AUC by more than 150-fold over free drug. Daunorubicin equivalents (daunorubicin-eq) include parent drug and fluorescent metabolites.

β-galactosidase (X-gal) or alkaline phosphatase (BCIP), activated indoxyl intermediates are generated, which dimerize to form an insoluble blue-violet dye. When liposomes containing these preparations were tested in tissue culture, formation of a compartmentalized, intense blue-violet color was noted within the tumor cell cytoplasm. Formation of similar pigmented regions in cells treated with the corresponding free substances was not observed. These observations are consistent with cellular uptake of DSPC:cholesterol liposomes by endocytosis followed by lysosomal fusion and release of liposome contents within lysosomes.

More recent studies take advantage of daunorubicin's fluorescence and its self-quenching properties while entrapped within liposomes.[92] Liposome entrapment quenches about 99% of daunorubicin's fluorescent signal; this signal increases to full intensity, however, as the drug is released from the vesicles upon their disruption, as illustrated in Figure 6.8. As with the in vitro investigations noted above, results from these fluorescence studies indicate that, following in vivo uptake, daunorubicin liposomes release their drug contents within the tumor cells over a prolonged period. This is illustrated in Figure 6.9 where data for DaunoXome in vivo delivery of daunorubicin to P1798 tumor is plotted both for total extractable drug and for in situ fluorescence. The extractable fluorescence is the sum of all daunorubicin-associated fluorescent species (parent drug and fluorescent

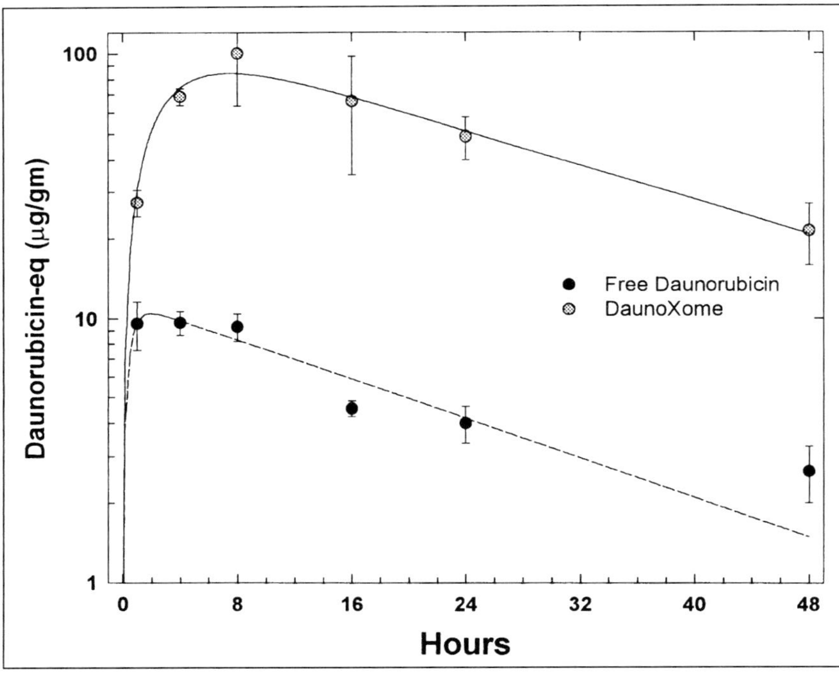

Fig. 6.6. Concentration of daunorubicin-eq in P1798 tumor tissue after administration of DaunoXome or free daunorubicin. Targeting manifests as a 10-fold increase in daunorubicin tumor AUC. Treatment was as noted in Figure 6.5.

Fig. 6.7. Inhibitory concentrations as a function of incubation time. P1798 lymphosarcoma tumor cells were incubated continuously in either DaunoXome or free daunorubicin. As expected, IC_{50} values were found to decrease for both free and liposomal daunorubicin with increasing incubation times. This rate of decrease, however, is more rapid for liposomal daunorubicin than for free drug, resulting in a greater cytotoxicity for DaunoXome with incubation times of 8 or more hours.

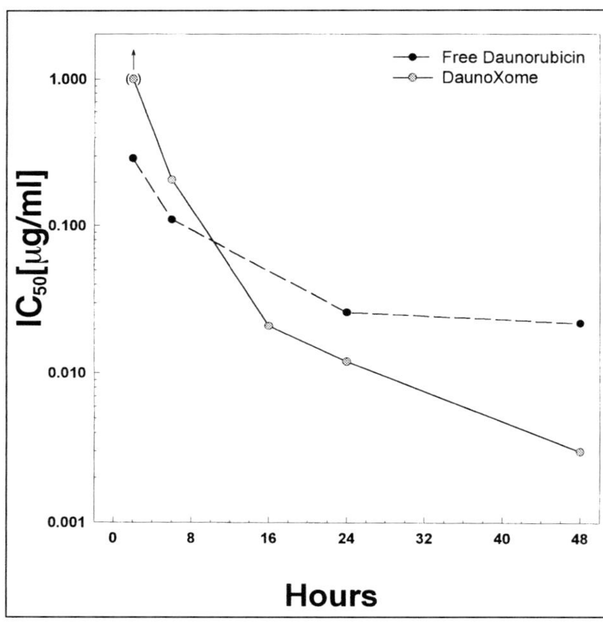

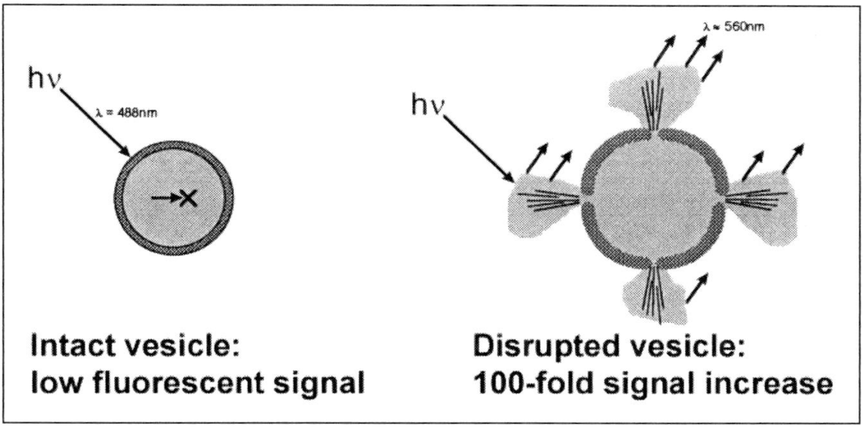

Fig. 6.8. DaunoXome (daunorubicin) fluorescence as a function of vesicle stability (fluorescence dequenching). While entrapped within the liposomes, the local concentration of daunorubicin is about 190 mM. At this high concentration, the fluorescent signal of daunorubicin is quenched nearly entirely. If the liposomes break down, releasing entrapped drug, the local concentration is diminished by about 100-fold. In diluted form (<2 mM), daunorubicin becomes fully fluorescent and its signal increases about 50- to 100-fold.

metabolites) whether free or still entrapped within the liposomes. On the other hand, in situ fluorescence (measured non-invasively by monitoring daunorubicin-associated fluorescence at the skin surface), is proportional to the amount of daunorubicin *released* from the vesicles. Differences in the peak times for these two profiles indicate the delay between delivery (8 h) and liposome release (30 h). In parallel to the in vitro findings, liposomal daunorubicin thus appears to be taken up by tumor cells *intact* and then breaks down slowly to release drug within the cells over a prolonged period, providing sustained high levels of active drug at the tumor site.

The improved tumor selectivity of DaunoXome, relative to free drug, is indicated by its 10-fold increased tumor AUC value, compared to only small differences (0.4- to 1.6-fold) observed for most normal tissues. With the exception of the spleen, DaunoXome AUC values in normal tissues are several-fold lower than for tumor. In contrast, the selectivity of free drug for tumor tissue is comparatively low, with only brain tissue demonstrating an AUC value below that for tumor. Although the pharmacokinetic data in this study did not include time points earlier than one hour, it is clear that peak tumor drug levels occur earlier for free drug (about 1-2 h, post injection) than for DaunoXome (8-9 h). The slower daunorubicin tumor accumulation for DaunoXome indicates the importance of maintaining high plasma levels of entrapped drug for enhanced tumor delivery. To achieve sustained plasma levels, it is necessary that drug-containing SUVs not only demonstrate minimal removal by non-tumor tissues but remain stable in the circulation. In the DaunoXome formulation, this stability is provided by the high phase transition temperature of DSPC and by the presence of 33 mole-percent cholesterol within the membrane bilayer. This composition reduces the loss of entrapped drug from interactions with serum proteins

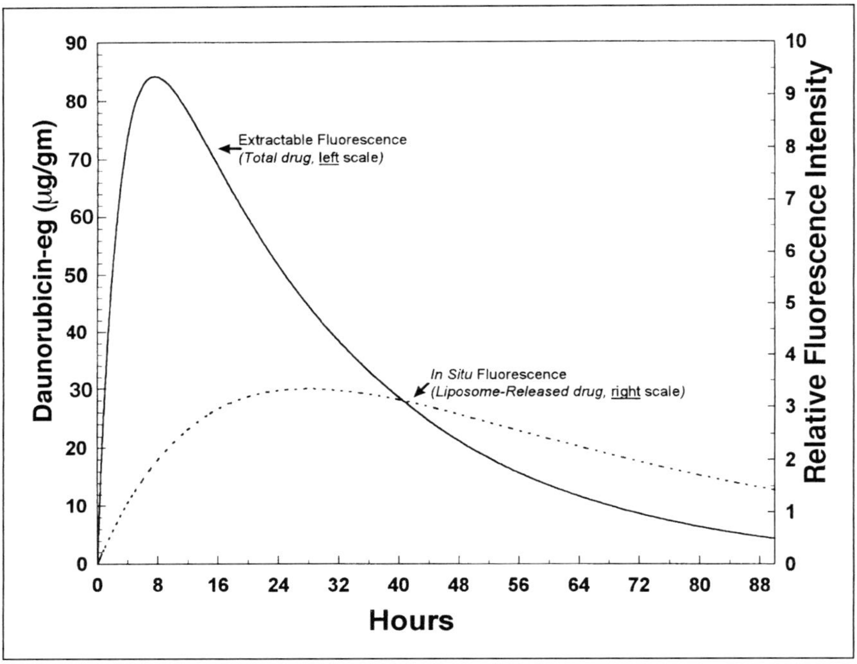

Fig. 6.9. In vivo tumor disposition of daunorubicin: profile of total vs. released drug (extracted vs. in situ drug). Total daunorubicin-associated fluorescence is based upon determination of all fluorescent species (parent drug and fluorescent metabolites) present in tissue extracts. This includes daunorubicin that has been released from the liposomes as well as material remaining entrapped. The in situ fluorescence, on the other hand, is based upon the signal produced in vivo from drug that has been released from liposome within the tumor (includes fluorescent metabolites); the fluorescence of material remaining entrapped within the liposomes is quenched and does not produce a signal under the conditions used.

at 37°C.[97-100] Comparison of the DaunoXome and free drug preclinical pharmacokinetic data thus suggests that, in the P1798 murine model, DaunoXome delivery operates on two levels: systemically as a *targeted* preparation and locally as a *sustained release* system.

DaunoXome Clinical Investigations

Due to the findings of high ^{111}In uptake by KS lesions in patients receiving the tumor imaging agent, patients with AIDS-related KS were included in the early Phase I/II dose escalation studies. In one study, involving 22 patients, response rates (partial, PR and complete responders, CR) of 55% were observed at doses of 50 and 60 mg/m^2 every two weeks.[101] Money-Kyrle et al[102] investigated the efficacy and toxicity of DaunoXome at 40 mg/m^2 every two weeks in 25 patients with advanced AIDS-associated KS of poor prognosis, achieving a partial response rate of 40% (4/10). For patients with no previous chemotherapy, the observed response rate was 57%.[47]

In these studies there was moderate myelosuppression, no observation of a significant decline in cardiac function, and an overall decrease in the frequency and severity of side effects commonly related to chemotherapy. DaunoXome human pharmacokinetics parallel those observed in preclinical investigations and are significantly different from those of the free drug. In patient studies, DaunoXome produces greater than 35-fold increases in plasma area under the curve (AUC) values, decreased total body clearance, greater peak plasma levels, and a smaller volume of distribution (Vd_{ss}).

KS has become an aggressive disease in HIV patients suffering severe immunosuppression.[103,104] The initial site of disease is usually cutaneous with nodular lesions potentially progressing rapidly to coalescence. This, in turn, may result in lymphatic obstruction and lymphedema. In addition, cutaneous disease may ulcerate, become infected, or produce local pain. Oral cavity and gastrointestinal tract involvement is common, and may involve as many as 50% of patients. Pulmonary involvement is also common, and can produce a range of symptoms including pulmonary insufficiency. In a few patients, visceral disease may precede cutaneous involvement and be the sole cause of symptomatology. KS can be the cause of death in many patients.

The incidence of AIDS patient deaths in Britain, attributable to KS, increased from 14% to 32% between 1986 and 1989.[105] Serious morbidity and disability can impair the quality of life in HIV patients even when KS is not the immediate cause of death. Epidemic KS afflicts a young population for whom preservation of quality of life and maintenance of performance status is especially important. Localized cutaneous disease can be reduced in a variety of ways, including radiation, cryotherapy, and intralesional chemotherapy with vinblastine. For more advanced disease, several chemotherapeutic agents have been used. Previously, interferon had been the only agent approved for the treatment of KS in the United States. It has been used typically for systemic disease that is less bulky and less aggressive. Clinical investigations have suggested that interferon can produce a response rate as high as 50%. This high rate has been limited primarily to patients retaining relatively intact immune systems. Patients with $CD4^+$ counts greater than 200 are most likely to achieve an objective response.[103] Unfortunately, however, most KS patients have low $CD4^+$ counts and advanced disease. The percent of responses in these patients may be no higher than 10%, and therefore interferon provides little benefit to these patients. Furthermore, interferon-α is associated with significant morbidity, including flu-like symptoms. These side-effects may impair quality of life and preclude long-term maintenance therapy.

Chemotherapy has become the principal treatment alternative for disseminated KS in patients who have failed interferon or who have low $CD4^+$ T cell counts. A wide range of cytotoxic agents has activity against KS; these include bleomycin, etoposide, doxorubicin, vinblastine and vincristine. The treatment of KS has been reviewed recently.[106] For single-agent therapy, reported response rates have varied widely and have been influenced greatly both by patient selection and by the choice of response criteria. In responding patients, chemotherapy can induce regression of cutaneous and visceral disease. Such tumor regression often is associated with improvement in tumor-associated symptomatology. Maintenance of responses requires continued therapy, however, and relapse without maintenance therapy is the rule. Combination chemotherapy regimens can result in high response rates. For example, a response rate exceeding 80% has been reported for the ABV combination of doxorubicin, bleomycin, and a vinca-alkaloid (vinblastine or vincristine).[103] Long-term therapy with aggressive combinations such as ABV is extremely difficult, however. The risks of cumulative cardiotoxicity with doxorubicin, pulmonary toxicity with bleomycin, and neurotoxicity with the vinca-alkaloids are superimposed on the possible increased risk of infectious complications and gastrointestinal toxicity observed with each treatment cycle. In addition, these agents are associated with significant myelosuppression, worsening the immunosuppression already present from the HIV infection.

Gill et al[107] conducted a comprehensive, randomized Phase III trial in 232 patients with AIDS-related KS to compare the therapeutic efficacy of DaunoXome to the standard regimen of doxorubicin, bleomycin, and vincristine (ABV). Treatments were administered every two weeks and consisted of either DaunoXome at 40 mg/m^2 or a combination of doxorubicin 10 mg/m^2, bleomycin 15 U, and vincristine 1 mg. Patients were kept on treatment until there was a complete response, unacceptable toxicity, or disease progression. Following strict criteria, the overall response rate was 25% (3 CR and 26 PR, of 116 treated) for DaunoXome and 28% (1 CR and 30 PRs of 111 treated) for ABV. There were no statistically significant differences between these treatments in terms of response rates, survival times, or median time to disease progression. There were, however, significant differences between the treatment groups in terms of toxicities. Patients treated with ABV had a significantly higher incidence of alopecia and neuropathy ($p < 0.0001$) while patients receiving DaunoXome experienced more grade 4 neutropenia. Cardiac functions were normal in all patients with none from either treatment group experiencing anthracycline-associated congestive heart failure.

Conclusions

The preclinical and clinical investigations presented above demonstrate that DaunoXome is efficacious within a safe and well-tolerated dose range. Preclinical studies, comparing DaunoXome and free drug, indicate that SUVs composed of DSPC:cholesterol (2:1) can increase the in vivo delivery of daunorubicin to solid tumors by 10-fold. Prolonged survival times and greater tumor growth inhibition indicate that the entrapped drug remains pharmacologically available and active. The clinical findings are consistent with a preferential tumor delivery of DaunoXome and suggest the use of DaunoXome at a dose of 40 mg/m^2 every two weeks for the treatment of KS in HIV positive patients.

Monitoring liposome integrity in vivo, using ^{111}In perturbed angular correlation, has shown that DSPC:cholesterol SUVs are highly stable in serum under physiological conditions.[23,98,108] However, once within tumors, these SUVs break down and release their entrapped contents.[26] Gamma-emitting ^{111}In becomes strongly tissue-bound and accumulates within tumors, thus enabling a variety of solid neoplasms in both animals and humans to be visualized by gamma-camera imaging.[24,34,36] In patients, the principal solid tumors demonstrating significant uptake of SUV-entrapped-^{111}In have included lung, breast and ovarian cancers, head and neck tumors,[36] as well as KS.[37] It is thus anticipated that DaunoXome, which is composed of similar DSPC:cholesterol SUVs may find potential applications for treatment of a similar range of solid neoplasms.

Metastasis is one of the more devastating aspects of cancer and a principal cause for treatment failures.[54] Systemic toxicity and inadequate tumor uptake of chemotherapeutic agents limit the effective treatment of disseminated malignant disease. Liposomes that target the increased permeability of tumor neovasculature have the potential for specifically delivering encapsulated agents to a wide range of disseminated solid neoplasms. A particular advantage to targeting this attribute of tumor physiology, is that it will be less likely to be downregulated without denying oxygen and essential nutrients to the tumor.

DaunoXome offers an effective and safe treatment for KS in HIV-positive patients.[37,109] It is associated with a significant rate of tumor regression and has been effective in patients who have failed prior chemotherapy. Side effects have generally been mild with sleepiness and loss of appetite most common. There have been no reports of more serious complications such as hepatotoxicity.[110] At the recommended dose of 40 mg/m^2, manageable myelosuppression associated with DaunoXome allows for the concomitant use of anti-HIV therapies, including AZT.[107] As a result of its efficacy and safety profile in Phase III clinical investigations, DaunoXome is the only approved first line therapy for KS in the United States and in several European countries. In ongoing Phase III studies, DaunoXome is demonstrating encouraging activity against several other neoplasms including breast cancer and lymphomas.[111,112] At the time of this writing, DaunoXome has been approved in 16 countries including: Canada, Denmark, France, Germany, Russia, the United Kingdom and the United States.

References

1. Kirby C, Gregoriadis G. The effect of lipid composition of small unilamellar liposomes containing melphalan and vincristine on drug clearance after injection into mice. Biochem Pharmacol 1983; 32:609-15.
2. Juliano RL, Stamp D. Pharmacokinetics of liposome-encapsulated anti-tumor drugs. Studies with vinblastine, actinomycin D, cytosine arabinoside, and daunomycin. Biochem Pharmacol 1978; 27:21-7.
3. Freise J, Mueller WH, Magerstedt P, Schmoll HJ. Pharmacokinetics of liposome encapsulated cisplatin in rats. Arch Int Pharmacodyn Ther 1982; 258:180-92.
4. Allen TM, Everest JM. Effect of liposome size and drug release properties on pharmacokinetics of encapsulated drug in rats. J Pharmacol Exp Ther 1983; 226:539-44.
5. Patel HM. Transcytosis of drug carriers carrying peptides across epithelial barriers. Biochem Soc Trans 1989; 17:940-2.
6. Postmes TJ, Hukkelhoven M, van den Bogaard AE, Halders SG, Coenegracht J. Passage through the blood-brain barrier of thyrotropin-releasing hormone encapsulated in liposomes. J Pharm Pharmacol 1980; 32:722-4.
7. Tökés ZA, St.Peteri AK, Todd JA. Availability of liposome content to the nervous system. Liposomes and the blood-brain barrier. Brain Res 1980; 188:282-6.
8. D Silva JB, Notari RE. Drug stability in liposomal suspensions: hydrolysis of indomethacin, cyclocytidine, and p-nitrophenyl acetate. J Pharm Sci 1982; 71:1394-8.
9. Fiore S, Serhan CN. Phospholipid bilayers enhance the stability of leukotriene A4 and epoxytetraenes: stabilization of eicosanoids by liposomes. Biochem Biophys Res Commun 1989; 159:477-81.
10. Kurosaki Y, Kimura T, Muranishi S, Sezaki H. The use of liposomes as enzyme carriers. I. Dependence of enzyme stability on the method of preparation. Chem Pharm Bull (Tokyo) 1981; 29:1175-8.
11. Forssen EA, Tökés ZA. In vitro and in vivo studies with adriamycin liposomes. Biochem Biophys Res Commun 1979; 91:1295-301.
12. Rahman A, Kessler A, More N, Sikic B, Rowden G, Woolley P, Schein PS. Liposomal protection of adriamycin-induced cardiotoxicity in mice. Cancer Res 1980; 40:1532-7.
13. Forssen EA, Tökés ZA. Use of anionic liposomes for the reduction of chronic doxorubicin-induced cardiotoxicity. Proc Natl Acad Sci USA 1981; 78:1873-7.

14. Forssen EA, Tökés ZA. Attenuation of dermal toxicity of doxorubicin by liposome encapsulation. Cancer Treat Rep 1983; 67:481-4.
15. Patel KR, Baldeschwieler JD. Mouse Lewis lung carcinoma and hepatoma ascites treatment by combination of liposome chemotherapy and non-specific immunotherapy. Int J Cancer 1984; 34:717-23.
16. Mauk MR, Gamble RC. Preparation of lipid vesicles containing high levels of entrapped radioactive cations. Anal Biochem 1979; 94:302-7.
17. Mauk MR, Gamble RC. Stability of lipid vesicles in tissues of the mouse: a gamma-ray perturbed angular correlation study. Proc Natl Acad Sci USA 1979; 76:765-9.
18. Hwang KJ, Mauk MR. Fate of lipid vesicles in vivo: a gamma-ray perturbed angular correlation study. Proc Natl Acad Sci USA 1977; 74:4991-5.
19. Meares CF, Bryant RG, Baldeschwieler JD, Shirley DA. Study of carbonic anhydrase using perturbed angular correlations of gamma radiation. Proc Natl Acad Sci USA 1969; 64:1155-61.
20. Meares CF, Sundberg MW, Baldeschwieler JD. Perturbed angular correlation study of a haptenic molecule. Proc Nat Acad Sci USA 1972; 69:3718-3722.
21. Goodwin DA, Meares CF, Song CH. The study of 111 In-labeled compounds in mice, using perturbed angular correlations of gamma radiations. Radiology 1972; 105:699-702.
22. Meares CF, Westmoreland DG. The study of biological macromolecules using perturbed angular correlations of gamma radiation. Cold Spring Harb Symp Quant Biol 1972; 36:511-6.
23. Wallingford RH, Williams LE. Is stability a key parameter in the accumulation of phospholipid vesicles in tumors? J Nucl Med 1985; 26:1180-5.
24. Proffitt RT, Williams LE, Presant CA, Tin GW, Uliana JA, Gamble RC, Baldeschwieler JD. Tumor-imaging potential of liposomes loaded with In-111-NTA: biodistribution in mice. J Nucl Med 1983; 24:45-51.
25. Proffitt RT, Williams LE, Presant CA, Tin GW, Uliana JA, Gamble RC, Baldeschwieler JD. Liposomal blockade of the reticuloendothelial system: improved tumor imaging with small unilamellar vesicles. Science 1983; 220:502-5.
26. Williams LE, Proffitt RT, Lovisatti L. Possible applications of phospholipid vesicles (liposomes) in diagnostic radiology. J Nucl Med Allied Sci 1984; 28:35-45.
27. Hwang KJ, Luk KFS, Beaumier PL. Volume of distribution and transcapillary passage of small unilamellar vesicles. Life Sci 1982; 31:949-55.
28. Ogihara-Umeda I, Sasaki T, Nishigori H. Active removal of radioactivity in the blood circulation using biotin-bearing liposomes and avidin for rapid tumour imaging. Eur J Nucl Med 1993; 20:170-2.
29. Ogihara-Umeda I, Sasaki T, Toyama H, Oda K, Senda M, Nishigori H. Rapid tumor imaging by active background reduction using biotin-bearing liposomes and avidin. Cancer Res 1994; 54:463-7.
30. Gordon K, Tajuddin A, Guitart J, Kuzel T, Eramo L, VonRoenn J. Hand-Foot syndrome associated with liposome-encapsulated doxorubicin therapy. Cancer 1995; 75:2169-2173.
31. Mauk MR, Gamble RC, Baldeschwieler JD. Vesicle targeting: timed release and specificity for leukocytes in mice by subcutaneous injection. Science 1980; 207:309-11.
32. Leipert TK, Baldeschwieler JD, Shirley DA. Applications of gamma ray angular correlations to the study of biological macromolecules in solution. Nature 1968; 220:907-9.
33. Williams LE, Duda RB, Proffitt RT, Beatty BG, Beatty JD, Wong JY, Shively JE, Paxton RJ. Tumor uptake as a function of tumor mass: a mathematic model. J Nucl Med 1988; 29:103-9.

34. Turner AF, Presant CA, Proffitt RT, Williams LE, Winsor DW, Werner JL. In-111-labeled liposomes: dosimetry and tumor depiction. Radiology 1988; 166:761-5.
35. Patel KR, Tin GW, Williams LE, Baldeschwieler JD. Biodistribution of phospholipid vesicles in mice bearing Lewis lung carcinoma and granuloma. J Nucl Med 1985; 26:1048-55.
36. Presant CA, Proffitt RT, Turner AF, Williams LE, Winsor D, Werner JL, Kennedy P, Wiseman C et al. Successful imaging of human cancer with indium-111-labeled phospholipid vesicles. Cancer 1988; 62:905-11.
37. Presant CA, Blayney D, Proffitt RT, Turner AF, Williams LE, Nadel HI, Kennedy P, Wiseman C et al. Preliminary report: imaging of Kaposi's sarcoma and lymphoma in AIDS with indium-111-labeled liposomes. Lancet 1990; 335:1307-9.
38. McNeil BJ. Guidlines for evaluating new tests. In: Rocha AFG, Harbert JC, eds. Textbook of Nuclear Medicine: Clinical Applications. Philadelphia: Lea and Febiger, 1979:473-484.
39. Kubo A, Nakamura K, Sammiya T, Katayama M, Hashimoto T, Hashimoto S, Kobayashi H, Teramoto T. Indium-111-labelled liposomes: dosimetry and tumour detection in patients with cancer. Eur J Nucl Med 1993; 20:107-13.
40. Dictor M, Bendsoe N, Runke S, Witte M. Major basement membrane components in Kaposi's sarcoma, angiosarcoma and benign vascular neogenesis. J Cutan Pathol (Denmark) 1955; 22:435-441.
41. Albini A, Barillari G, Benelli R, Gallo R, Ensoli B. Angiogenic properties of human immunodeficiency virus type 1 Tat protein. Proc Natl Acad Sci USA 1995; 92:4838-4842.
42. Schwartz RA. Kaposi's sarcoma: advances and perspectives. J Am Acad Dermatol 1996; 34:804-14.
43. Morris CB, Gendelman R, Marrogi AJ, Lu M, Lockyer JM, Alperin-Lea W, Ensoli B. Immunihistochemical detection of Bcl-2 in AIDS-associated and classical Kaposi's sarcoma. Am J Pathol 1996; 148:1055-63.
44. Noel JC, Hermans P, Andre J, Fayt I, Simonart T, Verhest A, Haot J, Burny A. Herpesvirus-like DNA sequences and Kaposi's sarcoma: relationship with epidemiology, clinical spectrum, and histologic features. Cancer 1996; 77:2132-6.
45. Brown LF, Tognazzi K, Dvorak HF, Harrist TJ. Strong expression of kinase insert domain-containing receptor, a vascular permeability factor/vascular endothelial growth factor receptor in AIDS-associated Kaposi's sarcoma and cutaneous angiosarcoma. Am J Pathol 1996; 148:1065-74.
46. Wu NZ, Klitzman B, Rosner G, Needham D, Dewhirst MW. Measurement of material extravasation in microvascular networks using fluorescence video-microscopy. Microvasc Res 1993; 46:231-53.
47. Presant CA, Scolaro M, Kennedy P, Blayney DW, Flanagan B, Lisak J, Presant J. Liposomal daunorubicin treatment of HIV-associated Kaposi's sarcoma. Lancet 1993; 341:1242-3.
48. Kennedy KA, Teicher BA, Rockwell S, Sartorelli AC. The hypoxic tumor cell: a target for selective cancer chemotherapy. Biochem Pharmacol 1980; 29:1-8.
49. Jain RK. Delivery of molecular medicine to solid tumors. Science 1996; 271:1079-80.
50. Netti PA, Baxter LT, Boucher Y, Skalak R, Jain RK. Time-dependent behavior of interstitial fluid pressure in solid tumors: implications for drug delivery. Cancer Res 1995; 55:5451-8.
51. Less JR, Skalak TC, Sevick EM, Jain RK. Microvascular architecture in a mammary carcinoma: branching patterns and vessel dimensions. Cancer Res 1991; 51:265-73.
52. Folkman J. Angiogenesis in cancer, vascular, rheumatoid and other disease. Nat Med 1995; 1:27-31.

53. Martiny-Baron G, Marme D. VEGF-mediated tumour angiogenesis: a new target for cancer therapy. Curr Opin Biotechnol 1995; 6:675-80.
54. Jahroudi N, Greenberger JS. The role of endothelial cells in tumor invasion and metastasis. J Neurooncol 1995; 23:99-108.
55. Weiss L, Orr FW, Honn KV. Interactions between cancer cells and the microvasculature: a rate-regulator for metastasis. Clin Exp Metastasis 1989; 7:127-67.
56. Bikfalvi A. Significance of angiogenesis in tumour progression and metastasis. Eur J Cancer 1995; 31A:1101-4.
57. Rak JW, St Croix BD, Kerbel RS. Consequences of angiogenesis for tumor progression, metastasis and cancer therapy. Anticancer Drugs 1995; 6:3-18.
58. Chaplain MA, Giles SM, Sleeman BD, Jarvis RJ. A mathematical analysis of a model for tumour angiogenesis. J Math Biol 1995; 33:744-70.
59. Skinner SA, Tutton PJ, PE OB. Microvascular architecture of experimental colon tumors in the rat. Cancer Res 1990; 50:2411-7.
60. Byrne HM, Chaplain MA. Mathematical models for tumour angiogenesis: numerical simulations and nonlinear wave solutions. Bull Math Biol 1995; 57:461-86.
61. Strugar JG, Criscuolo GR, Rothbart D, Harrington WN. Vascular endothelial growth/ permeability factor expression in human glioma specimens: correlation with vasogenic brain edema and tumor-associated cysts. J Neurosurg 1995; 83:682-9.
62. Warren BA. The vascular morphology of tumors. In: Peterson H-I, ed. Tumor Blood Circulation: Angiogenesis, Vascular Morphology and Blood Flow of Experimental and Human Tumors. Boca Raton, Florida: CRC Press, 1979:1-47.
63. Hwang KJ, Padki MM, Chow DD, Essien HE, Lai JY, Beaumier PL. Uptake of small liposomes by non-reticuloendothelial tissues. Biochim Biophys Acta 1987; 901:88-96.
64. Poste G, Bucana C, Raz A, Bugelski P, Kirsh R, Fidler IJ. Analysis of the fate of systemically administered liposomes and implications for their use in drug delivery. Cancer Res 1982; 42:1412-22.
65. Papahadjopoulos D, Gabizon A. Targeting of liposomes to tumor cells in vivo. Ann NY Acad Sci 1987; 507:64-74.
66. Kerr DJ, Kaye SB. Aspects of cytotoxic drug penetration, with particular reference to anthracyclines. Cancer Chemother Pharmacol 1987; 19:1-5.
67. Forssen EA. Chemotherapy with anthracycline liposomes. In: Gregoriadis G, ed. Liposomes as Drug Carriers: Recent Trends and Progress. Chichester: J Wiley and Sons, 1988:355-364.
68. Forssen EA, Coulter DM, Proffitt RT. Selective in vivo localization of daunorubicin small unilamellar vesicles in solid tumors. Cancer Res 1992; 52:3255-61.
69. Kaye SB, Boden JA, Ryman BE. The effect of liposome (phospholipid vesicle) entrapment of actinomycin D and methotrexate on the in vivo treatment of sensitive and resistant solid murine tumours. Eur J Cancer 1981; 17:279-89.
70. Ganapathi R, Krishan A, Wodinsky I, Zubrod CG, Lesko LJ. Effect of cholesterol content on antitumor activity and toxicity of liposome-encapsulated 1-beta-D-arabinofuranosylcytosine in vivo. Cancer Res 1980; 40:630-3.
71. Rustum YM, Dave C, Mayhew E, Papahadjopoulos D. Role of liposome type and route of administration in the antitumor activity of liposome-entrapped 1-beta-D-arabinofuranosylcytosine against mouse L1210 leukemia. Cancer Res 1979; 39:1390-5.
72. Gabizon A, Meshorer A, Barenholz Y. Comparative long-term study of the toxicities of free and liposome-associated doxorubicin in mice after intravenous administration. J Natl Cancer Inst 1986; 77:459-69.
73. Storm G, Roerdink FH, Steerenberg PA, de Jong WH, Crommelin DJ. Influence of lipid composition on the antitumor activity exerted by doxorubicin-containing liposomes in a rat solid tumor model. Cancer Res 1987; 47:3366-72.
74. van Hoesel QG, Steerenberg PA, Crommelin DJ, van Dijk A, van Oort W, Klein S, Douze JM, de Wildt DJ et al. Reduced cardiotoxicity and nephrotoxicity with pres-

ervation of antitumor activity of doxorubicin entrapped in stable liposomes in the LOU/M Wsl rat. Cancer Res 1984; 44:3698-705.
75. Forssen EA, Tökés ZA. Improved therapeutic benefits of doxorubicin by entrapment in anionic liposomes. Cancer Res 1983; 43:546-50.
76. Gabizon AA, Barenholz Y, Bialer M. Prolongation of the circulation time of doxorubicin encapsulated in liposomes containing a polyethylene glycol-derivatized phospholipid: pharmacokinetic studies in rodents and dogs. Pharm Res 1993; 10:703-8.
77. Gabizon A, Catane R, Uziely B, Kaufman B, Safra T, Cohen R, Martin F, Huang A et al. Prolonged circulation time and enhanced accumulation in malignant exudates of doxorubicin encapsulated in polyethylene-glycol coated liposomes. Cancer Res 1994; 54:987-92.
78. Allen TM. Long-circulating (sterically stabilized) liposomes for targeted drug delivery. Trends Pharmacol Sci 1994; 15:215-20.
79. Forssen EA, Ross ME. Daunoxome treatment of soild tumors: preclinical and clinical investigations. J Liposome Res 1994; 4:481-512.
80. Weiss RB, Bruno S. Daunorubicin treatment of adult solid tumors. Cancer Treat Rep 1981; 4:25-8.
81. Von Hoff DD, Rozencweig M, Slavik M, Muggia FM. Activity of daunomycin in solid tumors (letter). JAMA 1976; 236:1693.
82. Von Hoff DD. Use of daunorubicin in patients with solid tumors. Semin Oncol 1984; 11:23-7.
83. Dano K. Cross-resistance between Adriamycin, Daunomycin and vincristine in Ehrlich Ascites Tumor in vivo. In: Carter SK et al, eds. International Symposium on Adriamycin. Berlin, Heidleberg: Springer-Verlag, 1972:90-95.
84. Bachur NR, Steele M, Meriwether WD, Hildebrand RC. Cellular pharmocodynamics of several anthrocycline antibiotics. J Med Chem 1976; 19:651-4.
85. Bosanquet AG. Stability of solutions of antineoplastic agents during preparation and storage for in vitro assays. II. Assay methods, adriamycin and the other antitumour antibiotics. Cancer Chemother Pharmacol 1986; 17:1-10.
86. Iwasaki H, Liu YP, Nojyo Y, Ueda T, Nakamura T. Quantitative description of morphologic changes effected by antileukemic agents in L1210 leukemia cells. Anticancer Res 1995; 15:133-9.
87. Arcamone F. In: F. Arcamone, ed. Doxorubicin. New York: Academic Press, 1981:354.
88. Gheuens E, van der Heyden S, Elst H, Eggermont A, Van Oosterom A, De Bruijn E. Multidrug resistance in rat colon carcinoma cell lines CC531, CC531mdr+ and CC531rev. Jpn J Cancer Res 1993; 84:1201-8.
89. Nagasawa K, Natazuka T, Chihara K, Kitazawa F, Tsumura A, Takara K, Nomiyama M, Ohnishi N et al. Transport mechanism of anthracycline derivatives in human leukemia cell lines: uptake and efflux of pirarubicin in HL60 and pirarubicin-resistant HL60 cells. Cancer Chemother Pharmacol 1996; 37:297-304.
90. Michieli M, Michelutti A, Damiani D, Pipan C, Raspadori D, Lauria F, Baccarani M. A comparative analysis of the sensitivity of multidrug resistant (MDR) and non-MDR cells to different anthracycline derivatives. Leuk Lymphoma 1993; 9:255-64.
91. Dorr RT, Von Hoff DD, eds. Cancer Chemotherapy Handbook. 2nd ed. Norwalk, CT: Appleton and Lange, 1994:1020.
92. Forssen EA, Malé-Brune R, Adler-Moore JP, Lee MJ, Schmidt PG, Krasieva TB, Shimizu S, Tromberg BJ. Fluorescence imaging studies for the disposition of daunorubicin liposomes (DaunoXome) within tumor tissue. Cancer Res 1996; 56:2066-75.
93. Corbett TH, Griswold DPJ, Roberts BJ, Peckham JC, Schabel FMJ. Biology and therapeutic response of a mouse mammary adenocarcinoma (16/C) and its potential as a model for surgical adjuvant chemotherapy. Cancer Treat Rep 1978; 62:1471-88.

94. Casazza AM. Experimental evaluation of anthracycline analogs. Cancer Treat Rep 1979; 63:835-44.
95. Gibaldi M, Perrier D. Swarbrick J, ed. In: Drugs and the Pharmaceutical Sciences. Pharmacokinetics. Vol. 1. New York: Marcel Dekker, Inc., 1975; 329.
96. Tsou KC, Lo KW, Ledis SL, Miller EE. Indigogenic phosphodiesters as potential chromogenic cancer chemotherapeutic agents. J Med Chem 1972; 15:1221-5.
97. Ipsen JH, Karlstrom G, Mouritsen OG, Wennerstrom H, Zuckermann MJ. Phase equilibria in the phosphatidylcholine-cholesterol system. Biochim Biophys Acta 1987; 905:162-72.
98. Papahadjopoulos D, Cowden M, Kimelberg H. Role of cholesterol in membranes. Effects on phospholipid-protein interactions, membrane permeability and enzymatic activity. Biochim Biophys Acta 1973; 330:8-26.
99. Papahadjopoulos D, Jacobson K, Nir S, Isac T. Phase transitions in phospholipid vesicles. Fluorescence polarization and permeability measurements concerning the effect of temperature and cholesterol. Biochim Biophys Acta 1973; 311:330-48.
100. Kirby C, Clarke J, Gregoriadis G. Effect of the cholesterol content of small unilamellar liposomes on their stability in vivo and in vitro. Biochem J 1980; 186:591-8.
101. Gill PS, Espina BM, Muggia F, Cabriales S, Tulpule A, Esplin JA, Liebman HA, Forssen E et al. Phase I/II clinical and pharmacokinetic evaluation of liposomal daunorubicin. J Clin Oncol 1995; 13:996-1003.
102. Money-Kyrle JF, Bates F, Ready J, Gazzard BG, Phillips RH, Boag FC. Liposomal daunorubicin in advanced Kaposi's sarcoma: a phase II study. Clin Oncol (R Coll Radiol) 1993; 5:367-71.
103. Gill PS, Naidu YM, Salahuddin SZ. Recent advances in AIDS-related Kaposi's sarcoma. Curr Opin Oncol 1990; 2:1161-6.
104. Krown SE. The role of interferon in the therapy of epidemic Kaposi's sarcoma. Semin Oncol 1987; 14:27-33.
105. Kaplan SB. Neoplastic complications in HIV disease. Infections in Med 1992; 9:20.
106. Northfelt DW, Kahn JO, Volberding PA. Treatment of AIDS-related Kaposi's sarcoma. Hematol Oncol Clin North Am 1991; 5:297-310.
107. Gill PS, Wernz J, Scadden DT, Cohen P, Mukwaya GM, von Roenn JH, Jacobs M, Kempin S et al. Randomized phase III trial of liposomal daunorubicin (DaunoXome) versus doxorubicin, bleomycin, vincristine, (ABV) in AIDS-related Kaposi's sarcoma. J Clin Oncology 1996; 14:2353-2364.
108. Ogihara-Umeda I, Kojima S. Cholesterol enhances the delivery of liposome-encapsulated gallium-67 to tumors. Eur J Nucl Med 1989; 15:612-7.
109. Schurmann D, Dormann A, Grunewald T, Ruf B. Successful treatment of AIDS-related pulmonary Kaposi's sarcoma with liposomal daunorubicin. Eur Respir J 1994; 7:824-5.
110. Hengge UR, Brockmeyer NH, Rasshofer R, Goos M. Fatal hepatic failure with liposomal doxorubicin [letter]. 1993; Lancet 341:383-4.
111. Erdkamp FLG, Hupperets PSGJ, Ten Bokkel-Huinink WW, Nuyts GD, Eestermans GH. Phase II study of liposomal encapsulated daunorubicin (DaunoXome®) in advanced breast cancer. A Phase II pilot trial. (Abstract) 18th Annual San Antonio Breast Cancer Symposium, Dec. 10-13, 1995.
112. Tulpule A, Rarick MU, Kolitz J, Bernstein J, Traynor A, Myers A, Harvey-Buchanan L, Vergel de Dios-Salvosa M, Espina BM, Mukwaya G, Ross M, Levine AM. Liposomal encapsulated daunorubicin (DaunoXome®) has activity in relapsed/refractory low grade and intermediate grade non-Hodgkin's Lymphoma (NHL). Blood 88:Suppl 1:92a.

Chapter 7

Enhancement of Doxorubicin by Encapsulation in Long Circulating Thermosensitive Liposomes Combined with Local Hyperthermia

Kazuo Maruyama

Introduction

The therapeutic efficacy of antitumor drugs is restricted by dose-limiting toxicities to normal tissues in vivo, but would be greatly improved if the drugs could be selectively directed to the tumor sites and away from sensitive tissues. Liposomes have been proposed to be useful drug carriers for targeted drug delivery systems, and are under investigation in several therapeutic fields.[1,2] In order to achieve maximum targeting, liposomes should remain in the systemic circulation for a long time. However, formulations of liposomes used in the past were rapidly removed from the circulation by the reticuloendothelial system (RES).[3,4] It has been demonstrated recently that newly developed liposomes, containing either monosialoganglioside G_{M1} (G_{M1})[5] or amphipathic polyethylene glycol (PEG),[6-8] are not readily taken up by the macrophages in the RES and hence stay in the circulation for a relatively long period of time. More importantly, these long circulating liposomes, having small size (100-200 nm in mean diameter), showed significantly greater accumulation in solid tumors compared with the corresponding conventional liposomes.[9,10] These reports indicate that long circulating liposomes offer considerable potential for drug targeting to solid tumor.

It is known that the lamellar structure of the liposome membrane at the gel-to-liquid crystalline transition phase is somewhat loose and porous. Therefore, it is possible to design liposomes which can release entrapped drugs preferentially in an anatomical region subjected to local hyperthermia (thermosensitive liposomes, TSL) at their liquid-crystalline phase transition temperature.[11,12] It is well known, however, that the residence time of such liposomes in the blood circulation is very short, due to their rapid removal by liver and spleen, and this is obviously disadvantageous for local drug release by hyperthermia. If the thermosensitive liposomes can be endowed with a prolonged circulation lifetime, it may be possible to achieve much higher drug levels in tumors subjected to hyperthermia than could be obtained with ordinary thermosensitive liposomes.

Long Circulating Liposomes: Old Drugs, New Therapeutics, edited by Martin C. Woodle and Gerrit Storm.

For the study of long circulating, thermosensitive liposomes, doxorubicin (DXR) was selected to be encapsulated for the following reasons:

1) DXR, an anthracycline aminoglycoside, is a potent anticancer chemotherapy agent with a broad spectrum of antitumor activity.
2) Encapsulation of DXR into liposomes markedly reduces the cardiotoxicity of the drug.[12-14]
3) DXR exhibits strikingly enhanced cyotoxicity against tumor cells when combined with 42-43°C hyperthermia.[15]
4) DXR can be efficiently entrapped in liposomes by the pH gradient method.[16]

Thus, DXR is a good candidate for encapsulation in long circulating, thermosensitive liposomes, and such a formulation should be experimentally and clinically useful.

The purpose of this study was to establish the optimum conditions for preparing long circulating, thermosensitive liposomes from the viewpoint of effective drug release in response to hyperthermia. The drug biodistribution and chemotherapeutic activity of entrapped DXR were evaluated in colon 26 tumor-bearing mice subjected to local hyperthermia at the tumor site.

Methods

Materials

Dipalmitoylphosphatidylcholine (DPPC), distearoylphosphatidylcholine (DSPC) and distearoylphosphatidylethanolamine (DSPE) were kindly donated by NOF Corp., Tokyo. Monomethoxypolyethylene glycol succinimidyl succinate (PEG-OSu), with average molecular weights of 2000, was also provided by NOF Corp. Ganglioside G_{M1}; and cholesterol (Chol) were purchased from Funakoshi, Tokyo. [^{3}H]Cholesteryl hexadecyl ether, ^{67}Ga and ^{51}Cr were obtained from New England Nuclear. Doxorubicin (DXR) was a kind gift from Kyowa Hakko, Tokyo. Deferoxamine was purchased from Ciba Geigy, Kobe.

Synthesis of Amphipathic PEG

An amphipathic PEG was synthesized with a combination of DSPE, and PEG-OSu with 2000 Da in average molecular weight.[7] An aliquot of PEG-OSu in $CHCl_3$ was added to a solution of DSPE in $CHCl_3$/methanol (3:1, v/v), followed by addition of triethylamine with 3:1:3.5 in molar ratio of PEG-OSu:DSPE:triethylamine. The reaction mixture was stirred overnight at room temperature and $CHCl_3$ was evaporated. Full conversion of the primary amino group in DSPE was confirmed by the negative ninhydrin reactivity after the products were separated by thin layer chromatography. Phospholipid phosphorus assay showed the appearance of a new phosphate-positive spot with a higher R_f value than DSPE. A small amount of water was added to the evaporated reaction residues to form micelles. DSPE-PEG micelles were dialyzed against water using dialysis bags (Spectra-Por CE 300000 MWCO, Spectrum Medical) for 5 d and lyophilized.

Liposome Preparation and DXR Encapsulation

Liposomes were prepared with DPPC, DSPC, Chol and an appropriate amount of DSPE-PEG2000 derivative by the reverse-phase evaporation method. The lipid mixture (5 mg in total lipids) was dissolved in 600 µl of isopropyl ether/chloroform (1:1, v/v) and 300 µl of 300 mM citric acid (pH 4.0) were added. The mixture was sonicated to give a w/o emulsion with a bath-type sonicator (Iuchi VS-150), and the organic solvent in the emulsion was gradually evaporated at 60°C to form LUV suspension. Liposomes were extruded through Nuclepore polycarbonate filters to control size. Liposome size was measured by a Nicomp 370 submicron particle analyzer (HIAC Pacific Scientific). An appropriate amount of [^{67}Ga]DF or [^{3}H]cholesteryl hexadecyl ether was added for estimation of biodistribution or integrity of liposome in vivo, respectively.

The encapsulation of DXR into liposomes was done by employing the pH gradient method.[16] Briefly, the pH of the liposome suspension, initially at pH 4.0, was raised to pH 7.8 with 1 M NaOH. The liposome preparation was subsequently heated to 60°C for 5 min. These liposomes were then mixed with a preheated (60°C) DXR solution dissolved in distilled water at a DXR/lipid weight ratio of 0.2. This mixture was incubated with periodic mixing for 10 min at 60°C. The resulting preparation was finally passed through a Sephadex G-50 column. The amount of entrapped DXR was determined with a fluorescence spectrometer (Hitachi F-3000) by diluting liposomes with 0.3 N HCl-50% ethanol and measuring the fluorescence intensity (Ex: 470 nm, Em: 590 nm).

Thermosensitivity of Liposomes

Temperature-dependent release profiles of DXR from TSL were estimated by incubation of the liposomes at various temperatures. The DXR-liposomes were diluted with saline or 20% (v/v) fetal bovine serum. One ml of each sample was placed in a water bath (HAAKE F2) and heated at 36-43°C. The released DXR was separated from the liposomal suspension by ultrafiltration and assayed by fluorescence spectrometry as described under doxorubicin encapsulation. A Centrisart (molecular weight cut-off 20,000, Sartorius, 3000 rpm, 20 min) was used for this procedure. The absorption of DXR to membrane was less than 2%.

Biodistribution and Stability of DXR-liposomes In Vivo

Mouse colon carcinoma 26 cells (1 x 10^5 cells) were inoculated into the hind foot of male BALB/c mice (male, 8 weeks old, and weighing 22-25 g), and the tumor was allowed to grow for approximately 8 d, when the mean of its length and width was 8 mm. DXR-liposomes labeled with [^{67}Ga]DF or [^{3}H]cholesteryl hexadecyl ether were injected via the tail vein into the colon 26 solid tumor-bearing mice at a dose of 5 mg DXR/kg. At different times after injection, mice were anesthetized, bled via the retroorbital sinus and killed by cervical dislocation. Organs were excised and examined for ^{67}Ga using a gamma counter. Scintillation counting was done by the procedure described previously.[12] The weight of total blood was assumed to be 7.3% of the body weight.[17] Background blood concentration in each organ was corrected by examining the distribution of ^{51}Cr-labeled erythrocytes.[12]

In Vivo Studies with Hyperthermia

The colon 26 solid tumor-bearing mice were anesthetized with Nembutal. The free DXR or liposomal DXR was injected intravenously via the tail vein at a dose of 5 mg DXR/kg. Local hyperthermia using a radiofrequency oscillator (RF-hyperthermia HEH-100, Omron, Kyoto) with a 42°C heating temperature was started at 5 min after drug administration and was continued for 20 min. The hyperthermia apparatus used in this study was modified to accommodate small animals. At 5 min after completion of hyperthermia treatment, blood was collected from the retroorbital sinus and major organs were excised. A 0.1 g portion of tissue, or the whole tissue if it weighed less than 0.1 g, was used for measurement of DXR concentration. Samples were homogenized and extracted with butanol/toluene (1:1, v/v), then the extracts were subjected to HPLC assay according to the previous method.[12]

Tumoricidal Effect

Tumor-bearing mice were assigned at random into groups of 6-10. Treatment was started when the tumor had reached a diameter of 8 mm (= Day 0). The dose per injection was 5 mg DXR/kg body weight. Local hyperthermia at 42°C for 20 min was done at 5 min or 3 h after injection under light anesthesia. Tumor volumes were determined by a reported method.[18] Survival time was recorded in days after treatment with local hyperthermia.

Characterization

Encapsulation of DXR into Liposomes by pH Gradient Method

The encapsulation of DXR into liposomes was done under the following fixed conditions: ΔpH 3.8, a DXR/lipid weight ratio of 0.2, and a standing time of 10 min at 60°C at the loading step. DXR was entrapped into liposomes with 90-98% efficiency under the fixed conditions, even in the presence of 3 mol% of PEG2000 or 3-6 mol% of G_{M1} in the lipid components (Table 7.1). Though the ordinary method described in our previous report resulted in only 20% trapping efficiency,[19] the pH gradient method employed in this study was more effective for the encapsulation of DXR into liposomes. Furthermore, this process was not affected by liposomal size. However, the inclusion of 6 mol% DSPE-PEG derivatives or 3 mol% PEG2000 resulted in decreased trapping efficiency (data not shown). TSL lacking PEG precipitated during storage in a refrigerator for 1 d, owing to aggregation. On the other hand, no aggregation was observed, even after 5 d, in TSL prepared using PEG derivatives. The improved dispersion of these liposomes was considered to be due to the increased hydrophilicity of the liposome surface.

Thermosensitivity

Figure 7.1 shows the temperature-dependent release of DXR from liposomes. 58% or 45% of entrapped DXR was released from liposomes containing 3 mol% of PEG or

Table 7.1. Effect of PEG on pH-dependent DXR uptake into liposomes

Liposomal Composition	Molar Ratio	Diameter (nm(SD))	Trapping Efficiency (%)
DPPC/DSPC	9:1	182.9(44)	>98
PEG2000/DPPC/DSPC	0.31:9:1	190.6(60)	95
GM1/DPPC/DSPC	0.31:9:1	205.0(82)	>98
	0.61:9:1	198.5(80)	>95
DPPC/DSPC/Chol	8:1:1	135.5(55)	>95
PEG2000/DPPC/DSPC/Chol	0.31:8:1:1	135.2(61)	>90

6 mol% of G_{M1} with around 200 nm in size, by incubation with 20% serum for 5 min. Smaller sized liposomes (PEG2000/DPPC/DSPC/Chol, 135.2 nm) showed somewhat low release (40% at 43°C). These results indicate that DXR was released from liposomes in a temperature-dependent manner, and inclusion of 3 mol% of PEG or 6 mol% of G_{M1} did not interfere DXR release. However, the degree of DXR release was relatively low compared with that of cisplatin.[20] This relatively low efflux of DXR is due to the basic and amphipathic nature of the drug. Since DXR is in ionic form at pH 4.0, the hydrophobic interaction between DXR and acyl residues of DPPC and DSPC would be less effective. On the other hand, since DXR is in nonionic form in neutral or basic conditions, DXR can then interact with the acyl residues when it penetrates through the lipid layers. As described by Magin and Weinstein,[21] it has been considered that drugs to be encapsulated in thermosensitive liposomes should generally be hydrophilic, with a low affinity for the lipid bilayer. These liposomes were less leaky at temperatures below 37°C. These results suggest that the gel-to-liquid-crystalline phase transition of DXR-PEG2000/DPPC/DSPC (0.31:9:1), DXR-PEG2000/DPPC/DSPC/Chol (0.31:8:1:1) or DXR-G_{M1}/DPPC/DSPC (0.61:9:1) occurred at approximately 40-42°C.

In Vivo Behavior

Biodistribution of Long Circulating TSL Encapsulating DXR

DXR-TSL, DXR-PEG-TSL and DXR-GM1-TSL with 180-200 nm in mean size, respectively, and DXR-PEG-TSL (SUV) with 130-150 nm in mean size were labeled with [^{67}Ga]DF and injected iv into mice via the tail vein. The biodistribution was measured up to 3 h after injection. The results are summarized in Figure 7.2. TSL lacking PEG or G_{M1} were rapidly cleared from the blood circulation and concomitantly accumulated in the RES (liver), suggesting that liposomes of this lipid composition and size range are readily taken up by the RES. On the other hand, the inclusion of 3 mol% of DSPE-PEG2000 derivatives or 6 mol% of G_{M1} significantly increased the blood levels of TSL over a long period and decreased RES uptake.

Small-sized DXR-PEG-TSL (SUV) (135 nm average diameter), which showed greater prolonged circulation, accumulated most efficiently in the colon 26 solid tumor. Generally, extravasation of circulating molecules from blood vessels is a function of both local blood flow and microvascular permeability. Normal tissues outside the RES are generally known to have continuous and nonfenestrated vascular endothelia, and extravasation of macromolecules or liposomes is greatly limited. On the other hand, the permeability of the endothelial barrier in newly vascularized tumors is increased compared to that in normal tissues. Under physiological tumor conditions, only liposomes with small size and prolonged circulation half-life will have a sufficient chance to encounter the leaky vessels of the tumor.[10]

DXR Distribution in Mice

The blood and tissue-associated DXR levels were measured after injection of DXR-long circulating TSL, with or without local hyperthermia. As shown in Figure 7.3, DXR levels in blood after administration of DXR-long circulating TSL were significantly increased as compared with the other formulations. These results were due to the high blood level and low RES uptake of the long circulating TSL. After administration as a solution, DXR disappeared from the blood very rapidly. DXR-TSL was also cleared rapidly from blood due to high RES uptake. When hyperthermia

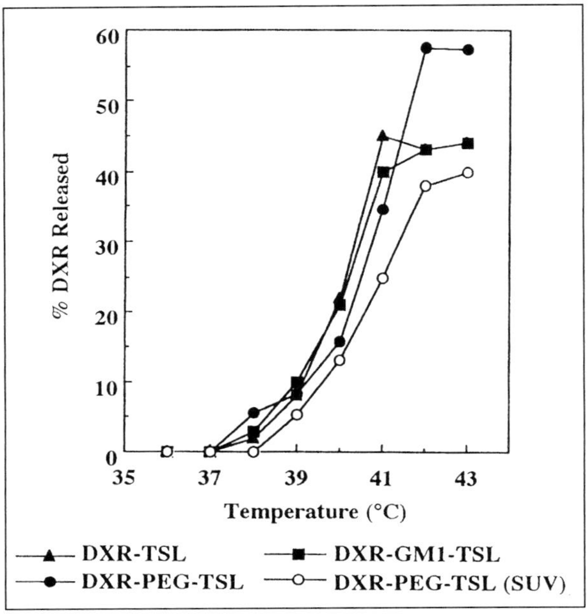

Fig. 7.1. Temperature dependent release of DXR from liposomes.

was applied, the blood levels of DXR were decreased, indicating that DXR was released from long circulating TSL in vivo in a thermosensitive manner and released DXR was cleared rapidly from the blood. There were no marked differences in tissue levels of DXR between groups with and without hyperthermia. For DXR-TSL, high levels of DXR were observed in liver (52.5 μg/g) and spleen (98.5 μg/g), presumably as a result of accumulation of TSL therein (Fig. 7.2). On the other hand, the distribution of DXR to liver and spleen after injection of DXR-long circulating TSL was significantly decreased. Thus, the blood level and the tissue distribution characteristics of DXR reflected those of the liposomes themselves. There was no marked difference in the accumulation of DXR in the heart among free drug and liposomal formulations.

Tumoricidal Effect

Accumulation of DXR in Tumor Tissue by Local Hyperthermia

The tumor-associated DXR levels with or without hyperthermia are summarized in Figure 7.4. Without hyperthermia, there was no marked difference in tumor accumulation of DXR at 30 min after injection among liposomal DXR formulations. However, for DXR-PEG-TSL (SUV), relatively high DXR levels in solid tumor were obtained at 3 h after injection without hyperthermia. Based on this result and those shown in Figure 7.2, we can see that intact DXR-PEG-TSL (SUV) passed fairly readily through the leaky tumor endothelium by passive convective transport. On the other hand, the system of DXR-long circulating TSL, and hyperthermia gave a value of 1.8-fold or 1.9-fold higher than that of DXR-TSL and hyperthermia, and 5.3-fold or 5.6-fold higher than that of free drug solution and hyperthermia at 30 min after injec-

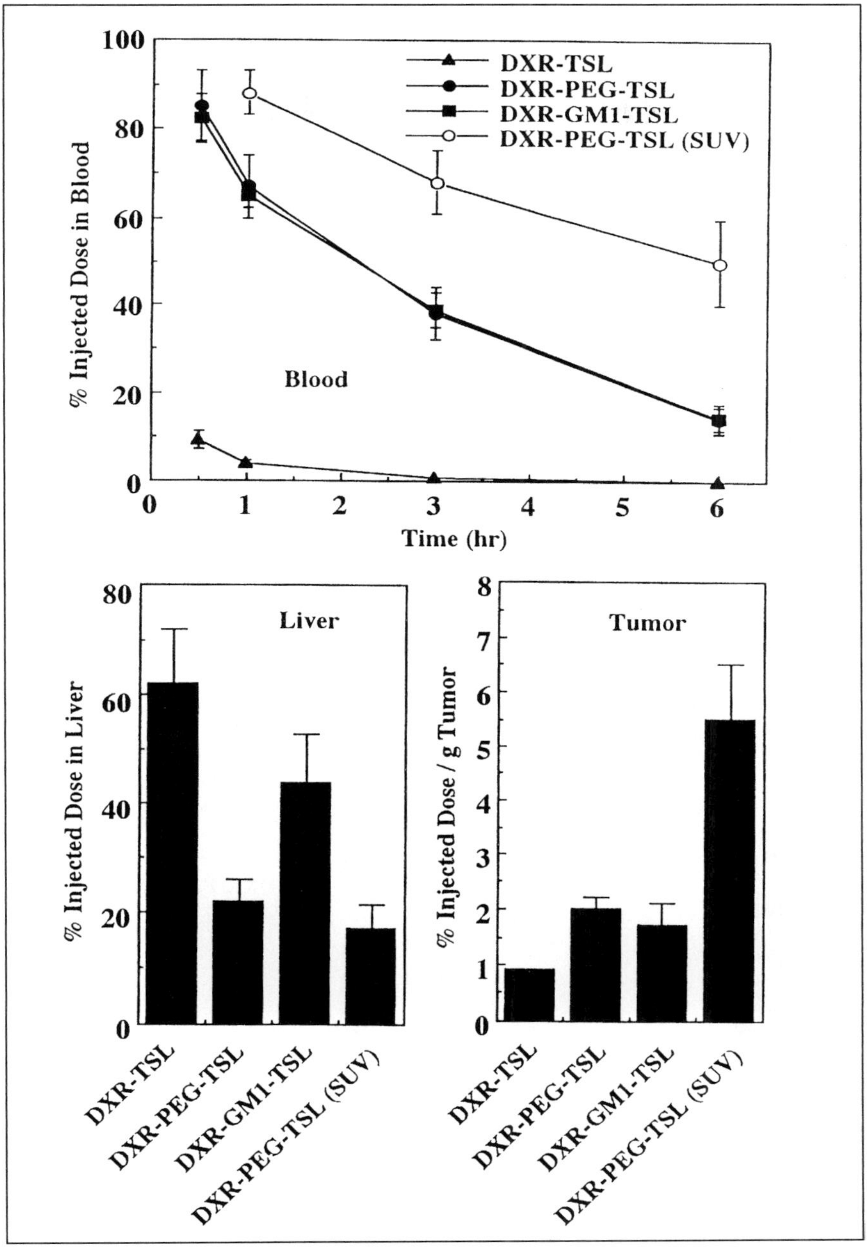

Fig. 7.2. Biodistribution of thermosensitive liposomes.

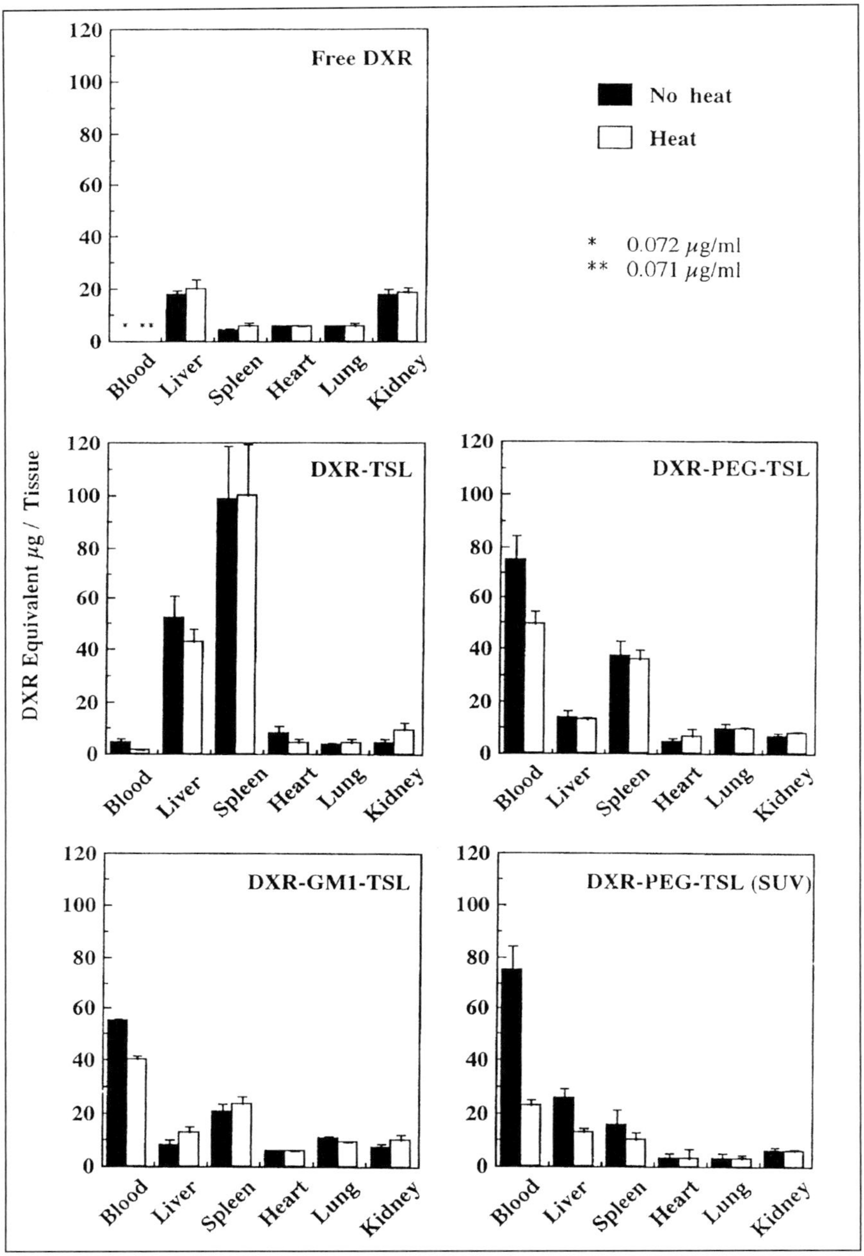

Fig. 7.3. Tissue DXR levels after administration of free DXR or DXR-thermosensitive liposomes to colon carcinoma 26 solid tumor-bearing mice with or without local hyperthermia. Local hyperthermia with 42°C heating temperature was applied at the tumor site on 5 min after drug administration and was continued for 20 min.

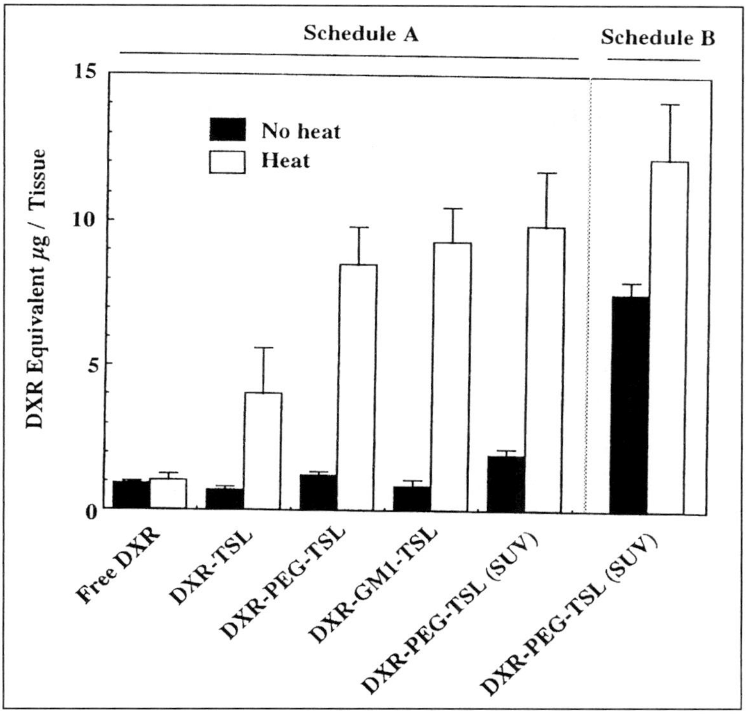

Fig. 7.4. Tumor DXR levels after administration of free DXR or DXR-thermosensitive liposomes to colon carcinoma 26 solid tumor-bearing mice with or without local hyperthermia at tumor site.

tion, respectively (Fig. 7.4, Schedule A). These levels with the liposomal systems were, of course, much higher than those obtained without hyperthermia. Since amphipathic PEG and G_{M1} prolonged the circulation time of liposomes, thereby increasing the amount of liposomes which pass the heating site, the amount of DXR released from long circulating TSL at the heating site is expected to be higher than that from TSL. Thus, we consider that the entrapped DXR was efficiently released from PEG-thermosensitive liposomes by hyperthermia at the tumor site and entered the tumor tissue by simple diffusion.

Antitumor Activity

The antitumor activities of free DXR and thermosensitive liposomal DXR with hyperthermia are summarized in Table 7.2 and Figure 7.5. Hyperthermia alone (without administration of drugs) caused tumor growth retardation but gave no improvement in survival time, compared with the control. The system of free drug and hyperthermia was ineffective, giving the same result as heating alone. On the other hand, both Schedule A and B (Fig. 7.4) which are the combinations of systemic liposomal DXR injection and local hyperthermia, effectively retarded tumor growth and increased survival time. These results show clearly that DXR encapsulated in long circulating TSL in combination with hyperthermia, causes a marked improvement in therapeutic efficacy, inhibiting the tumor growth.

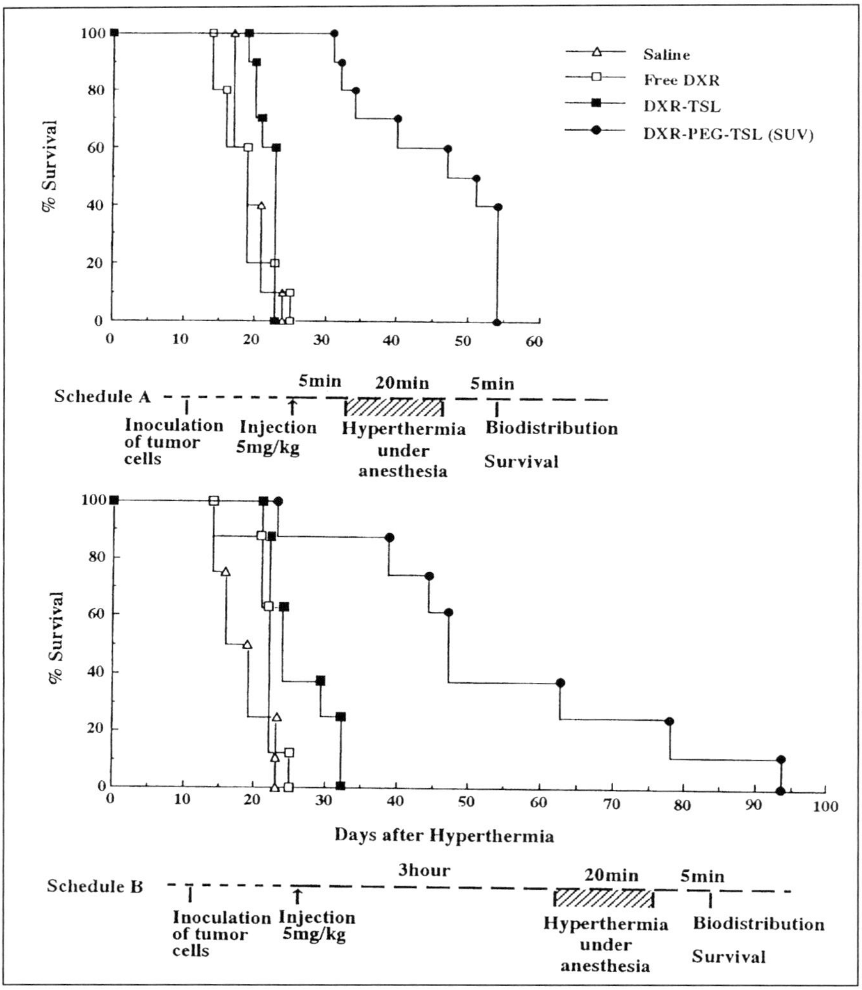

Fig. 7.5. Survival curve of colon carcinoma 26 solid tumor-bearing mice after iv injection of 5 mg/kg DXR/kg of free DXR or DXR-thermosensitive liposomes and hyperthermia with schedule A or B.

Conclusions

Weinstein et al[21] reported that thermosensitive liposomes injected iv leak their contents preferentially in an anatomic region subjected to local hyperthermia. Although thermosensitive liposomes could be useful for treatment of selected tumors in combination with local hyperthermia, they have drawbacks such as low stability and short half-life in the blood. The present results clearly indicated that the use of long circulating TSL containing DXR in combination with local hyperthermia could deliver a higher drug concentration to the solid tumor (Fig. 7.4) and resulted in superior therapeutic efficacy as compared with the PEG-lacking thermosensitive lipo-

Table 7.2. Effect of thermosensitive liposome encapsulated DXR with or without hyperthermia (42°C, 20 min) on tumor growth and survival of mice bearing colon carcinoma 26 tumor

Treatments	Tumor Volume Day 7 (cm³)	Tumor Volume Day 14 (cm³)	Survival Time Mean (days)	Survival Time T/C (%)
Control	3.0 ± 0.7	6.9 ± 1.5	29.4 ± 4.2	100
Free DXR	1.6 ± 3.9	3.9 ± 0.4	29.4 ± 5.2	100
DXR-TSL	1.4 ± 0.3	4.0 ± 1.0	25.2 ± 3.4	86
DXR-PEG-TSL	1.4 ± 0.2	3.3 ± 0.5	25.0 ± 3.6	85
DXR-GM1-TSL	1.4 ± 0.2	4.0 ± 0.7	26.6 ± 4.2	90
Heat	1.7 ± 0.4	4.8 ± 0.7	30.1 ± 4.5	102
Free DXR + Heat	1.6 ± 0.4	3.0 ± 1.0	31.5 ± 5.6	107
DXR-TSL + Heat	0.9 ± 0.2	1.3 ± 0.8	39.8 ± 8.4	135
DXR-PEG-TSL + Heat	0.6 ± 0.1	0.6 ± 0.2	56.0	190
DXR-GM1-TSL + Heat	0.5 ± 0.1	0.9 ± 0.1	47.4	162

Tumor-bearing BALB/c mice received 5 mg of DXR/kg iv in either free or liposomal DXR.
Data are the mean (SD) of 10 mice per group.
One mouse survived over 60 days.

somal system used in the past (Table 7.2 and Fig. 7.5). The inclusion of PEG or G_{M1} endowed TSL with the long circulation time to TSL and did not interfere with temperature-dependent DXR release from TSL. Although there was no difference in the release rate of DXR at 42°C between TSL and long circulating TSL, the drug concentration in the tumor provided by long circulating TSL was 2-2.5 times higher than that by ordinary liposomes. This process is shown in Figure 7.6. A much larger amount of long circulating TSL would pass the heating site during hyperthermia for 20 min compared with TSL lacking PEG. Therefore a larger amount of DXR was released in the capillary of the tumor mass from long circulating TSL by hyperthermia, and entered the tumor tissue by simple diffusion.

Small-sized PEG-TSL (SUV) were able to traverse the endothelium of blood vessels in tumors and extravasate into interstitial spaces. Since the capillary permeability of the endothelial barrier in nearly vascularized tumors is significantly greater than that of normal organs, it is possible that small-sized liposomes with prolonged circulation half-lives could predominantly pass through the leaky tumor endothelium by passive convective transport.[22,23] Therefore, the present results indicate that since a much larger amount of long circulating TSL (SUV) would extravasate into tumor tissue during circulation for 3 h after injection, encapsulated drug was released from extravasated liposomes in the interstitial spaces of solid tumor by local hyperthermia (Fig. 7.6).

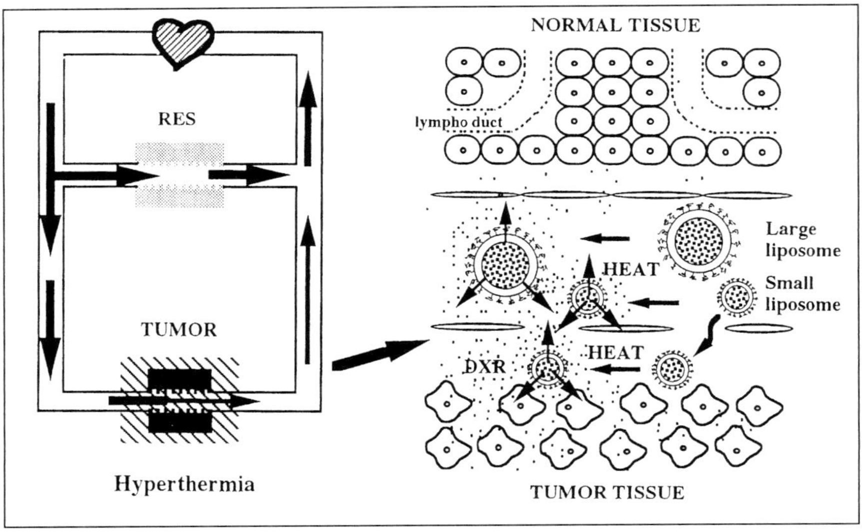

Fig. 7.6. DXR-long circulating thermosensitive liposomes and hyperthermia.

The mechanism by which PEG prolongs the circulation time of liposomes has not been fully clarified. However, increased hydrophilicity[24] and/or steric barrier on the liposomal surface[6] might be important for long circulating activity. A hydrophilic surface may prevent or reduce the interactions of liposomes with serum constituents, resulting in an enhanced stability of liposomes and a reduced rate of RES interaction. Liposomes with thermosensitivity and long circulating half-lives have a greater chance to release entrapped drug at a hyperthermic site. This system is expected to be clinically useful for delivering a wide range of chemotherapeutic agents in the treatment of solid tumors.

The combination of DXR-long circulating TSL and local hyperthermia resulted in a higher drug distribution to the tumor and more effective retardation of tumor growth than the ordinary TSL system. Thus, long circulating ability is effective for thermosensitive liposomal drug delivery in combination with hyperthermia. This system is expected to be clinically useful for delivering a wide range of chemotherapeutic agents in the treatment of solid tumor.

References

1. Young RC, Ozols RE, Mayers CE. The anthracycline antineoplastic drugs. New Engl J Med 1981; 305:139-153.
2. Lasic D. Application of liposomes in pharmacology and medicine. In: Lasic D, ed. Liposomes from Physics to Applications. Amsterdam: Elsevier, 1993:262-471.
3. Hwang KJ. Liposome pharmacokinetics. In: Ostro MJ, ed. Liposome from Biophysics to Therapeutics. New York: Marcel Dekker Inc. 1987:109-156.
4. Senior JH. Fate and behavior in vivo: a review of controlling factors. Critical Review in Drug Carrier Systems 1987; 3:123-139.
5. Allen TM, Chonn A. Large unilamellar liposomes with low uptake into the reticuloendotherial system. FEBS Lett 1987; 223:42-46.

6. Klibanov AL, Maruyama K, Huang L et al. Amphipathic polyethyleneglycols effectively prolong the circulation time of liposomes. FEBS Lett 1990; 268:235-237.
7. Maruyama K, Yuda T, Okamoto A et al. Prolonged circulation time in vivo of large unilamellar liposomes composed of distearoyl phosphatidylcholine and cholesterol containing amphipathic poly(ethylene glycol). Biochim Biophys Acta 1992; 1128:44-49.
8. Allen TM, Hansen C, Martin FL et al. Liposomes containing synthetic lipid derivatives of poly(ethylene glycol) show prolonged circulation half-lives in vivo. Biochim Biophys Acta 1991; 1066:29-36.
9. Yuan F, Leuning M, Huang SK et al. Microvascular permeability and interstitial penetration of sterically stabilized (stealth) liposomes in a human tumor xenograft. Cancer Res 1994; 54:3352-3356.
10. Unezaki S, Maruyama K, Hosoda J et al. Direct measurement of extravasation of polyethyleneglycol-coated liposomes into solid tumor tissue by in vivo fluorescence microscopy. Int J Pharm 1996; in press.
11. Maruyama K, Unezaki S, Takahashi N et al. Enhanced delivery of doxorubicin to tumor by long circulating thermosensitive liposomes and local hyperthermia. Biochim Biophys Acta 1993; 1149:209-216.
12. Maruyama K, Yuda T, Okamoto A et al. Effect of molecular weight in amphipathic polyethyleneglycol on prolonging the circulation time of large unilamellar liposomes. Chem Pharm Bull 1991; 39:1620-1622.
13. Olson F, Mayhew E, Maslow D et al. Characterization, toxicity, and therapeutic efficacy of adriamycin encapsulated in liposomes. Eur J Cancer Clin Oncol 1982; 18: 167-176.
14. Herman EH, Rahman A, Ferrans VJ et al. Prevention of chronic doxorubicin cardiotoxicity in beagles by liposome encapsulation. Cancer Res 1983; 43:54275432.
15. Hahn GM, Braun J, Har-Kedar I. Thermochemotherapy: Synergism between hyperthermia (42-43°C) and adriamycin (or Bleomycin) in mammalian cell inactivation. Proc Natl Acad Sci USA 1975; 72:937-940.
16. Mayer LD, Bally MB, Hope MJ et al. Uptake of antineoplastic agents into large unilamellar vesicles in response to membrane potential. Biochim Biophys Acta 1985; 816:294-302.
17. Wu MS, Robbins JC, Bugianesi RL et al. Modified in vivo behavior of liposomes containing synthetic glycolipids. Biochim Biophys Acta 1981; 674:19-29.
18. Corbett TH, Griswold D, Roberts BJ et al. Biology and therapeutic response of a mouse mammary adenocarcinoma (16/C) and its potential as a model for surgical adjuvant chemotherapy. Cancer Treat Rep 1978; 60:1471-1488.
19. Unezaki S, Matsumura M, Ohtani Y et al. Preparation and evaluation of temperature-sensitive liposomes containing doxorubicin by pH gradient method. Drug Delivery System 1992; 7:37-41.
20. Iga K, Hamaguchi N, Igari Y et al. Heat-specific drug release of large unilamellar vesicle as hyperthermia-mediated targeting delivery. Int J Pharm 1989; 57:241-251.
21. Weinstein JN, Magin RL, Yatvin MB et al. Liposomes and local hyperthermia: Selective delivery of methotrexate to heated tumor. Science 1979; 204:188-191.
22. Gabizon A, Papahadjopoulos D. Liposome formulations with prolonged circulation time in blood and enhanced uptake by tumors. Proc Natl Acad Sci USA 1988; 85:6949-6953.
23. Unezaki S, Maruyama K, Ishida O et al. Enhanced tumor targeting of doxorubicin by ganglioside GM1-bearing long circulating liposomes. J Drug Target 1993; 1:287-292.
24. Senior J, Delgardo C, Fisher D et al. Influence of surface hydrophilicity of liposomes on their interaction with plasma protein and clearance from the circulation: studies with poly (ethylene glycol)-coated vesicles. Biochim Biophys Acta 1991; 1062:77-82.

CHAPTER 8

Liposomal Localization and Chemotherapy for AIDS-Related Kaposi's Sarcoma

Mark Harrison, K.J. Harrington, J.S.W. Stewart

Introduction

Kaposi's sarcoma (KS) is a hallmark of the AIDS pandemic. Originally described over 100 years ago, the increased incidence of this tumor in a nontypical population led to the original hypothesis of AIDS in young, homosexual men. One of the three AIDS defining malignancies, KS is the most common tumor to affect HIV positive individuals. Within the AIDS population the disease is much more common amongst homosexual and bisexual men than heterosexual men. Tumor incidence, initially affecting up to 50% of HIV positive individuals at the start of the AIDS pandemic, now affects between 10 and 20% of HIV infected patients. This is thought to be due to changes in sexual practice, implying some form of sexual transmission of the agent that causes KS. Despite this decrease in incidence within HIV infected individuals, the increase in the worldwide incidence of HIV infection implies an increasing public health problem with advanced KS. For individuals affected, the average age at first diagnosis of KS is between 30 and 40 and the median survival period following diagnosis is 12 months.[1-3]

Classically KS affects the skin, with a wide spectrum of disease activity being seen. Lesions may be individual, offering little other than a cosmetic nuisance, or multiple with systemic involvement. Mucosal involvement is common, with the mouth being affected in up to 50% of patients. Other organs shown to be affected with KS include the lungs, gastrointestinal tract, liver, spleen and genitals. Clinically the lesions are brownish purple in color, elevated above the surrounding tissue, frequently nodular and may be surrounded by bruising and edema. Lesions may coalesce to form plaques, which in turn, may ulcerate. Advanced disease with limb edema, pulmonary and gastrointestinal symptoms, and frequently, superadded infection is life-threatening and requires urgent systemic therapy. Several clinical variants of KS have been identified and described; these include nodular, florid, infiltrative and lymphadenopathic varieties of disease. However, individuals, may show signs of all the potential clinical entities at some stage during their illness.[4-6]

Pathologically KS lesions consist of proliferative spindle shaped cells with slits between the cells associated with the presence of erythrocytes suggesting primitive

Long Circulating Liposomes: Old Drugs, New Therapeutics, edited by Martin C. Woodle and Gerrit Storm.

blood vessel development. The vascular spaces are thin walled and irregular and there is evidence of extrusion of erythrocytes with hemosiderin deposition. An inflammatory infiltrate with mononuclear cells is typically seen. The tumor thus shows evidence of proliferation, inflammation and angiogenesis. The cell of origin is thought to be the spindle shaped cell. The tissue of origin remains unclear, however, and is most likely vascular or lymphatic endothelium, though evidence suggesting smooth muscle origin has also been presented.[7-9]

The etiology of KS is beginning to be unravelled and evidence to date suggests the presence of a new herpes virus, termed Herpes Hominis Virus 8 (HHV8) or KSHV. Original work using PCR analysis showed the presence of unique viral DNA within the genomic DNA of several KS lesions. Continued work has included evidence of viral infection within a large number of KS affected individuals, both with and without HIV infection. Indirect immunoflouresence assays developed to specific KSHV antigens have shown the presence of KSHV antibodies in between 71-88% of serum samples in HIV positive KS affected patients from Italy, Uganda and the U.S. Serum samples from HIV negative but KS positive individuals showed a similar picture. Other groups have also produced data suggesting that HHV8 seropositivity correlates with the pattern expected for a sexually transmitted infection. The viral sequence has also been detected in other AIDS-related malignancies, especially lymphoma. The virus is thought to be under strict immunological control since evidence of viral infection does not correlate with the presence of KS lesions. Homology between the virus and HSV2 and Herpesvirus saimiri has helped confirm the phylogenetic place of this new herpesvirus. Further confirmation of the infective nature of this putative new virus includes evidence of B cell infection by DNA fragments from the virus. However, cell lines made from KS samples, though they behave in a proliferative manner similar to KS, have shown no evidence of viral presence. This suggests that other factors are important in the development of KS. A number of putative growth factors have been suggested, including the HIV *tat* gene which is capable of inducing KS-like lesions in the skin of transgenic mice. Interleukin-6, basic fibroblast growth factor, transforming growth factor β, and a 30 kDa protein secreted by HIV infected T cells known as Oncostatin M, have all been shown to be important in the development of KS. It would seem most likely that close interplay of all these factors are important in the generation of KS in the clinical setting.[10-18]

Standard Treatment of AIDS-Related Kaposi's Sarcoma

AIDS-related Kaposi's sarcoma is incurable and, therefore, the treatment intent is entirely palliative. In early stage disease, small, asymptomatic skin lesions in unexposed areas, which may remain clinically quiescent for prolonged periods, can be managed under a watch-and-wait policy. Active treatment is generally reserved until disease progression occurs or troublesome symptoms develop. This situation usually arises initially as a result of cutaneous disease causing cosmetic disfigurement or local symptoms, such as ulceration, edema and bleeding. Local treatment at this stage aims to secure durable local control of each individual lesion with optimal cosmesis at the cost of minimal toxicity. For patients with more generalized cutaneous or visceral disease, complete remissions are rarely achieved and the goal of treatment is to palliate symptoms without producing excessive treatment-related toxicity. Throughout the course of the illness, as the underlying immune deficiency progresses, the clinical behavior of the Kaposi's sarcoma may vary. Therefore, patients typically receive a number of different treatment modalities throughout the course of their illness. Treatment approaches may be either local or systemic, with many patients initially receiving local therapy to troublesome cutaneous lesions and later requiring systemic cytotoxic chemotherapy for progressive disease.

Local Treatment

Radiotherapy

Radiotherapy was used to treat AIDS-related Kaposi's sarcoma early in the disease epidemic and considerable experience has been gained by a number of centers, resulting in rapid progress in understanding the response of this disease to palliative radiotherapy. It is well-recognized that radiotherapy effectively palliates pain, cosmetic disfigurement, mass lesions, ulceration, bleeding and edema.

Like endemic KS, the lesions of AIDS-related Kaposi's sarcoma respond to relatively modest doses of radiotherapy. The optimal total radiation dose and fractionation schedule that should be employed remain contentious. Patients with poor performance status and a predicted survival of only a few months can be effectively palliated by the use of single fractions of radiotherapy. However, patients with a good performance status are not always best treated by single dose therapy and may benefit from prolonged, fractionated treatment. In randomized, prospective trials, conflicting results have been obtained. Higher dose, increased fractionation studies show that 40 Gy in 20 fractions (83% response rate) and 20 Gy in 10 fractions (79% response rate) are better than 8 Gy as a single fraction (50% response rate). Local control and cosmetic outcome were also significantly better for the fractionated regimens. However, conflicting results come from a study examining both response and normal tissue cosmesis in 596 lesions in 57 patients treated with either 8 Gy as a single fraction (86.7% response rate) or 16 Gy in 4 fractions (89.8% response rate). The differing results may, since such studies are relatively rare, reflect trans-Atlantic differences in assessment of cosmesis and response. In general, single fraction radiotherapy is an acceptable palliative treatment, except for larger, mature, symptomatic lesions in patients with good performance status who are likely to live long enough to develop troublesome local recurrence. In these patients, fractionated therapy should be considered.

Variations in the radioresponsiveness and local control rates of lesions in different areas of the body have been reported, with evidence that lesions on the face, eyelids, conjunctivae, hands and feet, and genitalia are particularly well controlled. Cosmetic outcome is also better in these areas, perhaps as a result of the patient presenting early for treatment because of cosmetic disfigurement.

Wide field radiotherapy may also be useful in palliating certain specific problems. Extended field irradiation has been employed for the treatment of extensive cutaneous disease, especially when associated with edema, although its use is limited by acute skin toxicity. Whole lung irradiation may occasionally play a role in patients with end-stage disease and symptomatic pulmonary involvement.

Patients receiving radiotherapy for AIDS-related Kaposi's sarcoma frequently develop severe acute reactions to treatment at doses which would be tolerated by patients without HIV infection. The basis of this apparent increased radiosensitivity remains unclear but it has been suggested that reduced systemic and intracellular glutathione levels are responsible. The increased radiosensitivity of patients with AIDS-related KS must be taken into account when planning treatment, especially when it involves the oral mucosa, and if necessary, treatment may be given as a split course. In addition, radiation recall phenomena have been documented in patients receiving cytotoxic chemotherapy.[19-33]

Intralesional Chemotherapy

Intralesional vinblastine injection is effective and easily performed for accessible lesions. Treatment may be associated with transient local pain and numbness at the site of injection. Scarring and alopecia in hair-bearing skin may preclude the use of intralesional chemotherapy in exposed areas. Response rates of greater than 90% have been reported. Intralesional vinblastine is considered to be the treatment of choice for small oral lesions with results similar to those obtained with palliative radiotherapy, but with considerably less mucosal toxicity.[31,34]

Other Local Modalities

Simple camouflage of unsightly lesions with cosmetics may be effective and is particularly useful for lesions at the hairline, where radiotherapy or intralesional chemotherapy may cause alopecia. Surgical removal of small troublesome lesions may also provide palliation with good cosmesis for selected sites. Otherwise, cryotherapy, laser or photodynamic therapy may have a role to play in certain selected situations.[31,35-37]

Systemic Therapy

Chemotherapy

The development of progressive or extensive skin disease, with or without visceral involvement, signals the need to switch from local to systemic therapy. The use of systemic therapy exposes the patient to the risks of systemic adverse effects, particularly, myelosuppression with resulting immune depression which can predispose the patient to life-threatening infective episodes. In addition, attention must be paid to the possible chronic effects of chemotherapy, since patients may require prolonged, repeated courses of treatment.

There exists an extensive literature on the use of cytotoxic chemotherapy in the treatment of AIDS-related Kaposi's sarcoma, both in terms of single agent and combination regimen therapy. The vinca alkaloids, the anthracyclines and bleomycin are most widely used, although other agents, such as etoposide, have been used. These agents have important dose limiting toxicities, such as lung, heart and neurological damage. Furthermore, patients with AIDS-related KS are likely to develop non-KS-associated disorders in the major organ systems, potentially lowering the threshold at which chemotherapy-induced toxicity may become manifest.

Single Agent Chemotherapy

Vinblastine, a vinca alkaloid, has been used in 38 patients at a median dose of 6 mg for a median duration of 72 weeks. Ten of the patients had a clinical response and a further 19 showed stable disease. An alternative vinca alkaloid, vincristine has also been used as a single agent to treat 23 patients with a dose of 2 mg. Initially, it is given on a weekly basis for up to 5 weeks and thereafter fortnightly. There were 11 responses in 18 evaluable patients (61%) with a median duration of response in excess of 4 months. The response rates of the various anthracyclines have been reported in small studies to vary from 0% for mitoxantrone and idarubicin, 14% for doxorubicin and 42% for epirubicin. Bleomycin 5 mg intramuscularly every 2 weeks was reported to have a response rate of 50% as a single agent in 38 patients. Response rates of 76% were reported with intravenous etoposide (150 mg/m² on days 1-3 of a four-week cycle) in 41 patients but this was achieved at the expense of considerable treatment-related toxicity. Tenoposide was evaluated in 25 patients treated with 350 mg/m² every 3 weeks, yielding a 40% response rate. Most recently, a 3 h infusion of single agent paclitaxel in a dose of 135 mg/m² (escalating to 175 mg/m²) was administered to 20 patients. A partial response rate of 65% was reported with neutropenia as the most frequent dose-limiting side effect.[38-47]

Combination Chemotherapy

The drugs with documented single agent activity have been combined in an attempt to increase the response rates. The greatest potential problem with such treatment combinations is the potential for overlapping toxicity to the bone marrow, causing severe immune suppression. This effect may be exacerbated by the continued use of antiretroviral therapy.

Alternating doses of vincristine 2 mg and vinblastine 0.1 mg/kg were employed in an attempt to circumvent the neurotoxicity of the former and reduce the myelotoxicity of the latter. Nevertheless, peripheral neurotoxicity was the major dose-limiting problem and the response rate was only 43%. The combination of adriamycin 40 mg/m² d 1, vinblastine 6 mg/m² d 1 and bleomycin 15 iU days 1 and 15, was initially investigated in 31 patients who received a median of 6 cycles. The response rate of 84% was a considerable improvement on the single agent data but was only achieved at the expense of considerable myelotoxicity, requiring dose reductions in more than half of the patients. In an attempt to reduce bone marrow toxicity, vincristine 2 mg was substituted for vinblastine and the bleomycin and doxorubicin doses were reduced to 10 mg/m² and 20 mg/m². A response rate of 8/10 patients was obtained, but there was still considerable immune suppression as demonstrated by the occurrence of bacterial infections in 5 patients. Further experience with this regimen was later reported to have a response rate of 79% in 33 patients.

In Europe anthracyclines are generally omitted from the first-line combination chemotherapy regimen. Most patients receive a combination of a vinca alkaloid and bleomycin as initial systemic therapy. A partial response rate of 56% was seen in 46 patients treated every 3-4 weeks with vincristine 2 mg and bleomycin 30 U, with manageable toxicity. The same regimen, but in a more intensive twice-weekly schedule, was associated with an increased response rate of 84% with 21% of patients experiencing episodes of neutropenia. Despite this fact, there were only 3 episodes of opportunistic infection during the study.

Sixty-one patients treated every two weeks with either low dose ABV (doxorubicin 20 mg/m², bleomycin 10 mg/m², vincristine 1.4 mg/m²) or single agent doxorubicin 20 mg/m² were investigated in a prospective, randomized trial. The combination chemotherapy yielded a significantly better response rate (88% vs. 48%). Disease free survival, but not median survival, was significantly better for ABV. Opportunistic infections developed in one quarter of evaluable patients, although there was no significant difference in the rates between the two regimens.[48-53]

Preclinical Data

Tumor Localization Studies

Liposomal accumulation in tumors by means of extravasation through leaky vasculature has been demonstrated by electron microscopic examination of C26 colon cancer xenografts and Kaposi's sarcoma-like lesions in mice. In the former study, staining studies revealed silver-enhanced gold particles in small blood vessels and in extracellular spaces around tumor cells in xenografts placed subcutaneously. In a similar study, extravasation and transcytosis of liposomes was demonstrated in Kaposi's-like dermal lesions in transgenic mice bearing the HIV *tat* gene.

These findings are supported by accumulating evidence showing deposition of radiolabeled liposomes in a range of solid tumors. In addition, deposition of small unilamellar vesicles labeled with ^{111}In has been demonstrated in patients with AIDS-related Kaposi's sarcoma and lymphomas. The ability of Stealth liposomes to target the lesions of Kaposi's sarcoma was also recently confirmed using ^{111}In[DTPA]-labeled Stealth liposomes.[54-56]

Therapeutic Studies in Xenograft Models

Preclinical studies have shown that cytotoxic drugs entrapped in liposomes are active against a range of tumor types, although, as yet, there have been no published reports of therapeutic models of Kaposi's sarcoma.

Increased efficacy and decreased toxicity of daunorubicin entrapped in small unilamellar vesicles, compared to the unencapsulated agent, was demonstrated in the treatment of murine lymphosarcoma P-1798 and murine MA16C mammary adenocarcinoma. In the former model system, 10-fold increased delivery of liposomal drug to the tumor was seen in comparison with the free agent. These results were correlated with the increased response rates for the liposomal preparation.

Comparison of the antitumor activity of epirubicin entrapped in either Stealth, conventional liposomes or as free drug was examined against mouse colon carcinoma C26. The pharmacokinetics of free and Stealth liposome-entrapped epirubicin were also assessed in rats. Free epirubicin was distributed to the tissues within minutes, whereas when given as the stealth liposome formulation, the distribution half-life of the drug was over 22 h. Stealth liposome-entrapped epirubicin was found to be more effective than either free epirubicin or epirubicin entrapped in conventional liposomes in inhibiting tumor growth and improving survival.

A mouse mammary cancer model was used to evaluate the therapeutic efficacy of doxorubicin either as a free drug or entrapped in conventional or Stealth liposomes. Both recently implanted and well-established tumors were treated, and the ability of treatment to prevent the development of spontaneous metastases from intramammary tumor implants was also assessed. The Stealth liposome preparation was significantly more effective than either the conventional liposome or the free drug in curing mice with recent tumor implants, in increasing the 8-week survival of mice with well-established tumors and in reducing the incidence of metastases from the intramammary tumor implants. In a further study, the activity of vincristine and doxorubicin was shown to be enhanced by entrapment within a Stealth liposomal preparation as compared with the unencapsulated drugs. The ability of doxorubicin in Stealth liposomes to impair the development of metastases from spontaneous mammary carcinoma was also demonstrated with a weekly dose of 6 mg/kg.

The activity of liposomal doxorubicin against a human prostate cancer xenograft in athymic nude Swiss mice was demonstrated and was shown to be associated with a 25-fold increase in the drug concentration at the tumor site, when compared to free drug.

These studies, and others that are ongoing in a wide range of tumor types, support the hypothesis that liposomal delivery of cytotoxic agents offers the potential for therapeutic benefits and reduced toxicity. Most clinical experience with these novel drug formulations has been gained in the arena of treating AIDS-related Kaposi's sarcoma, and these studies will now be reviewed in detail.[57-62]

DaunoXome

There have been a number of phase I, II and III clinical studies of DaunoXome in patients with advanced KS. The initial dose finding studies have now given way to larger prospective, randomized studies of DaunoXome versus conventional chemotherapy for advanced disease. Study patient numbers range from 3 to 232. All studies have required biopsy confirmed KS and the majority of patients, within reported studies, have had poor prognostic AIDS Clinical Trials Group (ACTG) criteria due either to previous AIDS-related infections or extensive systemic disease (Table 8.1). Patients have been predominantly male with ages in their thirties. Most have also received some other form of anti KS treatment prior to treatment with the liposomal anthracycline. The majority of published clinical studies have not mentioned CD4 counts but in two studies median CD4 counts were reported as 20 cells/µl and 29 cells/µl, implying advanced HIV disease. More advanced prognostic features such as viral load and copy number and HHV8 seropositivity have not, as yet, been reported in any of these patient populations. Reported intercurrent events in phase I and II studies appear similar to those seen in any group of patients with advanced AIDS. Infection rates in phase III studies show an increased incidence in the DaunoXome arm of the study but this is statistically nonsignificant. In those studies in which median survival duration was noted this appears to be between 9 and 12 months.

Clinical pharmacokinetic data regarding the optimal dose and scheduling of DaunoXome for treatment of KS was developed in a number of studies. Doses of 40, 50 and 60 mg/m^2 administered by intravenous bolus or short infusion two to three times weekly were seen to be clinically effective with acceptable toxicity. Prospective comparisons have used a dose of 40 mg/m^2. The median number of cycles of treatment varied between 7 and 11, depending upon patient numbers in the study and the dose given per treatment. Extended therapy to more than 20 cycles and a cumulative dose of up to 1660 mg/m^2 showed that the drug is tolerable and safe. Most studies required a good performance status with a Karnofsky score of at least 60%, an initial granulocyte count of greater than 1000/µl, a platelet count of greater than 75000/µl, a hemoglobin concentration greater than 10 g/dl and reasonable hepatic and renal function. Patients have, in general, continued routine prophylaxis against *Pneumocystis carinii*, and antiretroviral therapy where indicated.

The response rates have varied from 25-65.5%. Most responses were partial as judged by ACTG criteria (Table 8.2). Phase II studies have reported higher response rates than phase III comparisons. The median time from initiation of treatment to response was 4-6 weeks with a median duration of response of around 12 weeks. Within a phase III prospective randomized study median time to treatment failure was reported as 115 days. There were also reports of further response following reintroduction of the drug after drug cessation or dose decrease. A number or reports have commented on the effectiveness of DaunoXome after pretreatment with a "free" anthracycline. This suggests that there is a potential benefit in reversing multidrug resistance in cells that have become resistant to anthracyclines. Further investigations of this interesting observation are warranted.

Hematological toxicity has been the most common side effect associated with DaunoXome. Myelosuppression, predominantly manifesting as neutropenia, has occurred in up to 17% of cycles of chemotherapy administered. Phase I/II evaluations of DaunoXome report WHO grade 3 or 4 neutropenia complicating 49 of 293 cycles of treatment with two-thirds of patients being affected. In the phase III comparison of DaunoXome with ABV grade 3 or 4 neutropenia complicated 59 of 994 cycles of treatment with DaunoXome and 45 of 779 cycles of treatment with ABV. The lower incidence of toxicity within the randomized comparison is due to a lower, now standard, dose of DaunoXome (40 mg/m^2) being used for this study. The initial dose finding study had the majority of neutropenic toxicity within the group treated with 60 mg/m^2. Other studies have reported grade 3 or 4 neutropenia in 44% of patients at some stage during chemotherapy. Anemia with a hemoglobin less than 10 gm/dl was seen in between 24 and 28% of patients in all studies and thrombocytopenia with a platelet count of less than 100 x 10^9/l was reported in between 16 and 18% of patients receiving chemotherapy. Concurrent use of potentially myelosuppressive antiretroviral agents was continued in all studies reported. Dissecting out the overall role of DaunoXome when confounded by the use of antiretrovirals and AIDS-related neutropenia has proved difficult. Future, nonmyelosuppressive antiretrovirals should prove useful in this regard.

Nonhematological toxicities reported with liposomal daunorubicin include alopecia, nausea and vomiting and fatigue. Alopecia of grade 1 toxicity only has been reported in between 5 and 15% of patients. Nausea, without vomiting, is seen in 46% of patients (grade 1 and 2) in one phase I/II study and in 51% of patients treated in the phase III randomized comparison. This group also included 20% of patients who reported vomiting as a complication. Fatigue, increasingly being recognized as a problem, is reported in between 43-59% of patients. Indications of hepatic dysfunction have been minimal though 3 patients treated with 60 mg/m^2 of DaunoXome showed liver enzyme elevations greater than 5 times the normal values.

Cardiac tolerability has been closely examined. There have been no reports of cardiotoxicity, even with rigorous cardiac assessment, including measurement of left ventricular ejection fraction as measured by MUGA scan or echocardiography. This suggests that liposomal delivery may protect against this important side effect of the free drug.

There has been one, large randomized comparison of DaunoXome with conventional chemotherapy for AIDS-related KS. The study used a dose of 40 mg/m^2 of DaunoXome twice weekly versus doxorubicin 10 mg/m^2, bleomycin 15 U and vincristine 1 mg/m^2 twice weekly. An overall response rate of 25% was seen for DaunoXome and 28% for the combination treatment (ABV). In total, 232 patients from 13 centers were randomized into this study. Patients within this study had what would be considered quite severe advanced KS with almost two-thirds having associated lymphoedema, two-thirds having greater than 50 cutaneous lesions and one-third, some form of symptomatic visceral involvement. Both response rates were similar with the major difference being in the toxicity profiles of the two arms of the study. Patients receiving ABV had a higher incidence of alopecia and neuropathy, with patients receiving DaunoXome having more grade 4 neutropenia. Intercurrent infections also occurred more frequently with the DaunoXome arm with 36% of patients developing an opportunistic infection. Amongst patients treated with ABV, 26% developed infection at some stage during therapy. This difference was not statistically significant. The lower response rate than those previously reported for both DaunoXome and the combination therapy are almost certainly due to the changing population of KS affected patients. Improved therapies for HIV-related problems associated with patients having a longer survival have led to the disease being more advanced when the patients present for treatment. With improving therapies for HIV infection this would be expected to continue as a trend leading to an increased need for nontoxic therapies.

In conclusion, DaunoXome has shown itself to be an effective new therapy for the treatment of KS, particularly, AIDS-associated KS. The changing spectrum of HIV disease and time at which KS presents itself implies that this relatively nontoxic drug will have an important role in the management of advanced KS.[63-70]

Dox SL

There have been several phase I and II studies of Dox SL for AIDS-related KS. Additionaly, a number of phase III studies have now stopped recruiting and data have been presented in abstract form. Patient numbers within each study have ranged from 10 in initial dose finding studies to 238 within a randomized, prospective comparison. The majority of patients have, similarly to those in DaunoXome studies, poor prognostic features according to ACTG criteria. Median CD4 count, when reported, is low—between 25 and 30 cells/μl. All studies have required histological confirmation of the diagnosis and response assessments have followed ACTG criteria. Patients have, in the majority, continued routine antiretrovirals and other concomitant medications whilst receiving the chemotherapy. As expected, almost all patients have been male with a median age range in the mid to late thirties.

Initial phase I studies suggested that a dose of 20 mg/m^2 was associated with an optimal response and acceptable toxicity. Higher doses, such as 40 mg/m^2, are therapeutically effective but at the expense of quite marked toxicity, principally neutropenia. Several patients received 20 or more cycles of treatment without significant cumulative toxicity and there are now reports of patients receiving over 40 cycles of treatment. Response rates of between 30 and 92.5% have been reported. The median time to response was 6 weeks (2 cycles) with a median duration of response between 9 and 14 weeks. The majority of responses were partial according to ACTG criteria. A review of European patients treated with Dox SL suggested an overall response rate of 80.7% with a mean time to response of 2.3 cycles of therapy. The reported median survival is approximately 8 months.

Myelosuppression was the dominant toxicity with neutropenia of grades 3 or 4 being a continuous observation. This effect is dose-dependent. Neutropenia of grade 3 or higher was seen in between 34 and 72.5% of patients although the higher level was seen within the dose escalation study in which G-CSF was routinely used for grade 3 and 4 neutropenia. Neutropenia has also been reported as a progressive phenomena with increasing treatment cycle number, however, the number of patients within whom this was noted was small and it may not be significant. Within the European review of all patients treated with Dox SL, neutropenia of grade 3 or 4 complicated 281 of 2023 cycles of chemotherapy administered but affected greater than 50% of patients within the study. Four patients ceased therapy because of neutropenia. A prospective randomized study of Dox SL against a combination of bleomycin and vincristine showed leucopenia to have affected 87 of the 117 patients within the Dox SL arm of the study. Anemia is reported in between 15 and 20% of patients receiving Dox SL and thrombocytopenia is seen in between 5 and 10% of patients. These are similar levels to those seen with DaunoXome.

Nausea and vomiting, never exceeding WHO grade 2, was seen infrequently, affecting between 15 and 18% of patients. Stomatitis has been reported in between 5

Table 8.1. TIS staging system for AIDS-related Kaposi's sarcoma

Tumor (T)	
T0	Confined to the skin and/or lymph nodes and/or minimal oral disease
T1	Tumor-associated edema or ulceration
	Extensive oral disease
	Visceral involvement (other than lymph nodes)
Immune System (I)	
I0	CD4 T cells ≥ 200/ml
I1	CD4 T cells ≤ 200/ml
Systemic Disease (S)	
S0	No previous opportunistic infection or candidiasis
	No "B" symptoms
	Karnofsky Performance Status ≥ 70
S1	Previous opportunistic infection and/or candidiasis
	"B" symptoms (unexplained fever, night sweats, ≥ 10% weight loss)
	Karnofsky Performance Status ≤ 70

and 10% of the patients, particularly those treated with higher doses in initial dose escalation studies. Cardiac toxicity with large cumulative doses of Dox SL is not reported and studies of LVEF have revealed no deterioration. A single patient was reported as dying from cardiac failure after two cycles of Dox SL to a cumulative dose of only 76 mg. This was felt to be due to underlying HIV myocarditis rather than the study drug, although neither myocardial biopsy or post mortem examination were performed. Fatigue is seen in between 10 and 15% of patients.

An interesting side effect that appears to be unique to this drug is the development of a 'hand-foot' syndrome with increasing doses of Dox SL. Patients experience severe pain and erythema in the hands and feet, which may lead to loss of mobility and the need to discontinue treatment. This has been particularly noticed with the use of higher doses of Dox SL for more common, solid tumors. "Hand-foot" syndrome is not a common clinical problem within the doses required for the treatment of AIDS-related KS. Another unusual feature is the development of ulceration, which is frequently slow to heal, at the site of previous radiotherapy. This suggests that the presence of abnormal vasculature as a consequence of normal tissue damage by ionizing radiation may enhance the sequestration and transcytosis of liposomes at these sites with resulting local increases in drug deposition.

Comparative studies of Dox SL and conventional combination chemotherapy are nearing completion. Initial results, presented in abstract form, suggest a significant advantage for the drug over conventional chemotherapy for AIDS-related KS, i.e., bleomycin and vincristine or adriamycin, bleomycin and vincristine. Dox SL has a higher response rate, 58.7%, and a similar duration of response when compared with bleomycin and vincristine. This is not, however, necessarily associated with a decreased incidence of side effects. There is a significantly higher incidence of infection, particularly, oral candidiasis within the group treated with Dox SL and a trend for an increasing number of other infections within this arm.

Table 8.2. Response criteria for AIDS-related Kaposi's sarcoma

1. Complete response (CR)
Complete resolution of all existing lesions for a period of at least 4 weeks with no new lesions appearing in that time. Flat pigmented lesions must be biopsied to confirm the absence of residual disease. Endoscopic confirmation of complete response of previously detected visceral lesions must be obtained.
2. Clinical Complete Response (CCR)
Absence of clinically detectable residual disease for at least 4 weeks, but without biopsy or endoscopic confirmation.
3. Partial Response (PR)
One or more of the following in the absence of; (a) new skin lesions, (b) new oral/visceral lesions, (c) progressive disease-associated edema, (d) ≥ 25% increase in the product of the bidimensional diameters of any index lesion.

1. ≥ 50% decrease in the number of measurable cutaneous, oral or visceral lesions.
2. ≥ 50% decrease in the size of the lesions, defined by one of the following criteria.
   a. ≥ 50% decrease in the sum of the products of the largest bidimensional diameters ofthe index lesion.
   b. complete flattening of ≥ 50% of the lesions.
   c. ≥ 75% of nodular lesions become indurated plaques.

4. Stable Disease
Any response which fails to meet the above conditions.
5. Progressive Disease
Any of the following:

1. ≥ 25% increase in the product of the bidimensional diameter of any index lesion.
2. the appearance of new lesions.
3. ≥ 25% of previously flat lesions become raised.
4. the appearance of new or increased disease-associated edema.

Dox SL is an effective drug in AIDS related KS with the highest response rate seen for a single agent. The liposomal properties appear to have markedly changed the effectiveness of the drug when compared to "free" drug. Future studies comparing Dox SL with DaunoXome are important and should be informative as to which is the best agent. Based on presently available data, Dox SL has a superior effect though the toxicity seen with the drug is probably slightly more than with DaunoXome.[71-77]

References

1. Centers for Disease Control. Epidemiologic aspects of the current outbreak of Kaposi's sarcoma and opportunistic infections. N Engl J Med 1982; 306:248-52.
2. Beral V. The epidemiology of cancer in AIDS patients. AIDS 1991; 5:99-103 (Suppl 2).
3. Miles SA, Wang H, Elashoff R, Mitsuyasu RT. Improved survival for patients with AIDS related Kaposi's sarcoma. J Clin Oncol 1994; 12(9):1910-16.

4. Tappero JW, Conant MA, Wolfe SF et al. Kaposi's sarcoma. Epidemiology, pathogenesis, histology, clinical spectrum, staging criteria and therapy. J Am Acad Dermatol 1995; 28:371-395.
5. Buchbinder A, Friedman Kein AE. Clinical aspects of Kaposi's sarcoma. Curr Opin Oncol 1992; 4:867-874.
6. Krown SE, Myskowski PL, Paredes J. Medical management of AIDS patients: Kaposi's sarcoma. Med Clin North Am 1992; 76:235-252.
7. Miles SA. Pathogenesis of HIV related Kaposi's sarcoma. Curr Opin Oncol 1994; 6:497-502.
8. Ensoli BS, Nakamura S, Salahuddin SZ et al. AIDS Kaposi's sarcoma derived cells express cytokines with autocrine and paracrine growth effects. Science 1989; 243:223-226.
9. Karasek MA. Origin of spindle shaped cells in Kaposi's sarcoma. Lymphology 1994; 27:41-44.
10. Moore PS, Chang Y. Detection of herpesvirus-like DNA sequences in Kaposi's sarcoma in patients with and without HIV infection. N Engl J Med 1995; 332(18):1181-5.
11. Chang Y, Cesarman E, Pessin MS et al. Identification of herpesvirus-like DNA sequences in AIDS associated Kaposi's sarcoma. Science 1994; 266(5192):1865-9.
12. Su IJ, Hsu YS, Chang YC, Wang IW. Herpesvirus like DNA sequence in Kaposi's sarcoma from AIDS and non-AIDS patients in Taiwan. Lancet 1995; 345(8951):722-3.
13. Cohen J. Controversy: Is KS really caused by new herpesvirus? Science 1995; 268:1847-8.
14. Kedes DH, Operskalski E, Busch M et al. The seroepidemiology of human herpesvirus 8 (Kaposi's sarcoma-associated herpesvirus): Distribution of infection in KS risk groups and evidence for sexual transmission. Nat Med 1996; 2(8):918-924.
15. Gao SJ, KIngsley L, Li M et al. KSHV antibodies among Americans, Italians and Ugandans with and without Kaposi's sarcoma. Nat Med 1996; 2(8):925-928.
16. Zhong W, Wang H, Herndier B, Ganem D. Restricted expression of Kaposi's sarcoma associated herpesvirus (human herpesvirus 8) genes in Kaposi's sarcoma. Proc Natl Acad Sci USA 1996; 93(13):6641-6646.
17. Brown TJ, Rowe JM, Lui J, Shoyab M. Regulation of IL6 expression by oncostatin M. J Immunol 1991; 147:2175-2180.
18. Masood R, Lunardi Iskandar Y, Jean LFL et al. Inhibition of AIDS associated Kaposi's sarcoma cell growth by DAB389-interleukin 6. AIDS Res Hum Retroviruses 1994; 10:969-975.
19. Cooper JS, Fried PR, Laubenstein LJ. Initial observations of the effect of radiotherapy on epidemic Kaposi's sarcoma. JAMA 1984; 252:934-935.
20. Cooper JS, Steinfeld AD, Lerch I. Intentions and outcomes in the radiotherapeutic management of epidemic Kaposi's sarcoma. Int J Radiat Oncol Biol Phys 1991; 20:419-422.
21. Geara F, Le Bourgeois JP, Piedbois P, Pavlovitch JM, Mazeron JJ. Radiotherapy in the management of cutaneous epidemic Kaposi's sarcoma. Int J Radiat Oncol Biol Phys 1991; 21:1517-1522.
22. Berson AM, Quivey JM, Harris JW, Wara WM. Radiation therapy for AIDS-related Kaposi's sarcoma. Int J Radiat Oncol Biol Phys 1990; 19:569-575.
23. de Wit R, Smit WGJM, Veenhof KHN, Bakker PJM, Oldenburger F, Gonzalez D. Palliative radiation therapy for AIDS-associated Kaposi's sarcoma by using a single fraction of 800 cGy. Radiother Oncol 1990; 19:131-136.
24. Stelzer KJ, Griffin TW. A randomized prospective trial of radiation therapy for AIDS-associated Kaposi's sarcoma. Int J Radiat Oncol Biol Phys 1993; 27:1057-1061.

25. Harrison M, Harrington K, Tomlinson D, Stewart S. Response and cosmetic outcome of two fractionation regimes for AIDS-related Kaposi's sarcoma. Radiother and Oncology (In Press).
26. Vapnek JM, Quivey JM, Carroll PR. Acquired immunodeficiency syndrome-related Kaposi's sarcoma of the male genitalia: management with radiation therapy. J Urol 1991; 146:333-336.
27. Ghabrial R, Quivey JM, Dunn JP, Char DH. Radiation therapy of acquired immunodeficiency syndrome-related Kaposi's sarcoma of the eyelids and conjunctiva. Arch Ophthalmol 1992; 110:1423-1426.
28. Le Bourgeois JP, Frikha H, Piedbois P, Le Pechoux C, Martin L, Haddad E. Radiotherapy in the management of epidemic Kaposi's sarcoma of the oral cavity, the eyelid and the genitals. Radiother Oncol 1994; 30:263-266.
29. Meyer JL. Whole lung irradiation for Kaposi's sarcoma. Am J Clin Oncol 1993; 16:372-376.
30. Vallis KA. Glutathione deficiency and radiosensitivity in AIDS patients. Lancet 1991; 337:918-919.
31. Stewart S. AIDS-related cancer. In: Price P, Sikora K, eds. Treatment of Cancer. Third Edition. London: Chapman and Hall, 1995.
32. Nemechek PM, Corder MC. Radiation recall associated with vinblastine in a patient treated for Kaposi sarcoma related to acquired immune deficiency syndrome. Cancer 1992; 70:1605-1606.
33. Stelzer KJ, Griffin TW, Koh W-J. Radiation skin toxicity with bleomycin in a patient with Kaposi sarcoma related to acquired immune deficiency syndrome. Cancer 1993; 71:1322-1325.
34. Epstein JB. Treatment of oral Kaposi sarcoma with intralesional vinblastine. Cancer 1993; 71:1722-1725.
35. Northfelt D, Volberding P. AIDS-related Kaposi's sarcoma: clinical presentation, biology and therapy. Adv Oncol 1991; 7:9-17.
36. Langtry JA, Bottomley DM, Phillips RH, Staughton RC. The infra-red coagulator in the treatment of AIDS-related Kaposi's sarcoma and a comparison with radiotherapy. Clin Exp Dermatol 1994; 19:23-25.
37. Bernstein ZP, Wilson D, Summers K, Dougherty T, Brooks J, Mang TS. Pilot/Phase I study of photodynamic therapy (PDT) for treatment of AIDS-associated Kaposi's sarcoma (AIDS/KS). Proc Am Soc Clin Oncol 1995; 14:289 (abstract 285).
38. Volberding PA, Abrams DI, Conant M et al. Vinblastine therapy for Kaposi's sarcoma in the acquired immune deficiency syndrome. Ann Intern Med 1985; 103:335-338.
39. Mintzer DM, Real FX, Jovino L et al. Treatment of Kaposi's sarcoma and thrombocytopenia with vincristine in patients with acquired immunodeficiency syndrome. Ann Intern Med 1985; 102:200-202.
40. Chachoua A, Green M, Laubenstein L et al. Phase II study of oral idarubicin in patients with AIDS-associated Kaposi's sarcoma. Cancer Treat Rep 1987; 71:775-776.
41. Kaplan L, Volberding PA. Failure (and danger) of mitoxantrone in AIDS-related Kaposi's sarcoma. Lancet 1985; 2:396.
42. Goss PE, Shepherd FA, Burkes R et al. Phase II study of epirubicin in the treatment of Kaposi's sarcoma and AIDS. Proc Am Soc Clin Oncol 1989; 8:2.
43. Gill PS, Akil B, Colletti P et al. Pulmonary Kaposi's sarcoma: clinical findings and results of therapy. Am J Med 1989; 87:57-61.
44. Guermonprez G, Katlama C, Caumes et al. Bleomycin in treatment of AIDS-related Kaposi's sarcoma. Abstracts: VI International Conference on AIDS. San Francisco 1990; 3:213 (poster SB511).

45. Laubenstein LJ, Krigel RL, Odajnyk CM et al. Treatment of epidemic Kaposi's sarcoma with etoposide or a combination of doxorubicin, bleomycin and vinblastine. J Clin Oncol 1984; 2:1115-1120.
46. Schwartzsmann G, Sprinz E, Kronfeld M, Vinholes J, Sander E, Zampese M et al. Phase II study of tenoposide in patients with AIDS-related Kaposi's sarcoma. Eur J Cancer 1991; 27:1637-39.
47. Saville MW, Lietzau J, Pluda JM, Feuerstein I, Odam J, Wilson WH et al. Treatment of HIV-associated Kaposi's sarcoma with paclitaxel. Lancet 1995; 346:26-28.
48. Kaplan L, Abrams D, Volberding P. Treatment of Kaposi's sarcoma in acquired immunodeficiency syndrome with an alternating vincristine-vinblastine regimen. Cancer Treat Rep 1986; 70:1121-1122.
49. Gill PS, Rarick M, McCutchan JA et al. Systemic treatment of AIDS-related Kaposi's sarcoma: results of a randomized trial. Am J Med 1991; 90:427-434.
50. Gill PS, Rarick MU, Espina B, Loureiro C, Bernstein-Singer M, Akil B, Levine AM. Advanced acquired immune deficiency syndrome-related Kaposi's sarcoma: Results of pilot studies using combination chemotherapy. Cancer 1990; 65:1074-1078.
51. Gompels MM, Hill A, Jenkins P et al. Kaposi's sarcoma in HIV infection treated with vincristine and bleomycin. AIDS 1992; 6:1175-1180.
52. Lipman MCI, Swaden LS, Sabin CA, Collis C, Johnson MA. Kaposi's sarcoma in HIV infection treated with vincristine and bleomycin. AIDS 1993; 7:592-593.
53. Gill PS, Miles SA, Mitsuyasu RT et al. Phase 1 AIDS Clinical Trials Group (075) study of adriamycin, bleomycin and vincristine chemotherapy with zidovudine in the treatment of AIDS related Kaposi's sarcoma. AIDS 1994; 8(12):1695-9.
54. Huang SK, Lee K-D, Hong K et al. Microscopic localization of sterically stabilized liposomes in colon carcinoma-bearing mice. Cancer Res 1992; 52:5135-5143.
55. Huang SK, Martin FJ, Jay G et al. Extravasation and transcytosis of liposomes in Kaposi's sarcoma-like dermal lesions of transgenic mice bearing the HIVtat gene. Am J Pathol 1993; 143:1-14.
56. Presant CA, Blayney D, Proffitt RT, Turner AF, Williams LE, Nadel HI, Kennedy P, Wiseman C, Gala K, Crossley RJ, Preiss SJ, Ksionski GE, Presant SL. Preliminary report: imaging of Kaposi sarcoma and lymphoma in AIDS with indium-111-labeled liposomes. Lancet 1990; 335:1307-1309.
57. Forssen EA, Coulter DM, Profitt RT. Selective in vivo localization of daunorubicin small unilamellar vesicles in solid tumors. Cancer Res 1992; 52:3255-3261.
58. Mayhew EG, Lasic D, Babbar S et al. Pharmacokinetics and antitumor activity of epirubicin encapsulated in long circulating liposomes incorporating a polyethylene glycol-derivatized phospholipid. Int J Cancer 1992; 51:302-309.
59. Vaage J, Mayhew E, Lasic D et al. Therapy of primary and metastatic mouse mammary carcinomas with doxorubicin encapsulated in long circulating liposomes. Int J Cancer 1992; 51:942-948.
60. Vaage J, Donovan D, Mayhew E et al. Therapy of mouse mammary carcinomas with vincristine and doxorubicin encapsulated in sterically stabilized liposomes. Int J Cancer 1993; 54:959-964.
61. Vaage J, Donovan D, Loftus T, Working P. Prevention of metastasis from mouse mammary carcinomas with liposomes carrying doxorubicin. Br J Cancer 1995; 72:1074-1075.
62. Vaage J, Barbera-Guillem E, Abra R et al. Tissue distribution and therapeutic effect of intravenous free or encapsulated liposomal doxorubicin on human prostate carcinoma xenografts. Cancer 1994; 73:1478-1484.
63. Krown SE, Metroka C, Wernz JC. Kaposi's sarcoma in the acquired immune deficiency syndrome: a proposal for uniform evaluation, response and staging criteria. AIDS Clinical Trials Group Oncology Comitee. J Clin Oncol 1989; 7(9):1201-7.

64. Gill PS, Espina BM, Mugggia F et al. Phase I/II clinical and pharmacokinetic evaluation of liposomal daunorubicin. J Clin Oncol 1995; 13(4):996-1003.
65. Money Kyrle JF, Bates F, Ready J et al. Liposomal daunorubicin in advanced Kaposi's sarcoma: a phase 2 study. Clin Oncol (R Coll Radiol) 1993; 5:367-71.
66. Presant CA, Scolaro M, Kennedy P et al. Liposomal daunorubicin treatment of HIV associated Kaposi's sarcoma. Lancet 1993; 341:1242-3.
67. Schurmann D, Dormann A, Grunewald T, Ruf B. Successful treatment of AIDS related pulmonary Kaposi's sarcoma with liposomal daunorubicin. Eur Respir J 1994; 7:824-5.
68. Girard PM, Bouchard O, Goetschel A et al. Phase II study of liposomal encapsulated daunorubicin in the treatment of AIDS associated mucocutaneous Kaposi's sarcoma. AIDS 1996; 10(7):753-757.
69. Uthayakumar S, Bower M, Money Kyrle J et al. Randomized cross over comparison of liposomal daunorubicin versus observation for early Kaposi's sarcoma. AIDS 1996; 10(5):515-519.
70. Gill PS, Wernz J, Scadden DT et al. Randomized phase III trial of liposomal daunorubicin versus doxorubicin, bleomycin and vincristine in AIDS related Kaposi's sarcoma. J Clin Oncol 1996; 14(8):2353-2364.
71. Bogner JR, Kronawitter U, Rolinski B et al. Liposomal doxorubicin in the treatment of advanced AIDS related Kaposi's sarcoma. J AIDS 1994; 7(5):463-8.
72. Harrison M, Tomlinson D, Stewart S. Liposomal entrapped doxorubicin: an active agent in AIDS related Kaposi's sarcoma. J Clin Oncol 1995; 13(4):914-920.
73. Simpson JK, Miller RF, Spittle MF. Liposomal doxorubicin for the treatment of AIDS related Kaposi's sarcoma. Clin Oncol, R Coll Radiol 1993; 5(6):372-4.
74. Northfelt DW, Martin FJ, Working P et al. Doxorubicin encapsulated in liposomes containing surface bound polyethylene glycol: pharmacokinetics, tumor localization, and safety in pateints with AIDS related Kaposi's sarcoma. J Clin Pharmacol 1996; 36:55-63.
75. Goebel FD, Goldstein D, Goos M et al. Efficacy and safety of Stealth liposomal doxorubicin in AIDS related Kaposi's sarcoma. Br J Cancer 1996; 73:989-94.
76. Harrison M, Harrington KJ, Stewart JSW. Liposomal chemotherapy for AIDS related Kaposi's sarcoma. Expert Opin Invest Drugs 1995; 4:1197-1203.
77. Gordon KB, Tajuddin A, Guitart J et al. Hand foot syndrome associated with liposome encapsulated doxorubicin therapy. Cancer 1995; 75(8):2169-73.

CHAPTER 9

Pegylated-Liposomal Doxorubicin (Doxil®) in the Treatment of AIDS-Related Kaposi's Sarcoma

Donald W. Northfelt

Introduction

In 1981 reports of the development of Kaposi's sarcoma in previously healthy young homosexual men heralded the onset of the epidemic of acquired immunodeficiency syndrome (AIDS).[1-3] Kaposi's sarcoma is the most common neoplasm in persons infected with the human immunodeficiency virus (HIV), having been the initial ("AIDS-defining") diagnosis in 15% of the patients with AIDS in the United States.[4] It has been estimated that AIDS-related Kaposi's sarcoma (AIDS-KS) has a prevalence of about 20,000 cases in this country.[5] While the overall incidence seems to be declining, the actual morbidity and mortality ascribed to AIDS-KS has increased, probably as a result of improved therapy of opportunistic infections without concurrent improvements in the treatment of AIDS-KS.[6,7] Several modalities have been employed successfully to treat AIDS-KS, but their toxicities, as well as the patients' underlying immunodeficiency, have limited their usefulness. Also, AIDS-KS produces visceral involvement, especially in the lungs, for which current treatments are less effective than for cutaneous disease.[8,9]

Recently, pegylated-liposomal doxorubicin (Doxil®, Sequus Pharmaceuticals, Inc., Menlo Park, CA) has been developed as a therapy for AIDS-KS. It was theorized that the histopathology of AIDS-KS and the altered distribution of doxorubicin caused by liposomal encapsulation would create circumstances in which the efficacy of the drug would be enhanced while the toxicities would be ameliorated. Preclinical and early clinical testing of pegylated-liposomal doxorubicin validated this theoretical construct, and recently-completed clinical trials have demonstrated the advantages of this new formulation over standard treatment programs for the disease.

This chapter will review the clinical features of AIDS-KS and the findings of studies which have established the value of pegylated-liposomal doxorubicin as treatment for the disease.

Clinical Presentation of Kaposi's Sarcoma

Kaposi's sarcoma was a rare malignancy prior to the HIV pandemic: the annual incidence was approximately 21 per 1 million persons in the United States.[10] Originally

Long Circulating Liposomes: Old Drugs, New Therapeutics, edited by Martin C. Woodle and Gerrit Storm.

described by Moritz Kaposi in 1872, the "endemic" or "classical" form of Kaposi's sarcoma is characterized initially by an indolent multicentric skin tumor, usually affecting the lower extremities of elderly men of Jewish and Mediterranean ancestry.[11]

Clinically, endemic Kaposi's sarcoma appears as multiple subcutaneous plaques or nodules which have a typical violaceous hue; sometimes local edema occurs. Histopathologic examination shows interlacing spindle cells, endothelial cells and vascular spaces, which are interspersed with extravasated erythrocytes, infiltrating lymphocytes, and hemosiderin-laden macrophages. Endemic Kaposi's sarcoma is rarely life-threatening and visceral involvement is unusual.

More recently, other forms of Kaposi's sarcoma have been recognized. These include an endemic form in equatorial Africa[12] and a form that is presumed to be a complication stemming from iatrogenic immunosuppression occurring during renal transplantation.[13] Of note, withdrawal of immunosuppressive therapy has been associated with spontaneous remission of up to 29% of patients developing Kaposi's sarcoma in the posttransplant setting.[14] Besides being a nodular cutaneous disease, endemic African Kaposi's sarcoma can also become florid and infiltrating. Both the endemic African and immunosuppression-related forms of the disease involve visceral sites more commonly than does the classical variety.

AIDS-KS tends clinically to be more varied and aggressive than other forms of Kaposi's sarcoma. Most patients with AIDS-KS have cutaneous nodular disease, characterized by subcutaneous nodular lesions that vary in size from several millimeters to several centimeters in diameter. The nodules have the classical violaceous pigmentation and are generally not pruritic or painful. Nodular AIDS-KS does not usually cause necrosis of the overlying skin and rarely involves underlying bony structures; large exophytic lesions with superficial necrosis, bleeding, and bacterial superinfection do occasionally occur.

Tumor plaques of AIDS-KS tend to occur on the thighs and plantar surfaces of the feet, and commonly have underlying or surrounding lymphedema. Lymphedema may occur as a relatively isolated finding, seemingly out of proportion to the extent of visible cutaneous tumors. The face, penis, scrotum and lower extremities are frequently affected in this manner. Lymphadenopathic AIDS-KS affects the peripheral lymph nodes, sometimes causing massive nodal enlargement, but often leaving the skin unaffected. Oral AIDS-KS lesions can cause gingival bleeding, dental displacement and pain. The mucosa over the hard palate is frequently involved. Visceral involvement, particularly of the lungs and gastrointestinal tract, is common in patients with AIDS-KS. Such visceral involvement can result in respiratory insufficiency and gastrointestinal obstruction or malabsorption. Because the gastrointestinal and pulmonary lesions of AIDS-KS are usually submucosal, however, they do not often produce spontaneous hemorrhage at these sites. Pulmonary AIDS-KS is associated with a poor prognosis.[8,9]

Treatment of AIDS-Related Kaposi's Sarcoma

There is presently no cure for AIDS-KS; indeed, a cure may be impossible until a means of alleviating HIV-induced immunodeficiency is found. In the past, AIDS-KS was rarely considered to be life-threatening, as most patients with the disease ultimately died of opportunistic infections occurring as a result of their profound immunodeficiency. However, the morbidity and mortality attributable to AIDS-KS may be increasing. This trend has developed as significant improvements in treatment and prevention have substantially reduced *P. carinii* pneumonia as a cause of competing morbidity and mortality,[6,15] whereas no such advances have occurred in the management of AIDS-KS. Therefore, palliative therapy of AIDS-KS is frequently required for elimination or reduction of cosmetically unacceptable lesions, reduction of painful or unsightly edema or lymphadenopathy, and shrinkage of symptomatic oral lesions. Effective treatment is also essential for relief of respiratory symptoms caused by pulmonary involvement, and for relief of obstruction, pain, and hemorrhage caused by gastrointestinal involvement. As with any palliative therapy, treatment for AIDS-KS must be designed in such a way that it helps to curtail tumor progression but does not exacerbate the underlying immunodeficiency or significantly impair the patient's clinical status.

Based upon these goals, several strategies are currently employed to treat the various clinical problems encountered in AIDS-KS. Depending on the clinical situation, single- and multi-agent cytotoxic chemotherapy, treatment with interferon α, radiotherapy, and other local therapies have all been used successfully. The most appropriate treatment plan for an individual patient may combine several of these modalities in combination or sequentially.

A herpesvirus was recently identified within tissue and blood of persons with all varieties of Kaposi's sarcoma.[16] This new herpesvirus, human herpesvirus-8 (HHV-8) or Kaposi's sarcoma-associated herpesvirus (KSHV) may operate under conditions of HIV-induced immunodeficiency, by mechanisms which are not yet understood, to cause Kaposi's sarcoma and certain forms of non-Hodgkin's lymphoma. Intensive preclinical investigations are underway to determine whether any specific form(s) of antiviral therapy can affect the growth of HHV-8 and Kaposi's sarcoma, and it is anticipated that clinical testing of antiviral therapy for AIDS-KS will develop from these efforts. At the time of this writing, however, the value of antiviral therapy for AIDS-KS remains unproven.

For patients with widespread, aggressive AIDS-KS, combination cytotoxic chemotherapy is usually the therapy of choice.[17] The two most commonly employed regimens are the combination of bleomycin and vincristine ("BV"), and doxorubicin, bleomycin and vincristine ("ABV"). These regimens are currently considered to be first-line chemotherapy for advanced AIDS-KS. ABV has been shown to be superior to single agent therapy.[18] While these regimens control AIDS-KS in many patients, the disease will often eventually progress despite ongoing therapy. In addition, some patients cannot tolerate these regimens for a variety of reasons, including myelosuppression, neuropathies, nausea, and vomiting.

Rationale for Use of Liposome-Encapsulated Cytotoxic Chemotherapy in the Treatment of AIDS-KS

Encapsulation in liposomes has been shown in animal models and early human trials to reduce certain toxic effects of cytotoxic chemotherapeutic agents, including doxorubicin. It is believed that rapid uptake of doxorubicin-loaded liposomes by fixed macrophages (residing primarily in the liver), followed by slow release from those sites reduces peak plasma concentrations of doxorubicin.[19] Although this liver deposition phenomenon is useful in reducing side effects, the intrinsic activity of doxorubicin against systemic malignancies is not favorably influenced by encapsulation in conventional liposomes. In contrast, pegylated liposomes have been shown to circumvent uptake by hepatic macrophages and to circulate for prolonged periods in the bloodstream of rodents and dogs.[20] Radioactively-labeled pegylated liposomes have been shown to selectively enter implanted animal tumors and a direct correlation between blood circulation half-life and tumor uptake has been established.[21] In a preclinical model of Kaposi's sarcoma, pegylated liposomes containing colloidal gold particles were shown to selectively accumulate in both early and late stage cutaneous lesions.[22]

Following intravenous injection of pegylated-liposomal doxorubicin into tumor-bearing mice, doxorubicin levels measured in tumors were substantially higher than those seen in animals receiving comparable doses of standard doxorubicin.[23] Based on these observations, it was hypothesized that treatment with pegylated-liposomal doxorubicin could selectively deliver doxorubicin to tumor sites, including AIDS-KS lesions. If this hypothesis could be validated in patients with AIDS-KS, pegylated-liposomal doxorubicin therapy might therefore be expected to provide disease palliation at relatively lower doses than required with standard doxorubicin therapy by distributing a greater proportion of an administered dose of doxorubicin to AIDS-KS lesions.

Pharmacokinetics and Tumor Localization of Pegylated-Liposomal Doxorubicin in Patients with AIDS-KS

The first clinical trial conducted with pegylated liposomal doxorubicin in patients with AIDS-KS was designed to determine standard pharmacokinetic parameters of the drug in these patients, and to evaluate the ability of the pegylated liposome to deliver doxorubicin to disease sites in the manner hypothesized above.[24]

Eighteen patients with biopsy-proven AIDS-KS were randomly assigned to receive either standard doxorubicin or pegylated-liposomal doxorubicin. Consecutive participants were entered at three dose levels (10, 20 and 40 mg/m^2) in ascending fashion. Plasma samples were obtained at successive timepoints over the ensuing 96 h and analyzed for doxorubicin concentration; these data were used to derive pharmacokinetic parameters. In addition, a previously untreated AIDS-KS skin lesion was removed from each participant 72 h after dosing and analyzed for doxorubicin concentration.

The results of this trial are shown in Table 9.1. Pegylated-liposomal doxorubicin clearance ranged from 0.034-0.108 $l/h/m^2$, with a volume of distribution of approximately 6-7 l and an area under the concentration-time curve (AUC) ranging from 185 µg/hr/l at 10 mg/m^2 to 627 µg/hr/l at 40 mg/m^2. The distribution half-life of pegylated-liposomal doxorubicin was approximately 34-44 h. In contrast, clearance of standard doxorubicin at comparable dose levels has been reported to be much faster (24-73 l/h), with a substantially larger apparent volume of distribution (> 1500 l), smaller AUCs (range = 0.4-1.2 µg/hr/l), and distribution half-life of 5-10 minutes. Seventy-two hours after administration, doxorubicin levels achieved in AIDS-KS lesions of participants receiving pegylated-liposomal doxorubicin were 5.2-11.4 times greater than those in participants given comparable doses of standard doxorubicin. Pegylated-liposomal doxorubicin and standard doxorubicin were roughly equipotent in producing toxicities, and no toxicities were noted with pegylated-liposomal doxorubicin administration that would not have been expected with standard doxorubicin.

It was concluded that encapsulation in pegylated liposomes significantly limited the distribution and elimination of doxorubicin, enhanced accumulation of the drug in AIDS-KS lesions in comparison to standard doxorubicin administration, and might therefore improve the drug's efficacy and therapeutic index in the treatment of AIDS-KS.

Table 9.1. **Pharmacokinetics and tumor localization of pegylated liposomal doxorubicin**

Dose (mg/m^2)	N	C0 (mg/ml)	$t_{1/2}$ (h)	AUC (mg/h/ml)	clearance (l/h)	Vd (l)
10	3	3.0 ± 0.9	41.1 ± 6.8	151 ± 61	0.14 ± 0.05	7.9 ± 2.4
20	3	6.9 ± 2.7	43.5 ± 13.8	277 ± 163	0.17 ± 0.09	10.0 ± 4.4
40	3	17.6 ± 3.4	33.3 ± 8.4	580 ± 208	0.15 ± 0.06	6.5 ± 1.05
Pharmacokinetics of standard doxorubicin						
50	*	1-3	0.08	~ 1-5	10-40	~ 1000
Tissue Concentrations: Pegylated-liposomal doxorubicin vs. standard doxorubicin						
Dose (mg/m^2)				10	20	40
Tissue concentration of doxorubicin						
Patients treated with pegylated-liposomal doxorubicin, mg/g (SD):				2.06 (0.42)	1.61 (0.80)	7.71 (2.72)
Patients treated with standard doxorubicin, mg/g (SD):				0.18 (0.07)	0.31 (0.16)	0.82 (0.18)
Ratio				11.4	5.2	9.4

* values from literature sources

Effect of Pegylated-Liposomal Encapsulation on Distribution of Doxorubicin to AIDS-KS and Adjacent Normal Skin

Another small clinical trial was conducted to demonstrate the ability of the pegylated liposome to preferentially deliver doxorubicin to AIDS-KS lesions while restricting the ability of the drug to enter normal tissues.[25] Twenty-four patients with AIDS-KS were randomly assigned to receive pegylated-liposomal doxorubicin at doses of 10 or 20 mg/m². Each patient then underwent simultaneous biopsies of a representative AIDS-KS lesion and a nearby area of normal skin, and the tissue concentrations of doxorubicin were then determined. Patients were sequentially assigned to undergo the paired biopsies at either 48 h (eight patients at each dose level) or 96 h (four patient at each dose level) after receiving their dose of pegylated-liposomal doxorubicin.

The results of this study are presented in Table 9.2. Pegylated-liposomal doxorubicin was shown to deliver approximately 10-20 times more doxorubicin to AIDS-KS lesions than to adjacent normal skin. These findings provided further support for the hypothesis that treatment with pegylated-liposomal doxorubicin could selectively distribute drug to tumor tissue, with the consequent potential for greater efficacy and reduced toxicity.

Phase II Clinical Trials of Pegylated-Liposomal Doxorubicin in Patients with AIDS-KS

A number of investigators have reported the results of small phase II clinical trials of pegylated-liposomal doxorubicin therapy for AIDS-KS.[26-32] Their results are briefly summarized in Table 9.3. Partial responses were observed in 73-100% of patients treated in these studies. Toxicities were generally reported to be mild, even in patients with advanced HIV disease who had previously received chemotherapy treatments for AIDS-KS.

Efficacy of Pegylated-Liposomal Doxorubicin in the Treatment of AIDS-Related Kaposi's Sarcoma After Failure of Standard Chemotherapy

Enhanced delivery of doxorubicin to AIDS-KS lesions, promising results of small phase II studies, and an altered toxicity profile in preclinical studies raised the possibility that patients who experienced disease progression or toxicity with standard combination chemotherapy might benefit from single-agent salvage therapy with pegylated-liposomal doxorubicin. It was of particular clinical interest to test this

Table 9.2. Effect of pegylated-liposomal encapsulation on distribution of doxorubicin to AIDS-KS and adjacent normal skin

	48 hour biopsy		96 hour biopsy	
Dose (mg/m^2)	10 (n=8)	20 (n=8)	10 (n=4)	20 (n=4)
Normal skin*	0.36 (0.31)	0.84 (0.52)	0.95 (1.0)	0.27 (0.11)
AIDS-KS*	5.0 (7.8)	13.2 (8.5)	11.7 (16.6)	2.9 (1.4)

* mg doxorubicin/g tissue (SD)

Table 9.3. Phase II clinical trials of pegylated-liposomal doxorubicin in patients with AIDS-KS

Ref.	N	Dose*	Schedule	Response	Toxicity
26	20	20	every 2 weeks	PR 19	neutropenia common
27	16	20	every 2-3 weeks	PR 11 SD 5	grade 3/4 neutropenia in 50% of patients
28	15	10-20	every 2 weeks	PR 11	grade 4 neutropenia in 6% of cycles
29	39	10 (10 pts) 20 (27 pts) 40 (3 pts)	every 2 weeks x6	CR 3 PR 33 SD 3	stomatitis in 10% of patients alopecia in 10% of patients grade 4 neutropenia in 42% of patients
30	8	20	every 3 weeks x6	"significant regression" in all patients CR 1 PR 7	"minor"
31	3	10-20	every 2-4 weeks	PR 3	neutropenia common
32	34	20	every 3 weeks	CR 2 PR 23	grade 3/4 neutropenia 15% of cycles alopecia in 9% of patients nausea/vomiting in 17% of patients

* dose in mg/m^2. SD = stable disease.

hypothesis in patients who had already experienced disease progression on a regimen containing standard doxorubicin. A retrospective analysis was therefore performed to assess the efficacy of pegylated-liposomal doxorubicin treatment in a cohort of severely immunocompromised patients with advanced AIDS-KS who had experienced disease progression or unacceptable toxicity with combination systemic chemotherapy.[33]

Patients for the analysis were identified from a cohort of 383 patients enrolled in an open-label trial of pegylated-liposomal doxorubicin therapy for AIDS-KS which was carried out at 17 centers from March 1993 to December 1994. A panel of three reviewers independently assessed the information on patients' prior therapy. They identified patients within the larger study cohort who had received systemic chemotherapy with at least two concurrently administered agents for treatment of AIDS-KS. Patients experiencing disease progression, as defined by the AIDS Clinical Trials Group (ACTG) criteria,[34] or intolerable toxicity after two cycles of therapy with regimens including bleomycin and/or doxorubicin were specifically identified. The panel members were blinded to all other aspects of the patients' clinical status, including the results of their therapy with pegylated-liposomal doxorubicin. Only those patients identified by all reviewers as meeting these criteria were included in the analysis.

Responses to therapy were assessed using a modification of criteria established by the AIDS Clinical Trials Group (ACTG).[34] In addition, changes in disease-associated pain, edema, lesion color, and lesion flattening were assessed as a means of identifying treatment-induced clinical benefit not well-assessed by the ACTG criteria. Quantification of adverse events was based on investigator reports. Based on their clinical judgment, investigators evaluated the adverse events as mild, moderate or severe and whether the event was related to the study drug.

Patients received pegylated-liposomal doxorubicin at a dose of 20 mg/m^2 every three weeks. The dose could be reduced or cycle length increased if toxicity intervened. Assessment of response and clinical benefit were made every three weeks within 48 h prior to drug administration. Patients were followed for at least four weeks after discontinuation from the study. Use of myeloid colony-stimulating factors was permitted at the discretion of each investigator.

Of the 53 patients, 19 (36%) had a partial response and one patient (2%) had a complete clinical response (Table 9.4). Of the 28 patients who entered the trial having experienced disease progression on a combination regimen containing doxorubicin, 9 (32%) had a partial response. The median duration of response (i.e., time from initial response until development of progressive disease) for the entire group and for the doxorubicin treatment failures was 128 and 127 d, respectively. The median time to partial response was 109 d. The median time to treatment failure was 134 (range: 1–456+ days) for the 53 patients and 148 d (range: 1–251 d) for the 28 patients who had disease progression on prior doxorubicin therapy. For purposes of statistical analysis, patients who where found to have progressive disease on their first posttreatment evaluation were designated as treatment failures after one day on study.

The ability of pegylated-liposomal doxorubicin to induce clinical benefit was assessed by lesion flattening, color improvement, reduction of pain, and reduction in edema. Table 9.5 shows the percent of patients with a specific undesirable characteristic at baseline who showed a particular clinical benefit during treatment. Forty-eight percent of patients who had raised indicator lesions at baseline showed complete flattening of all indicator lesions; 56% of patients with red or purple indicator

Table 9.4. Efficacy of pegylated-liposomal doxorubicin in the treatment of AIDS-related Kaposi's sarcoma after failure of standard chemotherapy

Best Response	All Patients (N = 53)	Doxorubicin Failure* (N = 28)
Complete Clinical	1 (2%)	0
Partial	19 (36%)	9 (32%)
Stable	19 (36%)	14 (50%)
Progression	14 (26%)	5 (18%)
Median Time (days) to PR	109	109
Median Duration (days) of PR	128	127
Median Time (days) to Treatment Failure	134	148

* Patients whose AIDS-KS progressed on a combination regimen containing doxorubicin.

lesions at baseline had all the indicator lesions change to a less noticeable color; 45% of patients with moderate to severe pain at baseline had a reduction in pain to mild or absent without the aid of analgesics or antidepressants; of patients with edema at baseline, 83% had their edema disappear. Patients with a partial response were more likely to achieve each benefit than patients with stable disease or progressive disease. However, beneficial changes in lesion characteristics developed in some patients whose responses did not fulfill protocol criteria for partial response.

Twenty-six patients (49%) had more than one clinical benefit. One patient achieved clinical benefit in all four categories at one point during therapy. This patient's overall response to therapy was graded as a partial response. Nine patients (17%) achieved clinical benefit in three of the four categories at some point during study therapy. Seven of these patients achieved a partial response, and two had stable disease. Sixteen patients achieved clinical benefits in two of the four categories. The clinical benefits were durable, with median duration of 113 d for flattening, > 164 d for color improvement, > 111 d for pain reduction, and 150 d for edema reduction.

Table 9.6 lists the most common adverse events deemed by the investigators to be possibly or probably related to pegylated-liposomal doxorubicin. Overall, 76% of patients reported at least one adverse event which was possibly or probably related to pegylated-liposomal doxorubicin, and 30% reported at least one severe adverse event thought to be related to pegylated-liposomal doxorubicin.

Several adverse events occurred that were not expected from the usual toxicity profile of doxorubicin. Palmar-plantar erythrodysesthesia occurred in one patient, presenting as a severe erythematous rash. The rash resolved when therapy was withheld. One patient experienced an infusion-related reaction during his first infusion, characterized by the sudden onset of flushing of the entire body, associated with abdominal pain. The reaction subsided the same day, and the patient was

Table 9.5. Efficacy of pegylated-liposomal doxorubicin in the treatment of AIDS-related Kaposi's sarcoma after failure of standard chemotherapy: clinical benefits

	Complete Flattening	Color Improvement	Pain Reduction	Edema Reduction
N*	48	48	22	23
Benefit**				
Partial Response/ Complete Response	68%	82%	70%	100%
Stable Disease	38%	39%	25%	78%
Progressive Disease	31%	46%	25%	60%
All Patients	48%	56%	45%	83%

* Number of patients with the undesirable characteristic eligible for the benefit at baseline and, thus, with the potential to improve.
** Percentage of patients in each response category with the undesirable characteristic at baseline who experienced the benefit.

Table 9.6. Efficacy of pegylated-liposomal doxorubicin in the treatment of AIDS-related Kaposi's sarcoma after failure of standard chemotherapy: adverse events

N = 53	All Events* (% patients)	Severe (% patients)
EVENT		
Any Adverse Event	76%	30%
Leukopenia	40%	17%
Nausea and/or Vomiting	19%	0%
Alopecia	9%	0%
Asthenia	9%	2%
Fever	8%	2%
Diarrhea	6%	2%
Thrombocytopenia	6%	2%

* Adverse events occurring in ≥ 5% of patients, thought to be possibly or probably related to pegylated-liposomal doxorubicin

subsequently able to undergo a total of 12 cycles of pegylated-liposomal doxorubicin therapy without further incident when premedicated with prednisolone prior to dosing. No clinical evidence of cardiac toxicity or extravasation injury attributable to doxorubicin was seen in the study cohort.

In summary, the findings of this retrospective analysis demonstrated that pegylated-liposomal doxorubicin was effective treatment for patients who experienced disease progression or unacceptable toxicity with first-line, combination systemic chemotherapy for advanced AIDS-KS. The findings demonstrated that the pegylated-liposome encapsulation enhanced the efficacy of doxorubicin because patients who developed progressive disease while receiving standard doxorubicin subsequently responded to pegylated-liposomal doxorubicin. Use of doxorubicin as a single agent in the primary treatment of AIDS-KS has been reported to produce partial responses in 10% to 48% of patients.[18,35] In this study, the partial/complete response rate was 38% overall and 32% in patients who had already experienced disease progression while receiving doxorubicin. The response in these patients suggested that the longer half-life, greater AUC, and altered distribution characteristics of pegylated-liposomal doxorubicin enhanced the potency of doxorubicin, as had been observed in animal models.[23]

Objective responses were accompanied by important clinical benefits to patients. Patients remained active throughout the study as indicated by lack of decline in their Karnofsky performance scores. Pain was alleviated without the need to increase or add analgesics to patients' regimens and without the concurrent use of antidepressants. Lesions flattened, edema was reduced, and the color of lesions changed to more normal skin tones.

Pegylated-liposomal doxorubicin had no new or unexpected toxicities compared to those expected from conventional doxorubicin, with the exception of infrequent acute infusion-related reactions and skin eruptions that may be related to the altered delivery effected by the pegylated-liposome technology. The infusion reactions resolved with the cessation of the infusion. Palmar-plantar erythrodysesthesia was managed by withholding treatment until resolution and by resuming therapy at longer dosing intervals. Myelotoxicity was manageable with the use of colony-stimulating factors. The incidence of opportunistic infection was not significantly different than that seen for similar patients with AIDS receiving chemotherapy.[36,37]

Toxicities which often accompany doxorubicin-containing combination chemotherapy for AIDS-KS are alopecia, nausea and vomiting. While usually not of serious medical importance, these toxicities are very important to patients' quality of life. In the cohort described above, the incidence of alopecia was 9%. Nausea and vomiting occurred in 15% of patients and was not severe in any patient. In contrast, the incidence of nausea and vomiting reported with doxorubicin, ABV or BV ranged from 50% to 67% in patients with AIDS-KS, and alopecia was reported to occur in 55% to 67% of patients receiving these regimens.[18,36] Thus, pegylated-liposomal doxorubicin compared favorably with regard to these toxicities.

Pegylated-Liposomal Doxorubicin versus a Combination Regimen of Doxorubicin, Bleomycin and Vincristine in the Treatment of AIDS-Related Kaposi's Sarcoma: Results of a Randomized Phase III Clinical Trial

The demonstration of enhanced delivery of doxorubicin to lesions of AIDS-KS and the encouraging results from the analysis described above led to the hypothesis that pegylated-liposomal doxorubicin could provide an important new treatment option for AIDS-KS. A randomized comparative trial was therefore conducted to compare the efficacy and toxicities of pegylated-liposomal doxorubicin and the standard chemotherapy combination ABV in patients with advanced AIDS-KS.[38]

Patients were prospectively randomized to receive either pegylated-liposomal doxorubicin or the combination of doxorubicin, bleomycin and vincristine. They were enrolled at 25 sites from April 1993 to December 1994. Responses to therapy were assessed using the modified ACTG criteria described above. Pulmonary and gastrointestinal involvement with AIDS-KS was inferred at study entry from the appearance of patients' chest radiographs, respiratory and gastrointestinal symptoms, or reports of previous endoscopic evaluations. Endoscopic verification of visceral disease was not routinely obtained at study entry due to the reluctance of patients and investigators to permit invasive procedures when treatment decisions would not be altered by their outcome. Instead, patients were queried for symptoms of pulmonary and gastrointestinal AIDS-KS on each study visit. This information was not used in the formal assessment of response to study therapy, and the results were therefore reported separately. Standard World Health Organization (WHO) criteria were used to assess toxicities. Patients received pegylated-liposomal doxorubicin at a dose of 20 mg/m^2; doxorubicin, bleomycin and vincristine were administered at a dose of 20 mg/m^2 of doxorubicin (Adriamycin), 10 mg/m^2 of bleomycin, and 1 mg of vincristine. Treatments were administered every 14 d to a maximum of 6 cycles.

Two-hundred fifty-eight patients were enrolled in the trial. One-hundred thirty-three were randomized to receive pegylated-liposomal doxorubicin and 125 to receive ABV. The characteristics of study participants were typical of patients with advanced HIV disease. There were no significant differences between the groups. The median CD4+ T-lymphocyte count was 13 cells/mm^3 in those receiving ABV, and 12.5 cells/mm^3 in those receiving pegylated-liposomal doxorubicin.

Among patients receiving pegylated-liposomal doxorubicin, one had a clinical complete response and 60 had partial response (CCR + PR = 61/133, 45.9%; 95% confidence interval 37-54%). Among patients receiving ABV, 31 had partial response (PR = 31/125, 24.8%; 95% confidence interval 17-32%). This difference was statistically significant, $p < 0.001$ (Table 9.7). The median duration of response was 90 d with pegylated-liposomal doxorubicin and 92 d with ABV (Table 9.8). The median time to partial or complete response was 39 d with pegylated-liposomal doxorubicin and 50 d with ABV. The median time to treatment failure, defined as the time from initiation of therapy to the onset of progressive disease was 124 d and 128 d with pegylated-liposomal doxorubicin and ABV, respectively. The mean Karnofsky score did not change appreciably from baseline in either group.

Several additional parameters indicative of treatment efficacy were evaluated. The sum of the products of perpendicular diameters of indicator lesions of all patients decreased 24% from baseline to end of treatment in patients receiving pegylated-liposomal doxorubicin and 15% in patients receiving ABV ($p < 0.034$). Ninety-six percent of pegylated-liposomal doxorubicin patients had at least one indicator lesion raised at baseline, whereas 51% of patients had all their indicator lesions flattened at the end of treatment. Ninety-seven percent of ABV patients had at least one indicator lesion raised at baseline but only 28% of patients had all their indicator lesions flattened by the end of therapy. This difference in likelihood indicator lesion flattening with the respective therapies was statistically significant ($p < 0.001$). The median time required to achieve this benefit was 64 d in the pegylated-liposomal doxorubicin group and 115 d in the ABV group ($p < 0.001$).

Table 9.7. Pegylated-liposomal doxorubicin versus a combination regimen of doxorubicin, bleomycin and vincristine (ABV) in the treatment of AIDS-related Kaposi's sarcoma: results of a randomized phase III clinical trial

RESPONSE	Pegylated-Liposomal Doxorubicin	ABV	p-VALUE
	n = 133	n = 125	
Complete	0	0	
Complete Clinical	1 (0.8%)	0	
Partial	60 (45.1%)	31 (24.8%)	< 0.001*
Partial Response, 95% Confidence Interval	(37-54%)	(17-32%)	
Stable	70 (52.6%)	84 (67.2%)	
Progression	2 (1.5%)	10 (8.0%)	

* Fisher's exact test for between treatment differences for those patients having a complete/ partial response.

Table 9.8. Pegylated-liposomal doxorubicin versus a combination regimen of doxorubicin, bleomycin and vincristine (ABV) in the treatment of AIDS-related Kaposi's sarcoma: results of a randomized phase III clinical trial: response time variables

	Pegylated-Liposomal Doxorubicin n = 133	ABV n = 125	p
Median Time to Response	**39.0 days**	**50.0 days**	**0.014***
Median Duration of Response	**90.0 days**	**92.0 days**	**0.234***
Median Time to Treatment Failure	**124 days**	**128 days**	**0.259***

*** Log-Rank test**

Only 3% of pegylated-liposomal doxorubicin, and 4% of ABV patients had indicator lesions that were all brown at baseline. The balance of the patients had indicator lesions which were erythematous or violaceous and therefore considered to be more disfiguring. By the end of therapy 50% of pegylated-liposomal doxorubicin and 30% of ABV patients had indicator lesions that were all brown ($p < 0.002$). The mean time required to achieve this benefit was 68 d in the pegylated-liposomal doxorubicin group and 69 d in the ABV group.

Both regimens were associated with reduction in lesion-associated pain. At baseline, 40% of patients receiving pegylated-liposomal doxorubicin and 47% of patients receiving ABV had no pain; whereas, 33% and 31%, respectively, had moderate or severe pain. By the end of treatment 83% and 78% of pegylated-liposomal doxorubicin and ABV patients, respectively, were without lesion-associated pain.

Pain relief was achieved without increases or new additions of either analgesics or antidepressants. The median time required to achieve this benefit was 17 d in the pegylated-liposomal doxorubicin group and 15 d in the ABV group.

Seventeen percent of patients receiving pegylated-liposomal doxorubicin and 18% of patients receiving ABV had signs or symptoms of intestinal AIDS-KS at baseline. At the end of treatment 8% of pegylated-liposomal doxorubicin and 16% of ABV patients had such symptoms ($p \leq 0.05$). Nineteen percent of pegylated-liposomal doxorubicin and 23% of ABV patients had signs or symptoms of pulmonary AIDS-KS at baseline. These percentages did not change appreciably over the course of the study in either group.

The quality of life of patients enrolled in the trial was assessed at each study visit. A complete analysis of data obtained from two quality of life instruments was incomplete at the time of this writing. However, in all domains examined, quality of life among patients treated with pegylated-liposomal doxorubicin appeared to compare favorably with that experienced by patients receiving ABV.

Ninety patients receiving pegylated-liposomal doxorubicin (68%) and 43 patients receiving ABV (34%) completed the protocol. The mean number of cycles per patient was 5.2 ± 1.4 (SD) for patients receiving pegylated-liposomal doxorubicin and 3.8 ± 1.9 for patients receiving ABV. Thirty-seven percent of ABV patients and 11% of pegylated-liposomal doxorubicin patients discontinued their participation in the trial because of an adverse event ($p < 0.001$). Sixteen patients receiving pegylated-liposomal doxorubicin (12%) and 5 patients receiving ABV (4%) died during the trial. A Kaplan-Meier analysis indicated that the difference in death rates between the treatment groups was not statistically significant. Seven percent of ABV patients and no pegylated-liposomal doxorubicin patients withdrew because of disease progression. Other reasons for discontinuing study therapy included poor compliance (8 patients receiving ABV, 3 patients receiving pegylated-liposomal doxorubicin), lost to follow up (2 patients receiving ABV, 2 patients receiving pegylated-liposomal doxorubicin), intercurrent opportunistic infections (2 patients receiving ABV, 1 patient receiving pegylated-liposomal doxorubicin), and unspecified personal reasons (2 patients receiving ABV, 1 patient receiving pegylated-liposomal doxorubicin). One patient receiving ABV withdrew early due to lack of response, and one patient receiving pegylated-liposomal doxorubicin had therapy withheld for greater than 14 d.

Table 9.9 lists the most common adverse events experienced in the trial. Six episodes of acute infusion-related reactions occurred in the pegylated-liposomal doxorubicin group. These reactions were characterized by flushing, chest pain, dyspnea, difficulty swallowing, hypotension, and/or back pain. Three cases of palmar-plantar erythrodysesthesia were seen in the pegylated-liposomal doxorubicin group and one case was seen in the ABV group. Two other patients who received pegylated-liposomal doxorubicin developed rashes that were consistent with this syndrome but not explicitly designated as such by the investigator. Alopecia occurred in 42% of patients receiving ABV and in 11% of patients receiving pegylated-liposomal doxorubicin ($p < 0.001$).

In summary, the findings of this randomized comparative trial demonstrated that pegylated-liposomal doxorubicin was effective treatment in patients with advanced AIDS-KS and was superior to the combination of doxorubicin, bleomycin, and vincristine. Because doxorubicin delivered by pegylated liposomes was able to produce results superior to a proven combination regimen containing doxorubicin,

Table 9.9. Selected adverse Events*

Event	pegylated-liposomal doxorubicin n=133	ABV n=125	p-Value
leukopenia	56 (42%)	56 (45%)	NS
nausea and/or vomiting	46 (35%)	72 (58%)	<0.001
alopecia	15 (11%)	53 (42%)	< 0.001
peripheral neuritis	16 (12%)	35 (28%)	0.002
mucositis/stomatitis	24 (18%)	10 (8%)	0.026

* number of patients who experienced adverse event at any time during study treatment; advrese events judged by investigators to be directly related to study treatment.

it was concluded that pegylated-liposome encapsulation enhanced the therapeutic effect of doxorubicin.

Randomized Comparative Trial of Pegylated-Liposomal Doxorubicin versus Bleomycin and Vincristine in the Treatment of AIDS-Related Kaposi's Sarcoma

A randomized trial similar in design to that described above was also recently conducted in which treatments with pegylated-liposomal doxorubicin and the combination of bleomycin and vincristine were compared.[39] Partial results of this study have been reported and are shown in Table 9.10. These findings also demonstrated that pegylated-liposomal doxorubicin was superior to the standard combination regimen in its effectiveness. With respect to toxicity, pegylated-liposomal doxorubicin was found to cause significantly more neutropenia, but less nausea/vomiting, alopecia, and neuropathy.

Toxicity of Pegylated-Liposomal Doxorubicin in Patients with AIDS-KS

The primary toxicity of pegylated-liposomal doxorubicin has been found to be myelotoxicity with neutropenia in all clinical trials reported to date. Nausea, vomiting, alopecia, and neuropathy have all occurred with substantially lower frequency with pegylated-liposomal doxorubicin than is typically observed with standard chemotherapy treatments. Although definitive information is lacking, most investigators have considered the incidence of opportunistic infection to be no greater in patients treated with pegylated-liposomal doxorubicin than would be anticipated in patients with advanced HIV-related immunodeficiency.

Several unique aspects of the toxicity of pegylated-liposomal doxorubicin require specific discussion. There is an acute infusion reaction associated with pegylated-liposomal doxorubicin therapy. Symptoms of this reaction include flushing, facial edema, headache, back pain, rigors, hypotension, chest/throat tightness, and dyspnea. In the safety data base provided to the U.S. Food and Drug Administration (FDA) at the time of licensing of pegylated-liposomal doxorubicin, this infusion reaction was reported to have an incidence of 6.8% (48 of 705 patients treated with pegylated-liposomal doxorubicin). The reaction occurred during first cycle of therapy in all patients reported, and was self-limited, abating minutes to hours after

Table 9.10. Randomized comparative trial of pegylated-liposomal doxorubicin versus bleomycin and vincristine (BV) in the treatment of AIDS-related Kaposi's sarcoma

RESPONSE	Pegylated-Liposomal Doxorubicin n = 116	BV n = 102	p-VALUE
Complete	7 (5.8%)	1 (0.8%)	
Partial	63 (52.1%)	27 (22.5%)	< 0.001*

* Fisher's exact test for between treatment differences for those patients having a complete/partial response.

stopping the infusion, and usually not recurring with subsequent dosing. No specific preventative therapy has been recommended. In the author's experience, terminating the infusion has usually led to prompt resolution of the symptoms, allowing the infusion to be resumed at a slower rate without subsequent difficulty.

Palmar-plantar erythrodysesthesia has also been associated with pegylated liposomal doxorubicin therapy. Symptoms include painful macular erythema of palmar and plantar surfaces, similar to that seen with prolonged continuous infusions of doxorubicin and 5-fluorouracil. The incidence of this problem was 3.4% (24 of 705 patients) in the safety data base. This problem tended to occur after repeated dosing, especially at higher doses and shorter dosing intervals. Blistering, desquamation, and slow healing occur when treatment is stopped; the problem may recur with re-exposure to the drug.

Concern has been raised regarding the potential for extravasation injury associated with pegylated liposomal doxorubicin therapy because standard doxorubicin can cause significant local injury when it infiltrates into extravascular tissue. In contrast, extravasation pegylated-liposomal doxorubicin produces the histologic appearance of mild to moderate inflammation at subcutaneous injection sites in rabbits, as opposed to the severe inflammation and tissue necrosis caused by standard doxorubicin. At the University of California, San Francisco, we witnessed eight episodes of extravasation without significant clinical sequelae in patients receiving pegylated-liposomal doxorubicin for AIDS-KS.[40] Pegylated-liposomal doxorubicin is therefore considered an irritant, not a vesicant.

Therapy with any anthracycline raises concern over the potential for cardiotoxicity. In order to evaluate the cardiotoxicity caused by pegylated-liposomal doxorubicin, investigators identified 10 patients with AIDS-KS who had received in excess of 400 mg/m^2 of pegylated-liposomal doxorubicin and who were willing to undergo endomyocardial biopsy for histopathologic evaluation.[41] Control subjects matched for cumulative standard doxorubicin dose were identified from a data base of 131 patients who had undergone biopsies while enrolled in clinical trials using doxorubicin at Stanford University. Anthracycline-induced cardiotoxicity was scored using a well-established 6-point scale (0-3.0). Matched controls were available for 8 of the 10 pegylated-liposomal doxorubicin patients; 2 had received cumulative doses of 780 and 860 mg/m^2, but the scores of both these patients was 0.5, indicating no anthracycline-specific toxicity. In an unblinded comparison of the remaining 8 pegylated-liposomal doxorubicin patients with their matched controls, pegylated-liposomal doxorubicin patients had a lower median score (0.5 for pegylated-liposomal doxorubicin vs. 2.25 for standard doxorubicin) and the difference between the groups was highly significant (paired t-test, $p < 0.001$). These data suggest that pegylated-liposomal doxorubicin is less cardiotoxic than standard doxorubicin at comparable cumulative doses.

References

1. Friedman-Kein A, Laubenstein L, Marmor M et al. Kaposi's sarcom and pneumocystic pneumonia among homosexual men—New York City and California. MMWR 1981; 30:305-308.
2. Hymes KB, Cheung TL, Green JB et al. Kaposi's sarcoma in homosexual men: a report of 8 cases. Lancet 1981; 2:598-600.
3. Masur H, Michilis MA, Green JB et al. An outbreak of community-acquired Pneumocystis carinii pneumonia: initial manifestations of cellular immune dysfunction. N Engl J Med 1981; 305:1431-1438.
4. Beral V, Peterman TA, Berkelman RL et al. Kaposi's sarcoma among persons with AIDS: a sexually transmitted infection? Lancet 1990; 1:123-128.
5. Northfelt DW. AIDS-related Kaposi's sarcoma: still a problem, still an opportunity. J Clin Oncol 1994; 12:1109-1110.
6. Peters BS, Beck EJ, Coleman DG et al. Changing disease patterns in patients with AIDS in a referral centre in the United Kingdom: the changing face of AIDS. Br J Med 1991; 302:203-207.
7. Selik RM, Chu SY, Ward JW. Trends in infectious disease and cancers among persons dying of HIV infection in the United States from 1987 to 1992. Ann Intern Med 1995; 123:933-936.
8. Gill PS, Akil B, Colletti P et al. Pulmonary Kaposi's sarcoma: clinical findings and results of therapy. Am J Med 1989; 87:57-61.
9. Kaplan LD, Hopewell PC, Jaffe H et al. Kaposi's sarcoma involving the lung in patients with acquired immunodeficiency syndrome. J Acquir Immune Defic Synd 1988; 1:23-30.
10. Centers for Disease Control. Epidemiologic aspects of the current outbreak of Kaposi's sarcoma and opportunistic infections. N Engl J Med 1982:306:248-252.
11. Kaposi M. Idiopathisches multiples Pigmentsarkom der Haut. Arch Dermatol Syph 1872; 4:265-273.
12. Taylor JF, Templeton AC, Vogel CL et al. Kaposi's sarcoma in Uganda: a clinico-pathological study. Int J Cancer 1971; 3:122-135.
13. Harwood AR, Osoba D, Hofstader SL et al. Kaposi's sarcoma in recipients of renal transplants. Am J Med 1979; 67:759-765.
14. Penn I. Tumors of the immunocompromised patient. Ann Rev Med 1988; 39:63-73.
15. Lemp GF, Payne SF, Neal D et al. Survival trends for patients with AIDS. JAMA 1990; 263:402-406.
16. Moore PS, Chang Y. Detection of herpesvirus-like DNA sequences in Kaposi's sarcoma in patients with and those without HIV infection. N Engl J Med 1995; 332; 1181-1185.
17. Northfelt DW. Treatment of Kaposi's sarcoma: current guidelines and future perspectives. Drugs 1994; 48:569-582.
18. Gill PS, Rarick M, McCutchan JA et al. Systemic treatment of AIDS-related Kaposi's sarcoma: results of a randomized trial. Am J Med 1991; 90:427-433.
19. Gabizon A, Chisin R, Amselem S et al. Pharmacokinetic and imaging studies in patients receiving a formulation of liposome-associated Adriamycin. Br J Cancer 1991; 64:1125-1132.

20. Gabizon AA, Barenholz Y, Bialer M. Prolongation of circulation time of doxorubicin encapsulated in liposomes containing a polyethylene glycol-derivatized phospholipid: pharmacokinetic studies in rodents and dogs. Pharmaceut Res 1993; 10:703-708.
21. Huang SK, Mahew E, Gilani S. Pharmacokinetics and therapeutics of sterically stabilized liposomes in mice bearing C-26 colon carcinoma. Cancer Res 1992; 52:6774-6781.
22. Huang SK, Martin FJ, Jay G et al. Extravasation and transcytosis of liposomes in Kaposi's sarcoma-like dermal lesions of transgenic mice bearing the HIV tat gene. Am J Pathol 1993; 143:10-14.
23. Vaage J, Barbera-Guillem E, Abra R et al. Tissue distributions and therapeutic effects of intravenous free and liposomal doxorubicin against human prostatic carcinoma xenografts. Cancer 1994; 73:1478-1484.
24. Northfelt DW, Martin FJ, Working P et al. Doxorubicin encapsulated in liposomes containing surface-bound polyethylene glycol: pharmacokinetics, tumor localization, and safety of pegylated-liposomal doxorubicin in patients with AIDS-related Kaposi's sarcoma. J Clin Pharmacol 1996; 36:55-63.
25. Northfelt DW. Stealth® liposomal doxorubicin (SLD) delivers more doxorubicin (DOX) to AIDS-Kaposi's sarcoma (AIDS-KS) lesions than to normal skin. Proc ASCO 1994; abstract 5.
26. Hengge UR, Brockmeyer NH, Baumann M et al. Liposomal doxorubicin in AIDS-related Kaposi's sarcoma. Lancet 1993; 342:497.
27. Simpson JK, Miller RF, Spittle MF. Liposomal doxorubicin for treatment of AIDS-related Kaposi's sarcoma. Clin Oncol 1993; 5:372-374.
28. Bogner JR, Kronawitter U, Rolinski B et al. Liposomal doxorubicin in the treatment of advanced AIDS-related Kaposi sarcoma. J Acquir Immune Defic Synd 1994; 7:463-468.
29. James ND, Coker RJ, Tomlinson D et al. Liposomal doxorubicin (Doxil): an effective new treatment for Kaposi's sarcoma in AIDS. Clin Oncol 1994; 6:294-296.
30. Sturzl M, Zeitz C, Eisenburg B et al. Liposomal doxorubicin in the treatment of AIDS-associated Kaposi's sarcoma: clinical, histological and cell biological evaluation. Res Virol 1994; 145:261-269.
31. Wagner D, Kern WV, Kern P. Liposomal doxorubicin in AIDS-related Kaposi's sarcoma: long-term experiences. Clin Investig 1994; 72:417-423.
32. Harrison M, Tomlinson D, Stewart S. Liposome-entrapped doxorubicin: an active agent in AIDS-related Kaposi's sarcoma. J Clin Oncol 1995; 13:914-920.
33. Northfelt DW, Dezube BJ, Thommes JA et al. Efficacy of pegylated-liposomal doxorubicin in the treatment of AIDS-related Kaposi's sarcoma after failure of standard chemotherapy. J Clin Oncol 1997; 15:653-659.
34. Krown SE, Metroka C, Wernz JC. Kaposi's sarcoma in the acquired immunodeficiency syndrome: a proposal for uniform evaluation, response, and staging criteria. J Clin Oncol 1989; 7:1201-1207.
35. Fischl MA, Krown SE, O'Boyle KP et al. Weekly doxorubicin in the treatment of patients with AIDS-related Kaposi's sarcoma. J Acquir Immun Defic Synd 1993; 6:259-264.
36. Gill P, Rarick M, Bernstein-Singer M et al. Treatment of advanced Kaposi's sarcoma using a combination of bleomycin and vincristine. Am J Clin Oncol 1990; 13:315-319.
37. Rarick MU, Gill PS, Montgomery T et al. Treatment of epidemic Kaposi's sarcoma with combination chemotherapy (vincristine and bleomycin) and zidovudine. Ann Oncol 1990; 1:147-149.

38. Northfelt DW, Dezube BJ, Thommes JA et al. Randomized comparative trial of Doxil® vs. Adriamycin, bleomycin, and vincristine (ABV) in the treatment of severe AIDS-related Kaposi's sarcoma (AIDS-KS). 3rd Conference on Retroviruses and Opportunistic Infections 1996; abstract 379.
39. Stewart S, Jablonowski H, Goebel FD et al. Randomized comparative trial of Doxil® vs. bleomycin and vincristine in the treatment of AIDS-related Kaposi's sarcoma. XI International Conference on AIDS 1996; abstract LB.B.6026.
40. Madhavan S, Northfelt DW. Lack of vesicant injury following extravasation of liposomal doxorubicin. JNCI 1995; 87:1556-1557.
41. Berry G, Billingham M, Alderman E et al. Reduced cardiotoxicity of Doxil (pegylated liposomal doxorubicin) in AIDS Kaposi's sarcoma patients compared to a matched control group of cancer patients given doxorubicin. Proc ASCO 1996; abstract 843.

CHAPTER 10

DaunoXome® (Liposomal Daunorubicin) for First-Line Treatment of Advanced, HIV-Related Kaposi's Sarcoma

Geoffrey Mukwaya, Eric A. Forssen, Paul Schmidt, Michael Ross

Introduction

Kaposi's sarcoma, the most common tumor occurring in human immunodeficiency virus (HIV)-infected individuals and persons with the acquired immunodeficiency syndrome (AIDS), has become a more aggressive, increasingly difficult-to-treat tumor partly because of its rapid progression late in the course of AIDS.[1-4] Despite significant progress in understanding and treating Kaposi's sarcoma lesions cosmetically over the last 15 years, overall survival of patients with advanced, HIV-associated Kaposi's sarcoma has remained unchanged,[1] prompting continued research for effective and better tolerated therapies. By selectively enhancing drug delivery to tumor versus normal tissues, thereby enhancing the antitumor effects and limiting the toxicities of antineoplastic drugs, some liposomal delivery formulations have presented an exciting therapeutic advance in the treatment of advanced, HIV-associated Kaposi's sarcoma.

DaunoXome® (daunorubicin citrate liposome injection) (NeXstar Pharmaceuticals, Inc., San Dimas, California) is a unique liposomal formulation of the anthracycline antineoplastic agent, daunorubicin. Daunorubicin[5-7] has been used since the early 1970s for induction of remission in adults with acute nonlymphocytic leukemia and in adults and children with acute lymphocytic leukemia.[8-10] Its use, however, has been limited in part by its lack of specificity for tumor cells and resultant toxicity to normal tissues.

Formulation of daunorubicin (DaunoXome) is in the class of long circulating liposomes because its relatively small size and particular lipid bilayer composition combine to inhibit clearance by the mononuclear phagocyte system (MPS). DaunoXome has been shown to be highly selective for tumor tissues via enhanced extravasation through tumor microvasculature.[11-17] The liposomal anthracycline persists for prolonged periods within tumor cells and has significant efficacy with substantially less toxicities than free daunorubicin in the treatment of advanced, HIV-associated Kaposi's sarcoma.[18-20]

Long Circulating Liposomes: Old Drugs, New Therapeutics, edited by Martin C. Woodle and Gerrit Storm.

This chapter reviews the chemistry, clinical pharmacokinetics, efficacy, and safety and tolerability of DaunoXome. DaunoXome represents a significant advance in drug delivery for the selective targeting of tumor cells and offers a new first-line treatment option for advanced, HIV-associated Kaposi's sarcoma.

Chemistry

DaunoXome Formulation

DaunoXome® (daunorubicin citrate liposome injection) is a novel pharmaceutical formulation containing an aqueous solution of the citrate salt of the antineoplastic agent, daunorubicin, encapsulated within liposomes.[21] DaunoXome liposomes are composed of a single bilayer membrane of the chemically pure phospholipid distearoylphosphatidylcholine (DSPC) and cholesterol in a 2:1 molar ratio, and have a mean diameter of approximately 45 nm as determined by quasi-elastic light scattering. The lipid-to-drug weight ratio is 18.7:1 (total lipid:daunorubicin base), equivalent to a 10:5:1 molar composition of DSPC:cholesterol:daunorubicin.[21] DaunoXome injection is supplied in a single-use vial containing daunorubicin citrate equivalent to 50 mg daunorubicin base, at a concentration of 2 mg/ml. Storage is at 2°-8°C (36°-46°F) protected from light. The shelf life is 24 weeks when stored as recommended.[18]

Chemistry of Daunorubicin

Daunorubicin is an anthracycline glycoside antibiotic with antineoplastic activity that was originally obtained from *Streptomyces peucetius*; the drug is currently produced by *Streptomyces coeruleorubidus*. Daunorubicin consists of a 4-ring anthracycline moiety linked by a glycosidic bond to daunosamine, an amino sugar (Fig. 10.1). Daunorubicin differs structurally from doxorubicin in that daunorubicin contains an acetyl group instead of a hydroxyacetyl group in the 8-position. Chemically, daunorubicin is (8S-cis)-8-acetyl-10-[(3-amino-2,3,6-trideoxy-a-L-lyxo-hexopyranosyl)oxy]-7,8,9,10-tetrahydro-6,8,11-trihydroxy-1-methoxy-5,12-naphthacenedione.

Physical Properties of DaunoXome

The effects of a broad range of characteristics specific to the liposome formulation on liposome delivery to tumors have been evaluated, including composition and physical properties. The following properties have been found to be associated with optimal delivery of liposome contents to tumor tissues.[20]

1) Lipid phase transition temperature higher than physiological range (> 37°C).[22]
2) Presence of cholesterol within the membrane bilayer to minimize bilayer permeability and lipid exchange.[22-24]
3) Liposome size <100 nm in diameter, preferably 30-80 nm in diameter.[17-20]
4) Net neutral surface charge.[25,27]

Small unilamellar vesicles (single bilayer liposomes having diameters ≤100 nm) composed of DSPC and cholesterol in a 2:1 molar ratio have been shown to exhibit a particularly high affinity for solid tumors in vivo, capable of delivering high concentrations of imaging agents such as the radionuclide, In-111.[25-31] The lipid composition provides a strong bilayer that presents a barrier to plasma protein binding and leakage of contents. This and the neutral charge (zwitterion) of the surface lowers the affinity of such liposomes for cells of the MPS, extending circulation lifetimes

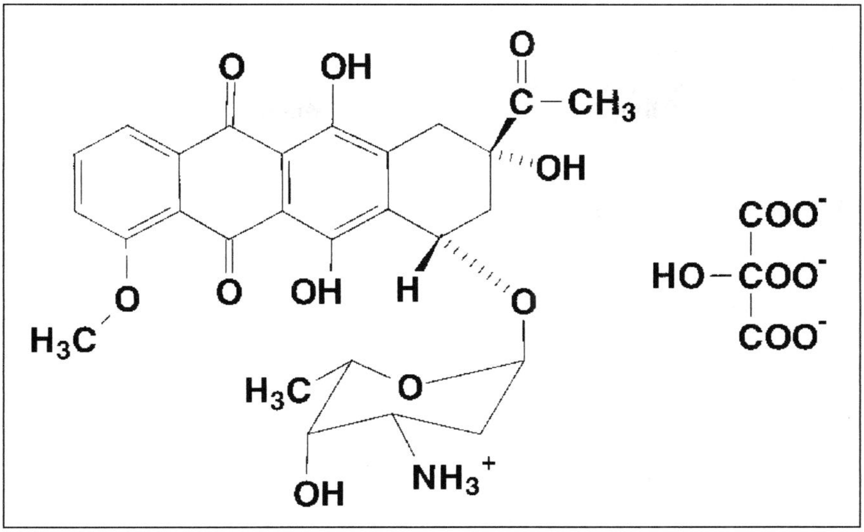

Fig. 10.1. Chemical structure of daunorubicin citrate.

and promoting passive transport into tumors through leaky vasculature. The imaging studies provided a basis for evaluating these liposomes as potential carriers of chemotherapeutic agents, including daunorubicin. The vesicles themselves are very well tolerated by humans and have been administered alone without associated drug in single doses >2 g/m² without significant adverse effects.

The liposomes in DaunoXome have been formulated to ensure optimal delivery of daunorubicin. The liposomes are unilamellar (one bilayer) and consist of DSPC and cholesterol, typical components of cellular membranes, in a ratio that provides significant physical stability. DSPC has a phase transition temperature of approximately 58°C, ensuring relative rigidity of the lipid side chains at physiological temperatures. Addition of cholesterol strengthens the bilayers of the liposomes, particularly in the presence of serum, by controlling loss of DSPC to high-density lipoproteins.[22,24,32] The net neutral surface charge also has been demonstrated to contribute to the stability of these liposomes in vivo.[25] Thus, the liposomes are especially stable and resistant to leakage of the entrapped daunorubicin, with documented leakage of daunorubicin from the liposomes noted to be less than 1% over 24 h in fetal calf serum at 37°C.[33]

The mechanism of action and preclinical pharmacology of DaunoXome are further detailed in chapter 6 of this volume.

Clinical Pharmacokinetics

Pharmacokinetic Profile

The plasma pharmacokinetics of DaunoXome differ significantly from the pharmacokinetics of free nonencapsulated daunorubicin hydrochloride.[7,18,20,34-36] Following iv administration of free drug at a dose of 80 mg/m² daunorubicin exhibits a low

Table 10.1. Pharmacokinetic parameters of free daunorubicin following a single dose of 80 mg/m² in patients with disseminated malignancies and of DaunoXome® following a single dose of 40 mg/m² in AIDS patients with Kaposi's sarcoma

Pharmacokinetic Parameter, Units	DaunoXome® Mean ± SD (n=30)	Free Daunorubicin Mean ± SD (n=4)
Peak plasma concentration, mg/ml	18.26 ± 3.03	0.4
Plasma clearance, ml/min	17.3 ± 6.1	236 ± 181*
Volume of distribution (steady state), l	6.4 ± 1.5	1006 ± 622
Distribution half-life, h	4.41 ± 2.33	0.77 ± 0.3
Elimination half-life, h	—	55.4 ± 13.7

*Calculated.

peak plasma concentration, a short initial distribution half-life and prolonged terminal elimination half-life, a large volume of distribution (reflecting extensive tissue distribution and binding), and rapid plasma clearance[7] (Table 10.1). After 12 h, plasma daunorubicin concentrations are considerably lower, resulting in low area under the plasma concentration-time curve (AUC) values; these observations are consistent with rapid removal of free daunorubicin from the circulation and extensive tissue uptake.

In contrast to free daunorubicin, DaunoXome plasma concentrations decline in a monoexponential fashion following a single dose of 40 mg/m², with a longer apparent elimination half-life (probably a distribution half-life), a smaller steady-state volume of distribution (probably much lower than for free drug because the liposomes are confined to vascular fluid volume), and a much slower plasma clearance.[21] The differences in volume of distribution and clearance result in a higher daunorubicin exposure (in terms of plasma AUC) from DaunoXome than with free daunorubicin. Following iv administration of DaunoXome at a dose of 80 mg/m², mean peak plasma concentrations are approximately 44 µg/ml, nearly 100-fold greater than free daunorubicin.[20] The AUC for DaunoXome at this dose is approximately 36-fold greater than free daunorubicin, despite the shorter observed terminal half-life.

The pharmacokinetics of a range of iv DaunoXome single doses (10, 20, 40, 60 and 80 mg/m²) were evaluated in patients with solid tumors, including some patients with Kaposi's sarcoma, in Phase I studies (Table 10.2).[20,35,36] In most patients, plasma concentrations of DaunoXome declined in a monoexponential fashion; however, in some patients a biexponential decline was observed. Peak plasma concentrations were dose-dependent and increased with dose in a nearly linear manner. AUC values also were dose-dependent and increased linearly with dose. Clearance rates decreased with increasing dose, but dose dependency was not strong.

Pharmacokinetics in Patients with AIDS-Related Kaposi's Sarcoma

Pharmacokinetic parameters for total daunorubicin (encapsulated drug plus free drug) following an iv dose of DaunoXome 40 mg/m² in 30 patients with AIDS-asso-

Table 10.2. Pharmacokinetic parameters of DaunoXome® [20]

DaunoXome® Single Dose, mg/m2	Peak Plasma Concentration, µg/mL	$t_{1/2}$, hr	AUC, µg ht/mL	Plasma Clearance, mL/min	Volume of Distribution, L
10 (n = 3)	5.9	2.8	16.9	15.7	3.75
20 (n = 4)	8.2	3.8	57.2	14.3	4.1
40 (n = 4)	18.15	4.0	120.1	10.5	3.7
60 (n = 2)	36.2	8.3	301.1	6.7	2.9
80 (n = 4)	43.6	5.2	375.3	6.6	2.9

ciated Kaposi's sarcoma are shown in Table 10.1. These data were consistent with the Phase I dose-ranging pharmacokinetic data, demonstrating that the plasma pharmacokinetics of DaunoXome in patients with AIDS-related Kaposi's sarcoma differ substantially from the pharmacokinetics of free daunorubicin. The AUC of the encapsulated daunorubicin represented 92% of the total daunorubicin (encapsulated drug plus free drug) AUC,[18] providing additional evidence that the DaunoXome is delivered intact to tumor cells with ensuing cytotoxicity, rather than by way of a slow release mechanism of daunorubicin into the systemic circulation. Similarly, the AUC of the major inactive (but potentially cardiotoxic) metabolite daunorubicinol, represented only 2% of the total daunorubicin AUC.[18]

Distribution

DaunoXome has a small steady-state volume of distribution (approximately 6.4 l), suggesting that it is confined mainly to the vascular fluid volume. In vivo studies in animals indicate that DaunoXome selectively results in greater accumulation and sustained levels of daunorubicin in tumor versus normal cells.[19,20,33]

Only limited data on the distribution of DaunoXome in humans are available. In a study of normal skin samples and Kaposi's sarcoma lesions taken 24 h after DaunoXome administration,[20] daunorubicin was not detected in normal skin samples of several evaluable patients. In contrast, following administration of DaunoXome, daunorubicin was detected in some samples from Kaposi's lesions at concentrations of 1.06 and 1.07 µg/g. These data are consistent with the results of the animal studies.

The selective tumor distribution results also have recently been supported by in situ laser-induced fluorescence studies in mice.[19] While comparable levels of daunorubicin were observed in normal and tumor tissues after administration of free daunorubicin, DaunoXome accumulated within tumor tissue more slowly than free daunorubicin, producing nearly 5-fold higher levels of drug in tumor versus normal tissues. Furthermore, drug levels persisted in tumor cells for several days following administration of DaunoXome, while daunorubicin levels decreased to baseline concentrations within a few hours following administration of free daunorubicin. The observed delay in achieving peak levels and slower elimination are consistent with accumulation of DaunoXome within the tumor, followed by vesicle breakdown and gradual release of daunorubicin. In summary, these data indicate that DaunoXome can be expected to produce high levels of daunorubicin in tumor versus normal tissues that persist for longer periods of time compared with administration of free daunorubicin.

Daunorubicin is approximately 63% bound to human serum albumin.[37] Conversely, DaunoXome minimizes protein binding of the encapsulated drug.

Elimination

Daunorubicin is extensively metabolized, mainly by cytoplasmic aldo-keto reductases, producing daunorubicinol, the major metabolite which has antineoplastic activity. Aglycones also may be formed and further metabolized, but their importance is not known.[38] Daunorubicinol is detected in low concentrations in plasma after iv administration of DaunoXome.[21]

Following administration of free drug, daunorubicin and its metabolites are excreted mainly in bile and to a lesser extent in urine. The in vivo metabolism and excretion of daunorubicin delivered by DaunoXome have not been evaluated. Studies to further evaluate the intracellular disposition of the liposomes and encapsulated daunorubicin are ongoing.

Special Populations

The pharmacokinetics of DaunoXome have not yet been evaluated in special populations, specifically in women, different ethnic groups, or individuals with hepatic and/or renal insufficiency.[21]

Clinical Efficacy

HIV-Associated Kaposi's Sarcoma

Kaposi's sarcoma, first described by Dr. Moritz Kaposi in 1872 as an unusual tumor afflicting males over the age of 50, is now the most common tumor occurring in HIV-infected individuals and persons with AIDS.[1-4] HIV-associated Kaposi's sarcoma is estimated to occur in 15% to 20% of all AIDS patients and is one of the diagnostic criteria for AIDS.[2] Homosexual and bisexual men with HIV are at much greater risk for developing Kaposi's sarcoma (up to 30%) than are other patient types with AIDS.[2] Despite the fact that the proportion of AIDS patients with Kaposi's sarcoma decreased across all risk groups from 1983 to 1989, partly because of the CDC expanding the definition of AIDS, the contribution of Kaposi's sarcoma to AIDS-related mortality has increased considerably over time (e.g., from 14% in 1984 to 32% in 1989 in the UK).[39] This increase in mortality may be due in part to its appearance later in the natural course of AIDS, and its more aggressive and rapid progression in patients with advanced AIDS.

Clinical Presentation

While classical Kaposi's sarcoma is a slow growing tumor primarily found on the skin of the lower extremities in elderly men, AIDS-related Kaposi's sarcoma is usually a much more aggressive tumor. The initial site of disease is commonly cutaneous with lesions appearing as asymptomatic, pink to deep purple or brown patches that are multifocal, widely disseminated, and symmetrical. Lesions are typically not painful, and they are often overlooked early in the course of the disease. In some cases, cutaneous lesions may ulcerate, become infected, or produce pain locally, but rarely do cutaneous lesions cause morbidity and mortality.[4]

Progression of AIDS-related Kaposi's sarcoma is generally more rapid than classical disease with frequent involvement of visceral sites, including the lungs, gastrointestinal (GI) tract, lymph nodes, and mucous membranes. When AIDS-related

Kaposi's sarcoma is responsible for death, extensive multiple organ involvement is commonly documented.[4]

The GI tract is a common site of progression of AIDS-related Kaposi's sarcoma. As many as 40% to 50% of patients with cutaneous lesions have GI involvement.[4,20] Many patients with GI involvement may be asymptomatic or have nonspecific signs. They may also present with manifestations that include rectal pain, bleeding, or masses; diarrhea or occult bleeding; and, infrequently, massive bleeding. Kaposi's sarcoma lesions can occur along the entire GI tract from the esophagus to the rectum and typically involve first the submucosa and later the mucosa.[4]

Pulmonary manifestations of AIDS-related Kaposi's sarcoma involvement carry a particularly poor prognosis, with pulmonary involvement noted postmortem in 20% to 50% of AIDS patients.[4] Evidence of tumor involvement is usually only discovered at necropsy, and varies from interstitial infiltrates in the alveoli and bronchovasculature to nodules in the underlying pulmonary tissue. Patients with pulmonary involvement generally present with nonspecific symptomatology, including fever, dyspnea and evidence of radiographic abnormalities (soft shadows, reticulonodular shadows, pleural effusions). Deficits in pulmonary function tests are common. However, lung biopsies are often necessary for a definitive diagnosis of Kaposi's sarcoma.[4]

Advanced, AIDS-related Kaposi's sarcoma is characterized by severe symptoms and a wide spectrum of disability. Cutaneous lesions in advanced disease may result in severe cosmetic problems that necessitate treatment.

Treatment for AIDS-Related Kaposi's Sarcoma

Treatment for AIDS-related Kaposi's sarcoma is individualized according to a particular patient's needs. Therapies are often effective but not curative, and they rarely induce complete remission. Ultimately, the optimal treatment for advanced disease depends on the extent of disease and the patient's underlying immune function.[35]

Numerous treatment options for AIDS-related Kaposi's sarcoma have been investigated (Table 10.3). Patients with minimal cutaneous disease are usually managed by observation alone. Locally symptomatic disease (e.g. painful lesions on extremities, oropharyngeal lesions threatening the airway) may be treated with radiation, laser therapy, intralesional vinblastine, or liquid nitrogen cryotherapy.[1] Interferon α, alone or in combination with zidovudine, has been shown to be active in patients with cutaneous disease, especially in those with minimal disease and high CD4 counts. Chemotherapy agents, alone or in combination, are generally utilized for patients with more progressive disease; a regimen of doxorubicin Adriamycin® (*Adriamycin* is a registered trademark of Pharmacia Adria), bleomycin and vincristine (ABV) has been shown to be more effective than doxorubicin alone in the treatment of advanced, AIDS-related Kaposi's sarcoma.[40]

Despite advances in the treatment of advanced, AIDS-related Kaposi's sarcoma with various single-agent and combination therapy regimens, subsequent improvements in patient survival rates have not been documented.[1] The search for more effective and better tolerated agents continues. New therapies for the treatment of advanced disease should do the following.

1) Alleviate tumor-associated symptoms–pain, edema, visceral symptoms.
2) Produce effective tumor responses–flattening and fading of lesions.
3) Be well tolerated alone and in combination with other medications.

Table 10.3. Traditional treatment options for AIDS-related Kaposi's sarcoma

Locoregional Therapy
Intralesional vinblastine
Cryotherapy
Irradiation
Laser therapy

Chemotherapy
Single agents
- interferon α
- vinca alkaloids
- bleomycin
- anthracyclines
- etoposide

Combination therapy
- bleomycin + vincristine (BV)
- doxorubicin (Adriamycin) + bleomycin + vincristine (ABV)

4) Be safe for long-term use.
5) Prevent the appearance of new disease.

Efficacy as First-Line Treatment of Advanced Disease

Phase II Studies

Six Phase II studies of DaunoXome 40 mg/m² administered iv every 2 weeks were conducted across the United States and Europe in patients with advanced, HIV-associated Kaposi's sarcoma.[18,20,41,42] Most patients received concomitant antiretroviral therapy and necessary medications for the prevention and treatment of opportunistic infections. Concomitant therapy with other local or systemic medications for Kaposi's sarcoma was not allowed. Efficacy was evaluated by determining overall response, duration of response, and effects on quality of life. Unlike the Phase III trial described below in which responses were evaluated by an independent central reviewer according to ACTG criteria, responses in the Phase II studies were assessed by each individual study investigator.

In Phase II studies, DaunoXome demonstrated significant antitumor activity in advanced, HIV-associated Kaposi's sarcoma.[20] In a pooled analysis of 91 evaluable patients, an overall response rate of 64.8% (59/91) with 2.2% complete responses was observed. The mean duration of DaunoXome treatment was 20 weeks or 9.6 cycles. Quality of life was assessed by analyzing patients' Karnofsky Performance Status as a function of the cumulative DaunoXome dose. No change in mean Karnofsky Performance Status was observed for cumulative doses up to 1000 mg/m² (25 cycles). In a Phase II study of 25 patients with advanced, HIV-associated Kaposi's sarcoma,[41] DaunoXome 40 mg/m² administered iv every 2 weeks produced an overall response rate of 57% when administered as first-line therapy. Partial responses were observed in two patients who had pulmonary Kaposi's sarcoma; one patient experienced a complete response documented by bronchoscopic evaluation of pulmonary lesions, but several cutaneous lesions persisted. Interestingly, response to DaunoXome was independent of CD4 counts, i.e., patients with CD4 counts < 50 cells/mm³ were as likely to have antitumor responses to DaunoXome (2/15 complete responses and 7/15 partial responses) as patients with CD4 counts > 50 cells/mm³ (0/7 complete responses and 5/7 partial responses).

Phase III Studies

An open-label, multicenter, randomized Phase III trial[47] was conducted in 232 patients with advanced HIV-associated Kaposi's sarcoma to compare the efficacy and safety of DaunoXome (40 mg/m²) and a combination therapy regimen of doxorubicin (Adriamycin®, 10 mg/m²), bleomycin (15 U), and vincristine (1.0 mg) (ABV) given every 2 weeks. The dose and schedule of ABV were chosen to provide an equivalent level of myelosuppression. Advanced disease was defined as the presence of ≥25 mucocutaneous lesions, the development of ≥10 mucocutaneous lesions in a 1-month period, symptomatic visceral involvement, and/or tumor-related edema. The patients with advanced HIV-associated Kaposi's sarcoma evaluated in this study are representative of those Kaposi's sarcoma patients who would require palliative therapy in a normal clinical setting.[18,21]

One hundred sixteen patients were treated with DaunoXome 40 mg/m² and 111 patients with ABV. Patient demographic and baseline tumor characteristics are shown in Tables 10.4 and 10.5.

Response to therapy was assessed by an independent central reviewer according to the AIDS Clinical Trials Group (ACTG) criteria[43] and required a minimum of one of the following for at least 28 d.

1) ≥50% decrease in number of total lesions.
2) ≥50% decrease in the sums of the products of the largest perpendicular diameters of bidimensionally measurable marker lesions.
3) Complete flattening of ≥50% of all previously raised lesions.

DaunoXome was comparable to ABV in the treatment of advanced, HIV-associated Kaposi's sarcoma with respect to tumor response, duration of response, and time to progression of the disease (Table 10.6). The overall response rate was 23% (27/116) with DaunoXome and 30% (33/111) with ABV. The 95% CI for difference in response rates (ABV-Daunoxome) was (-5%, 18%). The lower response rate for Phase III relative to Phase II is likely due to the different response assessment method for the trial designs. Twenty of 33 ABV responders and 11/27 DaunoXome responders had responses according to criteria other than flattening of lesions (i.e. shrinkage of lesions and/or reduction in the number of lesions). Photographic evidence of tumor response was comparable for DaunoXome and ABV across all anatomic sites (e.g. face, oral cavity, trunk, legs, and feet). Time to first evidence of response was rapid for both treatment groups, occurring at the first or second follow-up visit (4 or 8 weeks after initiation of treatment). A trend toward longer survival was observed with DaunoXome therapy (342 d) compared with ABV (291 d).

Patients treated with DaunoXome had significantly ($p < 0.0001$) less premedication use (antiemetic, antipyretic) than those treated with ABV. Dexamethasone (11.5% vs. 40.9%) and acetaminophen (12.8% vs. 35.2%) were required in a lower proportion of treatment cycles in DaunoXome patients than in ABV patients, respectively. Safety data from this trial are reviewed below in the section on safety and tolerability.[21]

Patient quality-of-life evaluations also demonstrated comparable effects of treatment with DaunoXome and ABV. Quality-of-life assessments included evaluation of

Table 10.4. Demographic characteristics at study entry

	DaunoXome® (n=116)	ABV (n=111)
Mean age, years	37.6	37.7
Mean Kartnofsky Performance Status, %	85.8	85.4
Median CD4 count, cells/ml	30	28

Table 10.5. Baseline tumor characteristics

	DaunoXome® (n=116)	ABV (n=111)
Skin lesions >50, %	64.7	65.6
Visceral involvement, %	31.0	30.0
Pulmonary, %	19.8	19.4
Gastrointestinal, %	11.2	9.2
Tumor-associated edema, %	56.0	58.6
CD4 count <100 cells/mL, %	72.4	74.9

Table 10.6. Efficacy results of first-line therapy for HIV-associated Kaposi's sarcoma[18,21]

	DaunoXome® (n=116)	ABV (n=111)
Response rate, %	23*	30
Median duration of response, days	110	113
Median time to progression, days	92	105
Survival, days	342	291

changes from baseline in Karnofsky Performance Status, quality-of-life scores based on patient-recorded diaries, and patient body weight over the study period. While patients who received ABV appeared to exhibit progressively greater decreases in weight and quality-of-life scores after five cycles, patients receiving DaunoXome had no changes in weight, quality-of-life scores, or Karnofsky Performance Status over 15 cycles of treatment.[18]

The results from this large multicenter, randomized, controlled Phase III trial demonstrating the efficacy of DaunoXome as first-line therapy for advanced, HIV-associated Kaposi's sarcoma[47] are supported by the results of previous Phase II studies[40,41] in similar patient populations.

Safety and Tolerability

Overview

Knowledge of the common side effects of free daunorubicin is important to any analysis of the safety and tolerability profile of DaunoXome, which contains daunorubicin encapsulated within a liposome. The dose-limiting side effect of free daunorubicin is acute myelosuppression, manifested primarily on the granulocytic series. In addition, free daunorubicin causes alopecia and nausea and vomiting in a significant number of patients. Chronic administration of daunorubicin therapy also has been associated with a cardiomyopathy manifested as congestive heart failure that appears to increase in incidence in adults after a cumulative dosage above 400-550 mg/m.[33] Extravasation of free daunorubicin can cause severe local tissue necrosis.

Daunorubicin has substantially altered pharmacokinetics and some differences in toxicity when encapsulated within a liposome and administered as DaunoXome. Administered as the liposome formulation, several common toxicities of DaunoXome appear to be less severe than those of free daunorubicin.

Phase III Safety Summary

Safety was evaluated in the open-label, randomized, controlled Phase III trial that compared the effects of DaunoXome and ABV as first-line therapy in 227 patients with advanced, HIV-associated Kaposi's sarcoma.[47] Patients with advanced, HIV-associated Kaposi's sarcoma are seriously ill and immunocompromised due to their underlying HIV infection. Because these patients are receiving several concomitant medications, including potentially toxic antiviral and antiretroviral agents, the contribution of the study drugs to the adverse experience profile is difficult to establish.

DaunoXome 40 mg/m^2 was generally well tolerated in the Phase III study (Tables 10.7 and 10.8). As with free daunorubicin, the most important acute toxicity of DaunoXome was myelosuppression, which was manifested primarily on the granulocytic series. The incidences of neutropenia were similar between the DaunoXome and ABV treatment groups (Table 10.7).

The incidence of alopecia was significantly ($p < 0.001$) lower in DaunoXome-treated patients (8%) than in ABV-treated patients (36%) (Fig. 10.2). In contrast with free daunorubicin, which causes alopecia in most patients,[44] the incidence of alopecia with DaunoXome is notably low. This is an important consideration for the AIDS patient, for whom alopecia can be particularly stigmatizing.

Similarly, the incidence of neuropathy was significantly ($p < 0.001$) lower in DaunoXome-treated patients (13%) than in ABV-treated patients (41%). This can be an important consideration for patients who may take other HIV/AIDS drugs associated with neuropathy.

Of particular importance is the lack of cardiotoxic effects with DaunoXome in the Phase III trial at cumulative doses that historically have been associated with significant cardiotoxicity with free daunorubicin. The results show minimal change from baseline in the LVEF at cumulative doses of DaunoXome surging from 100 to > 1000 mg/m^2 (Fig. 10.3). In contrast to free daunorubicin and other anthracyclines, reports of clinical cardiotoxicity have been rare in Phase II and III trials at cumulative

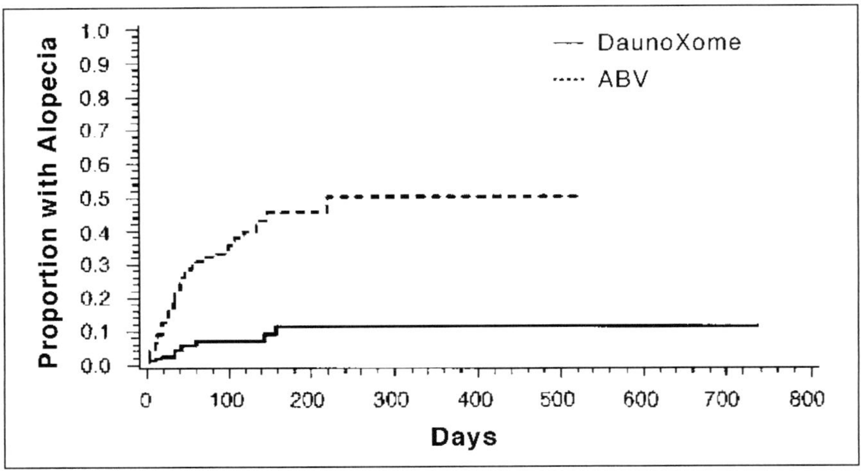

Fig. 10.2. Cumulative incidence of alopecia by treatment group in the Phase III comparative trial.

Table 10.7. Summary of key safety data from the Phase III comparative trial

	DaunoXome (n=116)	ABV (n=111)
Adverse events, % of pateints		
Alopecia (all grades)	8	36*
Neuropathy (all grades)	13	41*
Neutropenia (<1000 cells/mm^3)	36	35
Neutropenia (<500 cells/mm^3)	15	5
Number of cases with absolute 20%-25% reduction in ejection fraction†	3	1
Number of cases removed from therapy due to cardiac causes†	2	0
Opportunistic infections/illnesses	40	27
Median time to first opportunistic infections/illnesses (after enrollment)	214 day	412 days‡

*p < 0.001.
†The denominator is uncertain since there were many instances of missing repeat cardiac evaluations.
‡p = 0.21.

Table 10.8. Summary of all adverse events reported by ≥ 5% of DaunoXome patients from the Phase III comparative trial

	DaunoXome (n=116)		ABV (n = 111)	
	Mild-Moderate %	Severe, Life-threatening %	Mild-Moderate %	Severe, Life-threatening %
Nausea	51	3	45	5
Fatigue	43	6	44	7
Fever	42	5	49	5
Diarrhea	34	4	29	6
Cough	26	2	19	0
Dyspnea	23	3	17	3
Headache	22	3	23	2
Allergic reactions	21	3	19	2
Abdominal pain	20	3	23	4
Anorexia	21	2	26	2
Vomiting	20	3	26	2
Pain	19	3	15	1
Rigors	19	0	23	0
Back pain	16	0	8	0
Sweating increased	12	2	12	0
Neuropathy	12	1	38	3
Rhinitis	12	0	6	0
Edema	9	2	8	1
Chest pain	9	1	7	0
Depression	7	3	6	0
Malaise	9	1	11	1
Stomatitis	9	1	8	0
Alopecia	8	0	36	0
Dizziness	8	0	9	0
Sinusitis	8	0	5	1
Arthralgia	7	0	6	0
Constipation	7	0	18	0
Myalgia	7	0	12	0
Pruritus	7	0	14	0

Table 10.8. (continued)

	DaunoXome (n=116)		ABV (n = 111)	
	Mild-Moderate %	Severe, Life-threatening %	Mild-Moderate %	Severe, Life-threatening %
Insomnia	6	0	14	0
Influenza-like symptoms	5	0	5	0
Tenesmus	4	1	1	0
Abnormal vision	3	2	3	0

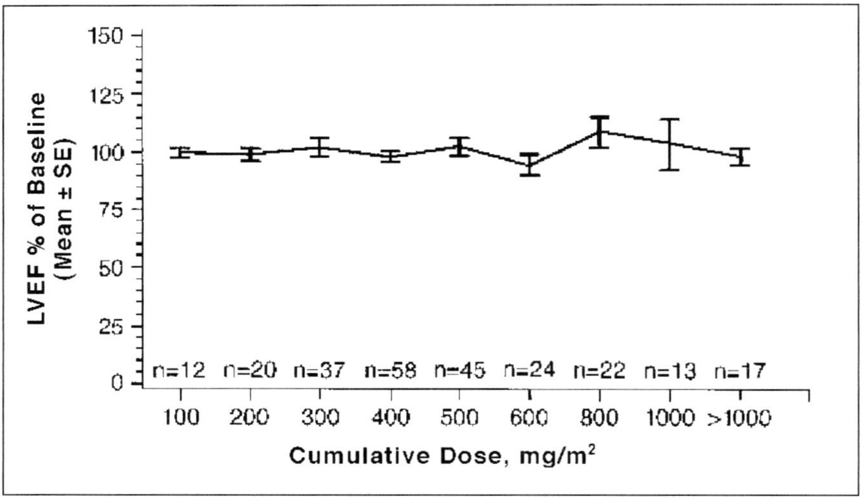

Fig. 10.3. Left ventricular ejection fraction (LVEF) in patients receiving DaunoXome during the Phase III comparative trial.

DaunoXome doses > 600 mg/m^2 (Table 10.9).[18,45] Although this increased tolerance of high doses of DaunoXome may be partly due to the specific dose schedule used in these clinical trials, the improved cardiac safety profile in this population following relatively high cumulative dose of the anthracycline should be noted.

Overall adverse events reported by at least 5% of patients in the Phase III comparative trial were generally comparable between patients who received DaunoXome and ABV (Table 10.8). Mild alopecia was reported in 6% of patients treated with DaunoXome and moderate alopecia in 2% of patients, while no instances of severe alopecia were reported. Nausea was reported to be mild in 35% of patients, moderate in 16% of patients, and severe in 3% of patients treated with DaunoXome. Mild

Table 10.9. Cardiac function in Phase II and III trials of DaunoXome[4]

Group	Number of Patients	Initial LVEF	Terminal LVEF	Change
All patients ≤ 600 mg/m²	53	59.9 ± 9.4	58.9 ± 7.8	–1.0 ± 9.2
600 mg/m² - 999 mg/m²	34	59.9 ± 10.6	59.1 ± 6.7	–0.8 ± 8.9
1000 mg/m² - 1499 mg/m²	12	60.1 ± 7.2	58.2 ± 5.2	–1.9 ± 9.5
≥ 1500 mg/m²	7	59.3 ± 7.0	59.3 ± 15.0	0.0 ± 11.1

and moderate vomiting were each reported by 10% of DaunoXome-treated patients, and severe vomiting by 3% of patients. Although grade 3 or 4 injection site inflammation was reported in two patients treated with DaunoXome, no instances of local tissue necrosis were observed with extravasation, a common occurrence with extravasation of free daunorubicin. Severe adverse events were relatively uncommon for both treatment groups.

Conclusions

On the basis of the Phase III efficacy and safety data, DaunoXome has been approved for first-line therapy of advanced AIDS-related Kaposi's sarcoma in the United States and 19 other countries. The efficacy and safety of DaunoXome in treatment of advanced AIDS-related Kaposi's sarcoma suggest that the liposomal drug will be useful for therapy of other cancers. Phase II clinical trials are underway to test DaunoXome as a single agent against lymphoma, leukemia, and solid tumors such as breast and prostate. There is evidence of activity from these early studies, e.g., non-Hodgkin's lymphoma,[46] and DaunoXome was well tolerated at a dose of 100 mg/m² every 3 weeks.[46]

References

1. Abrams DI. Acquired immunodeficiency syndrome and related malignancies. Semin Oncol 1991; 18(5, suppl 7):41-45.
2. Beral V. Epidemiology of Kaposi's sarcoma. Cancer Surv 1991; 10:5-22.
3. Beral V, Peterman TA, Berkelman RL, Jaffe HW. Kaposi's sarcoma among persons with AIDS: a sexually transmitted infection? Lancet 1990; 335:123-128.
4. Buchbinder A, Friedman-Kien AE. Clinical aspects of epidemic Kaposi's sarcoma. Cancer Surv 1991; 10:39-52.
5. Theologides A, Yarbro JW, Kennedy BJ. Daunomycin inhibition of DNA and RNA synthesis. Cancer 1968; 21:16-21.
6. Andersson B, Beran M, Peterson C, Tribukait B. Significance of cellular pharmacokinetics for the cytotoxic effects of daunorubicin. Cancer Res 1982; 42:178-183.
7. Alberts DS, Bachur NR, Holtzman JL. The pharmacokinetics of daunomycin in man. Clin Pharmacol Ther 1971; 12:96-104.
8. McEvoy GK, ed. Daunorubicin hydrochloride. In: AHFS Drug Information 96. Bethesda, MD: American Society of Hospital Pharmacists, 1996; 10:00.
9. Calabresi P, Chabner BA. Antineoplastic agents. In: Gilman AG, Rall TW, Nies AS, Taylor P, eds. Goodman and Gilman's The Pharmacological Basis of Therapeutics. 8th ed. Elmsford, NY: Pergamon Press, 1990:1209-1263.
10. Wang JJ, Chervinsky DS, Rosen JM. Comparative biochemical studies of adriamycin and daunomycin in leukemic cells. Cancer Res 1972; 32:511-515.

11. Warren BA. The vascular morphology of tumors. In: Peterson H-I, ed. Tumor Blood Circulation: Angiogenesis, Vascular Morphology, and Blood Flow of Experimental and Human Tumors. Boca Raton, Fl: CRC Press, Inc. 1979:1-47.
12. Hwang KJ, Padki MM, Chow DD et al. Uptake of small liposomes by non-reticuloendothelial tissues. Biochim Biophys Acta 1987; 901:88-96.
13. Less JR, Skalak TC, Sevick EM, Jain RK. Microvascular architecture in a mammary carcinoma: branching patterns and vessel dimensions. Cancer Res 1991; 51:265-273.
14. Poste G. Liposome targeting in vivo: problems and opportunities. Biol Cell 1983; 47:19-38.
15. Skinner SA, Tutton PJM, O'Brien PE. Microvascular architecture of experimental colon tumors in the rat. Cancer Res 1990; 50:2411-2417.
16. Weiss L, Orr FW, Honn KV. Interactions between cancer cells and the microvasculature: a rate-regulator for metastasis. Clin Exp Metastasis 1989; 7:127-167.
17. Poste G, Bucana C, Raz A et al. Analysis of the fate of systemically administered liposomes and implications for their use in drug delivery. Cancer Res 1982; 42:1412-1422.
18. Data on file. NeXstar Pharmaceuticals, Inc., San Dimas, CA.
19. Forssen EA, Malé-Brunë R, Adler-Moore JP et al. Fluorescence imaging studies for the disposition of daunorubicin liposomes (DaunoXome) within tumor tissue. Cancer Res 1996; 56:2066-2075.
20. Forssen EA, Ross ME. DaunoXome® treatment of solid tumors: preclinical and clinical investigations. J Liposome Res 1994; 4:481-512.
21. DaunoXome® (daunorubicin citrate liposome injection) Prescribing Information. NeXstar Pharmaceuticals, Inc., San Dimas, Calif.
22. Papahadjopoulos D, Jacobson K, Nir S, Isac T. Phase transitions in phospholipid vesicles: fluorescence polarization and permeability measurements concerning the effect of temperature and cholesterol. Biochim Biophys Acta 1973a; 311:330-348.
23. Kerr DJ, Kaye SB. Aspects of cytotoxic drug penetration, with particular reference to anthracyclines. Cancer Chemother Pharmacol 1987; 19:1-5.
24. Papahadjopoulos D, Cowden M, Kimelberg H. Role of cholesterol in membranes: effects on phospholipid-protein interactions, membrane permeability and enzymatic activity. Biochim Biophys Acta 1973b; 330:8-26.
25. Proffitt RT, Williams LE, Presant CA et al. Tumor-imaging potential of liposomes loaded with In-111-NTA: biodistribution in mice. J Nucl Med 1983a; 24:45-51.
26. Turner AF, Presant CA, Proffitt RT et al. In-111-labeled liposomes: dosimetry and tumor depiction. Radiology 1988; 166:761-765.
27. Williams LE, Proffitt RT, Lovisatti L. Possible applications of phospholipid vesicles (liposomes) in diagnostic radiology. J Nucl Med Allied Sci 1984; 28:35-45.
28. Williams LE, Duda RB, Proffitt RT et al. Tumor uptake as a function of tumor mass: a mathematic model. J Nucl Med 1988; 29:103-109.
29. Proffitt RT, Williams LE, Presant CA et al. Liposomal blockage of the reticuloendothelial system: improved tumor imaging with small unilamellar vesicles. Science 1983b; 220:502-505.
30. Presant CA, Proffitt RT, Turner AF et al. Successful imaging of human cancer with indium-111-labeled phospholipid vesicles. Cancer 1988; 62:905-911.
31. Presant CA, Blayney D, Proffitt RT et al. Preliminary report: imaging of Kaposi sarcoma and lymphoma in AIDS with indium-111-labelled liposomes. Lancet 1990; 335:1307-1309.
32. Gregoriadis G, Davis C. Stability of liposomes in vivo and in vitro is promoted by their cholesterol content and the presence of blood cells. Biochem Biophys Res Commun 1979; 89:1287-1293.

33. Forssen EA, Coulter DM, Proffitt RT. Selective in vivo localization of daunorubicin small unilamellar vesicles in solid tumors. Cancer Res 1992; 52:3255-3261.
34. Forssen E, Chan KK, Muggia FM et al. Clinical pharmacokinetics (PK) of liposomal daunorubicin (VS103). Proc Am Assoc Cancer Res 1990; 31:181 Abstract 1078.
35. Gill PS, Espina BM, Muggia F et al. Phase I/II clinical and pharmacokinetic evaluation of liposomal daunorubicin. J Clin Oncol 1995; 13:996-1003.
36. Guaglianone P, Chan K, DelaFlor-Weiss E et al. Phase I and pharmacologic study of liposomal daunorubicin (DaunoXome). Invest New Drugs 1994; 12:103-110.
37. Eksborg S, Ehrsson H, Ekqvist B. Protein binding of anthraquinone glycosides, with special reference to Adriamycin. Cancer Chemother Pharmacol 1982; 10:7-10.
38. Robert J, Gianni L. Pharmacokinetics and metabolism of anthracyclines. Cancer Surv 1993; 17:219-252.
39. Kaplan L. Neoplastic complications of HIV disease. Infect Med 1992; 9(suppl F):20.
40. Gill PS, Rarick M, McCutchan JA et al. Systemic treatment of AIDS-associated Kaposi's sarcoma: results of a randomized trial. Am J Med 1991; 90:427-433.
41. Presant CA, Scolaro M, Kennedy P et al. Liposomal daunorubicin treatment of HIV-associated Kaposi's sarcoma. Lancet 1993; 341:1242-1243.
42. Money-Kyrle JF, Bates F, Ready J et al. Liposomal daunorubicin in advanced Kaposi's sarcoma: a phase II study. Clin Oncol 1993; 5:367-371.
43. Krown SE, Metroka C, Wernz JC. Kaposi's sarcoma in the acquired immune deficiency syndrome: a proposal for uniform evaluation, response, and staging criteria. J Clin Oncol 1989; 7:1201-1207.
44. Cerubidine® (Daunorubicin HCL) for Injection Prescribing Information. Emeryville, CA: Chiron Therapeutics, January 1995.
45. Gill PS, Wernz J, Scadden DT et al. Lack of cardiac toxicity of liposomal encapsulated daunorubicin (DaunoXome®) after long term use in AIDS related Kaposi's sarcoma. Ann Oncol 1996; 7(suppl 1); 134.
46. Tulpule A, Rarick MU, Kolitz J et al. Liposomal encapsulated daunorubicin (DaunoXome®) has activity in relapsed/refractory low grade and intermediate grade non-Hodgkin's lymphoma (NHL). Blood 1996; 88(10, suppl 1):92a.
47. Gill PS, Wernz J, Scadden DT et al. Randomized phase III trial of liposomal daunorubicin versus doxorubicin, bleomycin, and vincristine in AIDS-related Kaposi's sarcoma. J Clin Oncol 1996; 14:2353-2364.

CHAPTER 11

Initial Clinical Evaluation of Pegylated-Liposomal Doxorubicin in Solid Tumors

Alberto A. Gabizon, Franco M. Muggia

Introduction

Recently, two liposomal anthracycline formulations have been approved by the U.S. Food and Drug Administration for clinical use (Doxil, DaunoXome).[1-3] Another formulation (D-99)[4] is in advanced clinical trials. As these products become available to an enlarged circle of clinicians and nurses, the need to recognize their toxicity profile becomes more urgent. When a drug is presented in a liposome formulation, the medical teams are faced with a new entity which does not fit the usual definition of a new drug. This new entity is more complex than a simple drug since it consists of a carrier, the liposome, and a drug, which may be an approved and well-known drug (e.g., doxorubicin or daunorubicin) or a newly-designed drug (e.g., annamycin[5]). Some clinicians tend to look at liposomal doxorubicin or liposomal daunorubicin as another drug analog among the many (epirubicin, mitoxantrone, idarubicin) that have become available in recent years. Moreover, many clinicians will refer to liposomal anthracyclines as a homogeneous group of formulations with relatively similar pharmacological and biological effects. These are dangerous oversimplifications. A learning process is required to grasp the novelty and complexity of liposome-based drug delivery systems.

In considering a liposome formulation, we should be aware that there are basically three variables that may affect its biological activity and toxicity profile: the lipid vesicle, the drug, and the form of interaction between drug and lipid vesicle. Let's briefly examine each of these variables.

(1) The liposome (lipid vesicle) is a supramolecular assembly, whereby amphipathic lipids, generally phospholipids, form a closed bilayer creating a vesicle with an entrapped water phase separated physically from the external medium.[6] The liposome serves as the drug carrier and may contain additional lipid ingredients, such as cholesterol, intercalated in the bilayer. Liposomes may differ in composition (e.g., fatty acid and headgroup composition of phospholipid, modification of lipids with special chemical groups) and physical properties (number of concentric bilayers, vesicle diameter, surface charge, surface hydrophilicity). The liposome chemical

Long Circulating Liposomes: Old Drugs, New Therapeutics, edited by Martin C. Woodle and Gerrit Storm.

composition and physical properties will affect significantly its pharmacokinetic properties and as such may be tailored in the desired direction. Liposomes are generally nontoxic inasmuch as their components are biocompatible and biodegradable. In fact, one basic principle of the liposome pharmaceutics field is that none of the manipulations done to the chemical and physical properties of liposomes should result in a toxic preparation. Thus, the aim is to engineer a carrier essentially devoid of inherent toxicity. In addition, liposomes are not expected to have any pharmacologic activity interfering with the active principle of the preparation.

(2) The drug is the active principle of the formulation. One should realize that, as long as the drug is still encapsulated in the liposome, the drug is not bioavailable. To become bioavailable the drug has to be released from the vesicles either by a process of drug leakage with preservation of the liposome integrity, or following vesicle breakdown as in the case of phagocytosis and lysosomal enzymatic attack.[7] In most cases, the drug is a compound approved already in humans for administration as free drug, as in the case of doxorubicin, daunorubicin, and amphotericin B. A baseline toxicity profile is therefore known for the free drug. The unknown is how liposome encapsulation will modify toxicity. From a regulatory standpoint, the safest approach would probably be to consider a liposome formulation of an old drug, as a new drug. However, the free drug should always be used as the baseline control against which a balanced evaluation of the toxicology of the liposome formulation is done.

(3) Interaction of drug with carrier: This will determine the control the liposome has over the bioavailability and the biodistribution of the drug. Liposomes may function as dual purpose carriers: providing a slow-release system in the intravascular compartment and tissue of distribution and/or redirecting the tissue distribution of the drug according to the liposome profile of biodistribution.[8] Drugs can be entrapped either in the water phase or in the lipid bilayer of the vesicle. Entrapment may be a passive process or driven actively by a chemical or electrochemical gradient. The method of entrapment will affect the rate of release of the drug with implications on the pharmacological activity.[9] Thus, it is possible to have two liposome formulations of the same drug with the same chemical characterization, but displaying a different pharmacologic and toxicologic profile because of the way the drug has been encapsulated.[10] Another important result of liposome encapsulation is a reduction of the drug renal clearance.[11] This is due to the fact that a small molecular weight drug is encapsulated in a nanoparticle which is far beyond the threshold of glomerular filtration. Thus, no renal excretion of the drug is possible as long as the drug has not been released from the liposomes.

With regard to doxorubicin, there were a handful of reasons that motivated investigators to search for liposomal formulations. This drug is a potent anticancer compound widely used in a broad spectrum of cancer types.[12] However, the toxicity of doxorubicin is substantial and imposes serious limits in our ability to use the drug repeatedly. In addition to acute toxicity consisting mainly of myelosuppression, doxorubicin also causes irreversible, cumulative-dose-dependent myocardial damage. As a result, the maximal cumulative dose of doxorubicin is generally limited to 550 mg/m^2.[13] Besides myelosuppression, other toxic effects related to administration of doxorubicin include nausea and vomiting, transient EKG disturbances, stomatitis, alopecia, local phlebitis and extravasation necrosis. Thus, a number of investigators have tried to formulate doxorubicin and other anthracyclines in liposomes to buffer its toxicity while preserving its antitumor activity. A vast amount of preclinical literature on this topic is available,[14] from which several pharmaceutical-grade formulations have emerged and are currently being tested in clinical trials (Table 11.1). In this report, we will concentrate on Doxil, a formulation of liposomal doxorubicin characterized by prolonged circulation time and high stability,[14] and with which we have extensive and direct experience at the clinical level. It also represents the formulation that differs most greatly from the free drug in pharmacokinetics, toxicity modifications, and in emerging therapeutic spectrum.

Clinical Toxicity

Doxil (Sequus Pharm. Inc., Menlo Park, CA) is a liposomal formulation of doxorubicin presented in liquid form and stored at 5°C. Each vial contains a 10 ml liposome suspension. The concentration of doxorubicin is 2 mg/ml. The drug is encapsulated in the water phase of the vesicles by an ammonium sulfate generated proton gradient.[15] Because of its high intraliposomal concentration, the drug appears to undergo a reversible gelification process which may help to stabilize its retention in the vesicles.[16] One critical component of Doxil is a polyethylene-glycol derivatized phospholipid which accounts for the long circulating properties of these liposomes.[17]

We will divide the toxicity observations into three categories according to standard criteria: Acute, Subacute and Chronic.

Acute Side-Effects

Acute reactions to Doxil infusion have been reported at an incidence of 10% of treated patients during the course of a Phase I study.[2] These reactions are of variable severity and are characterized by facial and neck flushing, breathing difficulties, and, in some cases, back pain, cyanosis and transient drop in blood pressure. They resolve usually within a few minutes after discontinuing the infusion. These reactions appear to be related to a fast rate of infusion. In many cases, resuming the

Table 11.1. Liposomal anthracyclines-formulations currently in clinical testing[a]

Name	Composition	Vesicle Size	Max. Tolerated Dose	Pharmacokinetics[e]
Doxil[b]	HPC-Ch-DSPE/MPEG	80-120 nm	50-60 mg/m^2 q4wk	$t_{1/2}$ (biphasic), 84 hr min/46 hr; Cl, 0.09 L/hr; V_{ss}, 5 L
DaunoXome[c]	DSPC/Ch	50-80 nm	60 mg/m^2 q2wk	$t_{1/2}$ (single), 4-8 hr; Cl, 0.6 L/hr; V_{ss}, 4 L
D-99[d]	PC/Ch	180 nm	75-90 mg/m^2 q3wk	$t_{1/2}$ (biphasic), 12 min/17 hr; Cl, 19 L/hr; V_{ss}, 160 L

a. Based on refs. 2-4, and 24. Abbreviations: HPC = hydrogenated phosphatidyl-choline; Ch = cholesterol; DSPE = distearoylphosphatidyl-ethanolamine; MPEG = methoxy-polyethylene glycol; DSPC = distearoylphosphatidyl-choline; PC = phosphatidyl-choline.
b. From SEQUUS Pharmaceuticals (Menlo Park, CA)
c. From NEXSTAR Pharmaceuticals (San Dimas, CA)
d. From The Liposome Company (Princeton, New Jersey)
e. Median values for dose levels of 50 mg/m^2 (Doxil), 40-60 mg/m^2 (DaunoXome), and 90 mg/m^2 (D-99). $t^1/_2$=half-life; CP=clearance; Vss=volume of Distribution at steady state

infusion at a slower rate does not result in toxicity. Acute back pain during infusion has also been reported for another formulation of liposomes containing a nonanthracycline drug.[18] The mechanism for this acute type of reaction to liposome infusion is still unclear, but it is probably related to an interaction of some blood component with the liposome surface. One possibility is activation of the alternate pathway of the complement system as described in a rat model.[19]

At Hadassah Hebrew University Medical Center, we currently administer Doxil by intravenous drip after prior dilution in 250 ml 5% Dextrose bags, and the infusion rate is typically 1-1.5 mg Doxil per minute. When a patient is treated with Doxil for the first time, it is advisable to start with a test dose of Doxil at 0.1-0.2 mg/min for 10 minutes to check for a possible acute reaction. If the patient feels normal, the infusion rate can then be raised to 1.0-1.5 mg/ml. We do not premedicate patients, except for those who have had an acute reaction in which case we administer 50 mg hydrocortisone iv immediately prior to Doxil infusion. With these precautions, we have seen only one acute reaction in our last series of 25 patients (author's unpublished results).

Toxicity to the site of injection in the form of local pain, phlebitis, or vein sclerosis has not been observed with Doxil and other liposomal anthracycline formulations. Regarding the vesicant damage to the skin that follows extravasation in the site of injection, there seems to be a protective effect of liposomal encapsulation. In fact, several episodes of Doxil extravasation that did not result in skin damage have been reported.[20] In a mouse study in which Doxil was injected intradermally, we found a mild inflammatory reaction with no ulceration or skin necrosis,[21] suggesting that the liposomes are removed from the site of injection by lymphatic drainage and scavenger cells before any significant drug leakage from the vesicles takes place. This is in contrast to free doxorubicin which causes severe skin damage with necrosis and ulceration.

Acute nausea and vomiting is seldom seen with Doxil treatment. Overall, gastrointestinal toxicity appears to be reduced with liposomal anthracyclines. As a rule, we do not use any intravenous antiemetics as premedication before Doxil administration, unless a particular patient is extremely susceptible.

Fever with or without chills has been observed at a significant incidence in patients receiving D-99,[4] and occasionally with DaunoXome[29] but not in the case of Doxil,[2] a finding that may be related to the difference in reticulo-endothelial system (RES) uptake between these formulations.[14] In one patient manifesting fever after Doxil, a similar occurrence had been noted following intraperitoneal delivery of a radioimmunoconjugate, suggesting the reaction could have represented radiation 'recall' (i.e., an inflammatory reaction due to drug effects in an area of previous radiation).

Subacute Side-Effects

The most important subacute toxicity of Doxil is mucositis or stomatitis. This is also the dose-limiting toxicity with regard to the maximal tolerated dose (MTD) of a single dose of Doxil. Out of 9 patients receiving 70-80 mg/m^2 Doxil, 3 developed Grade 3 or 4 stomatitis which lasted in some cases for 7 d or more.[2] Therefore the maximal recommended dose of Doxil should not exceed 70 mg/m^2. Some patients have complained of pain reminiscent of gastritis and esophagitis, occasionally extending to the upper neck and lasting 1 to 2 weeks after treatment. This is also likely to be the result of damage to the mucous membranes in the upper gastrointestinal tract. Stomatitis is also a significant side-effect of free doxorubicin, especially when administered as prolonged infusion or when dose-intensive regimens with hematopoietic growth factors are used.[22] However, for conventional bolus administration of doxorubicin at the standard 3-week schedule and without growth factor support, myelosuppression (leukopenia and granulocytopenia) is generally the dose-limiting toxicity.[12] In contrast, Doxil causes mild myelosuppression. Grade 4 leukopenia/granulocytopenia was observed in only one case out of 105 courses in 35 patients treated with a dose range of 60-80 mg/m^2. Mild leukopenia is commonly seen at doses of 40 mg/m^2 and greater but it is generally uneventful and not dose-limiting.[2] In patients who have had bone marrow ablation with autologous rescue, or who have had total abdominal radiation we have observed grade 4 thrombo- and neutropenia. As to the timing of leukopenia, there seems to be a slight delay as compared to free doxorubicin. When weekly blood counts are done, the nadir is usually seen at 2 weeks after injection but in a substantial number of patients we have observed the nadir 3 weeks after treatment (author's unpublished data). With other liposomal anthracycline formulations, granulocytopenia is a significant form of toxicity.[14] Bone marrow is a tissue rich in macrophages and with a fenestrated capillary system which generally results in a high uptake of liposomes. The fact that some of the longer circulating liposomes partly avoid the RES or are removed by the RES at a very slow pace may account for the differences in severity of myelosuppression between Doxil, a long circulating formulation and other formulations with shorter circulation times.

A unique feature of Doxil is the high incidence of cutaneous toxicity,[2] especially in the form of hand-foot syndrome[22] (also defined as 'palmar-plantar erythrodysesthesia'). Skin toxicity is seldom seen after a single course of Doxil. It becomes more frequent with repeated treatments, especially when short intervals (less than 4 weeks) are allowed. In fact, for multiple treatments with Doxil, skin toxicity is the dose limiting factor (Fig. 11.1). As a result a schedule of 4 weeks, rather than the standard 3-week schedule is becoming the recommended schedule for doses of Doxil greater than 50 mg/m^2, although more experience is needed before definitive statements can be made. Clearly, there is a cumulative damage in the skin from repeated treatments with Doxil. This appears to be caused by cytotoxic damage to proliferating keratinocytes of the basal layers of the epidermis. Skin repair may be a process too slow to enable retreatment at standard 3-week intervals. This, however, does not hold for low doses (20 mg/m^2) as in the cases of AIDS-related Kaposi's sarcoma.[1] Clinically, skin toxicity manifests generally by painful erythema and edema, followed by desquamation and finally re-epithelization of the damaged areas. Generally, it develops between 1-3 weeks after a second or further injection of Doxil, affecting primarily contact pressure areas such as palms of hands and soles of feet. Its severity may range from a minimal erythema to major skin breakdown, crippling temporarily the patient. Recovery lasts between 2-3 weeks, thus sometimes delaying further treatment with Doxil for up to 6 weeks. Interestingly, other liposomal anthracycline formulations have not been reported to cause this type of toxicity, despite administration of greater nominal doses of the drug. With regard to free doxorubicin, hand-foot syndrome has only been observed in a small fraction of patients receiving dose intensive regimes such as prolonged infusions (> 15 d) or high dose treatments.[23] This special feature of Doxil may have to do with its extremely long circulation time and small vesicle size (distribution half-life of ~2 d),[24] which may enable liposomes to localize in the skin, depositing a substantial fraction of the drug payload. Other skin manifestations that have been noted are pigmentation predominantly in intertrigenous areas, and in the nails. A striking reaction was 'radiation recall' noted in the skin in a patient with breast cancer. In addition, some patients have complained of pain over previously irradiated serosal surfaces.

Despite the skin toxicity problem, Doxil treatment (dose range: 40-80 mg/m^2) has a very low incidence of alopecia (~10%),[2] as compared to free doxorubicin which causes universal alopecia. Near total alopecia was only noted in two patients with severe liver involvement. In the case of D99 and DaunoXome, the reported incidence of alopecia is ~40% and 5%, respectively.[3,4]

Chronic Toxicity

By chronic toxicity of anticancer drugs, we refer to the permanent damage caused by these drugs to a specific tissue. This is the case of the anthracycline-induced cardiomyopathy. Free radical damage caused by doxorubicin and analogs to the mitochondria of cardiac myocytes appear to account for this severe, irreversible, and cumulative-dose related toxic effect.[12] Although electron microscopy can reveal the increasing damage of repeated doses of doxorubicin relatively soon after treatment,[25] the clinical manifestations may only develop many years later. Because scoring of cardiac damage with electron microscopy is not widely available, the clinical studies with liposomal anthracyclines published so far have relied for detection of cardiotoxicity on cardiac function measurements. In a phase I study of Doxil, 14 patients received cumulative doses equal or greater than 450 mg/m^2 without any significant loss of cardiac function as determined by the left ventricle ejection fraction.[2] This is encouraging because it points at a possible reduction of cardiotoxicity when Doxil is used to deliver doxorubicin. However, caution is needed because of the small patient numbers and the relatively short follow-up of patients treated with liposomal anthracyclines.

Other Potential Toxicities

One interesting observation from the reported clinical studies with liposomal anthracyclines is the lack of hepatic toxicity. Because of the high hepatic uptake of most types of liposomes, it was feared that liver toxicity may ensue liposome-based therapy. This has not been the case so far, at least with regard to liposomal anthracyclines. Another source of concern is the possibility of damage to Kupffer cells.[26] In fact, studies in rats showed that liposomal doxorubicin can exert major toxic effects on the macrophage population of the liver, including impaired phagocytic function for about two weeks.[26] More recently, though, this problem was not encountered in the same animal model when Doxil was administered in a regimen more closely resembling the clinical setting.[27] Although there are no direct clinical laboratory tests to assess Kupffer cell function, the clinical experience with Doxil and other liposomal anthracyclines does not point at a higher sensitivity of patients to infections. We have seen, however, four patients manifesting Herpes zoster rashes out of 100 patients treated. Even if we are overlooking a subtle damage to the RES, this may still be an acceptable toxicity price when the clinical outlook of patients with metastatic cancer is considered.

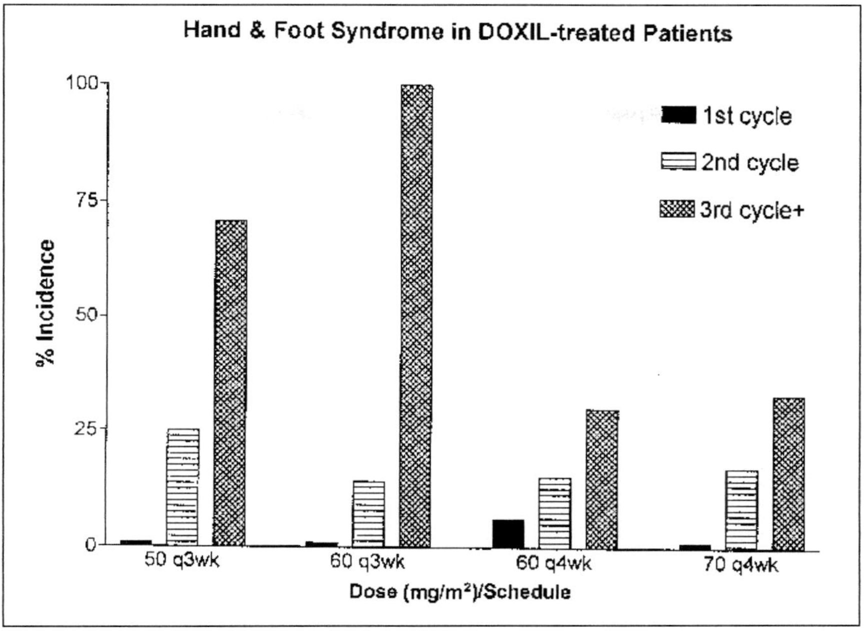

Fig. 11.1. Effect of dose-schedule and number of treatments on the skin toxicity of Doxil. Data obtained from a phase I study.[2]

Antitumor Activity

Kaposi's Sarcoma

The efficacy of both DaunoXome and Doxil against this neoplasm has been covered in chapters 9 and 10, respectively. Noteworthy features of this activity are an improved therapeutic index of either preparation as opposed to the three-drug free doxorubicin (Adriamycin)-based regimen, also including bleomycin and vinblastine, ABV; and in the case of Doxil, activity against ABV-resistant disease. This experience suggests that the spectrum of activity of Doxil might differ from that of the parent compound. Alternatively, improved delivery of Doxil to certain tumors might render them more susceptible to its antitumor effects relative to free doxorubicin at conventional doses.

Ovarian Cancer

At the University of Southern California, experience in platinum- and paclitaxel refractory ovarian cancer was obtained in the phase I study (8 patients entered)[2] and continued in a phase II study (35 patients entered together with the Sarah Cannon Cancer Center),[28] and further expanded in a broad phase II study for previously treated gynecologic malignancies. Consistent activity with objective responses in 20-30% of patients has been noted, and the median duration of response will be near one year—that is, considerably longer than what has been reported with high-dose epirubicin.[30] Responses have been maintained for as long as 30 months, in spite of dose reductions from 50 mg/m² every 4 weeks to usually 40 mg/m² as a maintenance dose and repeated in intervals from 4-6 weeks—the latter if required for healing of some skin ulcerations. It has not been unusual for responses to be documented objectively several months after treatment, and even for the CA-125 to show some delay in declining to after the third cycle of treatment. The responses have been almost uniformly partial, but a complete response has been registered, and maintained for over one year in spite of discontinuation of therapy. Future directions should not only compare Doxil to other drugs recently found useful in this setting, such as topotecan, but also seek to include Doxil in combinations with either taxanes or platinums.

Breast Cancer

As anticipated, during the phase I study activity was noted against previously treated breast cancer;[2] one patient has had control of bone metastases for 30 months. Two studies in this disease, one in England and one in Israel, have subsequently published their preliminary experiences in abstract form. The English study reduced the 60 mg/m² starting dose to 45 mg/m² after the first 13 patients experienced considerable plantar erythrodysesthesia. Among 43 patients entered, 3 CR and 13 PR had been documented, with 17 being too early.[31] The study at Hadassah Hebrew University Hospital reports 7 PR or improvement among 13 patients that have been entered in a randomized study of 60 mg/m² every 4 weeks versus 45 mg/m² every 3 weeks.[32] Because of its unique toxicity spectrum and its attenuated cardiotoxicity, a future role in combination with other drugs such as the taxanes or cyclophosphamide in a first-line or adjuvant setting must be explored. Conversely, combinations with vinorelbine and other drugs that are not used included in the initial regimens, may also prove of interest for palliative purposes.

Other Tumors

Following the phase I studies,[2,3,29] activity-seeking trials of Doxil and DaunoXome are ongoing in hepatocellular cancer, head and neck cancer, prostate cancer, nonsmall cell lung cancer, hepatocellular cancer, malignant lymphomas and soft-tissue sarcomas. Results are currently not available. A future role against these tumors awaits the results of these studies as well as eventual studies of drug combinations. Nevertheless, toxicologic considerations favor the use of a drug such as Doxil in the adjuvant setting for soft-tissue sarcomas over use of free doxorubicin. Similarly, our study in gynecologic cancer suggests activity against endometrial cancer: because these patients generally have poor tolerance of systemic chemotherapy, such activity should be pursued vigorously in future clinical trials.

Concluding Remarks

The potential of liposomal anthracyclines to buffer a number of undesirable side effects of doxorubicin is a clinical fact. Toxicity buffering stems mainly from the slow release of drug from liposomes and from the tissue distribution changes. There are, however, significant differences among the various liposome formulations which may have a major clinical weight. The unique toxicity of Doxil to skin underscores the contention that drastic pharmacokinetic changes may result in a compound with a qualitatively different toxicity profile. Ultimately, the extended use of liposomal drugs will depend on an assessment of their therapeutic index, i.e., the benefits/risks odds equation. For AIDS-related Kaposi's Sarcoma, Doxil at 20 mg/m^2 (see ref. 1) and DaunoXome[3] at 40-50 mg/m^2 offer an improved therapeutic index, judged to be sufficient by regulatory authorities for their approval. In the case of common forms of cancer, promising antitumor activity has been observed with Doxil in Phase I-II trials,[2,28] but we still have to wait for the results of definitive Phase III clinical trials that will determine the "benefits/risks" ratio, in reference to the standard forms of therapy currently available.

References

1. Harrison M, Tomlinson D, Stewart S. Liposomal-entrapped doxorubicin: An active agent in AIDS-related Kaposi's sarcoma. J Clin Oncol 1995; 13:914-920.
2. Uziely B, Jeffers S, Isacson R et al. Liposomal doxorubicin: antitumor activity and unique toxicities during two complementary phase I studies. J Clin Oncol 1995; 13:1777-1785.
3. Gill PS, Espina BM, Muggia F et al. Phase I/II clinical and pharmacokinetic evaluation of liposomal daunorubicin. J Clin Oncol 1995; 13:996-1003.
4. Cowens JW, Creaven PJ, Greco WR et al. Initial clinical (phase I) trial of TLC D-99 (doxorubicin encapsulated in liposomes). Cancer Res 1993; 53:2796-2802.
5. Zou Y, Priebe W, Stephens LC et al. Preclinical toxicity of liposome-incorporated Annamycin: Selective bone marrow toxicity with lack of cardiotoxicity. Clin Cancer Res 1995; 1:1369-1374.
6. Lasic DD. Liposomes: From Physics to Applications. Amsterdam: Elsevier, 1993.
7. Gregoriadis G. Liposomes as Drug Carriers: Recent Trends and Progress. Chichester: John Wiley & Sons, 1988.
8. Gabizon A. Liposomes as a drug delivery system in cancer chemotherapy. In: Roerdink FH, Kroon AM, eds. Drug Carrier Systems. Biochemical and Biophysical Basis and Medical Prospects. Horizons in Biochemistry and Biophysics. Chichester: Wiley, 1989; 9:185-211.
9. Barenholz Y, Cohen R. Rational design of amphiphile-based drug carriers and sterically stabilized carriers. J Liposome Res 1995; 5:905-932.
10. Goren D, Gabizon A, Barenholz Y. The influence of physical characteristics of liposomes containing doxorubicin on their pharmacological behavior. Biochim Biophys Acta 1990; 1029:285-294.
11. Gabizon A. Liposome circulation time and tumor targeting: Implications for cancer chemotherapy. Adv Drug Deliv Rev 1995; 16:285-294.
12. Young RC, Ozols, RF, Myers CE. The anthracycline antineoplastic drugs. N Engl J Med 1981; 305:139-153.
13. Von Hoff DD, Layard MW, Basa P et al. Risk factors for doxorubicin-induced congestive heart failure. Ann Int Med 1979; 91:710-717.
14. Gabizon A. Liposomal anthracyclines. Hematol. Oncol Clin North Am 1994; 8:431-450.
15. Haran G, Cohen R, Bar LK et al: Transmembrane ammonium sulfate gradients in liposomes produce efficient and stable entrapment of amphipathic weak bases. Biochem Biophys Acta 1990; 1025:143-151.
16. Lasic DD, Frederik PM, Stuart MC et al: Gelation of liposome interior—A novel method for drug encapsulation. FEBS Lett 1992; 312:255-258.
17. Woodle MC, Lasic DD. Sterically stabilized liposomes. Biochem Biophys Acta 1992; 1113:171-199.
18. Sculier JP, Coune A, Brassinne C et al. Intravenous infusion of high doses of liposomes containing NSC 251635, a water-insoluble cytostatic agent: A pilot study with pharmacokinetic data. J Clin Oncol 1986; 4:789-797.

19. Szebeni J, Wassef NM, Spielberg H et al. Complement activation in rats by liposomes and liposome-encapsulated hemoglobin: evidence for anti-lipid antibodies and alternative pathway activation. Biochem Biophys Res Com 1994; 205:255-263.
20. Madhavan S, Northfelt DW. Lack of vesicant injury following extravasation of liposomal doxorubicin. J Natl Cancer Inst 1995; 87:1556-1557.
21. Gabizon A, Pappo O, Goren D et al. Preclinical studies with doxorubicin encapsulated in polyethyleneglycol-coated liposomes. J Liposome Res 1993; 3:517-528.
22. Lokich JJ, Moore C. Chemotherapy-associated palmar-plantar erythrodysesthesia syndrome. Ann Int Med 1984; 101:798-800.
23. DeSpain JD. Dermatologic toxicity. In: Perry MC, ed. The Chemotherapy Source Book. Baltimore: Williams & Wilkins, 1992:531-547.
24. Gabizon A, Catane R, Uziely B et al. Prolonged circulation time and enhanced accumulation in malignant exudates of doxorubicin encapsulated in polyethylene-glycol coated liposomes. Cancer Res 1994; 54:987-992.
25. Billingham ME, Mason JW, Bristow MR et al. Anthracycline cardiomyopathy monitored by morphologic changes. Cancer Treat Rep 1978; 62:865-869.
26. Daemen T, Hofstede G, Ten Kate MT et al. Liposomal doxorubicin-induced toxicity: Depletion and impairment of phagocytic activity of liver macrophages. Int J Cancer 1995; 61:716-721.
27. Storm G, Ten Kate MT, Working PK et al. Doxorubicin entrapped in sterically stabilized liposomes: effects on bacterial blood clearance capacity of the mononuclear phagocyte system. Clin Cancer Res (submitted).
28. Muggia F, Hainsworth J, Jeffers S et al. Phase II study of liposomal doxorubicin (Doxil) in refractory ovarian cancer: antitumor activity and toxicity modification by liposomal encapsulation. J Clin Oncol 1997; 15:987-993.
29. Guaglianone P, Chan K, DelaFlor-Weiss E et al. Phase I and pharmacologic study of liposomal daunorubicin (DaunoXome). Invest. New Drugs 1994; 12:103-110.
30. Vermorken JB, Kobierska A, Chevallier B et al. Phase II study of high-dose epirubicin (HDE) in ovarian cancer patients previously treated with cisplatin. Proc Amer Soc Clin Oncol 1995; 14:276 (Abstr 772).
31. Ranson M, O'Bryne K, Carmichael J et al. Phase II dose-finding trial of Dox-SL (Stealth liposomal doxorubicin HCl) in the treatment of advanced breast cancer. Proc Am Soc Clin Oncol 1996; 15:124 (Abstr 161).
32. Gabizon A, Uziely B, Lotem M et al. Doxil in patients with pretreated metastatic breast cancer (MBC): a dose-schedule finding study with pharmacokinetics. Proc Am Soc Clin Oncol 1997; 16:147a (Abstr 516).

Section II

Current Medical Applications

B. Other Therapeutic Applications

CHAPTER 12

Antibacterial Therapy with Sterically Stabilized Liposome Formulations

Irma A.J.M. Bakker-Woudenberg, Gerrit Storm, Martin C. Woodle

Introduction

In the last few years a major breakthrough with respect to liposomal delivery of antibiotics has occurred in the development of liposome formulations with long circulation half-lives.[1-4]

Liposomes have been widely studied as carriers of antibiotics in a variety of models of intracellular infections.[5,6] In most studies "classical" liposomes encapsulating antibiotics were applied. As these liposomes are taken up primarily by cells of the mononuclear phagocyte system (MPS), they can be exploited to achieve the selective delivery of antibiotics in models of intracellular infections localized in MPS tissues, primarily liver and spleen. In these studies application of liposomes aims for site-specific drug delivery resulting in high and prolonged antibiotic concentrations in the infected tissues c.q. infected cells. Results have shown that in intracellular infections the therapeutic index of the antibiotic was substantially increased when administered in the liposome-encapsulated form.

Another rationale to apply liposomes as carriers of antibiotics is to achieve site-avoidance drug delivery, which is particularly important in the case of potentially toxic antibiotics. In this respect extensive studies have been performed with amphotericin B encapsulated in liposomes or bound to other lipid carriers, with the aim of reducing its toxicity compared to conventional amphotericin B desoxycholate, resulting in an increase in its therapeutic index. Animal studies showed promising results, and were followed by clinical studies using various amphotericin B lipid formulations.[7,8] Three industrially produced amphotericin B lipid formulations are now available and have been evaluated in compassionate use studies in immunocompromised patients with life-threatening fungal infections and a variety of underlying diseases who show intolerance to or failure on conventional amphotericin B. These compounds share the advantage of a low nephrotoxicity, but are different in terms of acute reactions. Their efficacy in this patient population has been clearly demonstrated. This is an impressive outcome as one should take into account that these patients are a selection of more serious cases which all failed on conventional amphotericin B therapy.

Long Circulating Liposomes: Old Drugs, New Therapeuitics, edited by Martin C. Woodle and Gerrit Storm.

Various clinical trials comparing the individual amphotericin B lipid formulations with conventional amphotericin B with respect to their clinical efficacy in patients with fungal infections are ongoing. However, the comparative efficacy of the various amphotericin B lipid formulations remain to be elucidated.

Safe and effective treatment with liposomal amphotericin B was also reported in patients with multidrug-resistant visceral leishmaniasis, although relapse may sometimes occur in AIDS patients. As the causative organism *Leishmania donovani* survives and multiplies primarily in liver macrophages, both site-specific delivery and site-avoidance delivery of amphotericin B is contributing to the increase in therapeutic index in the treatment of visceral leishmaniasis.

In summary, site-specific delivery of antibiotic in intracellular infections in macrophages in liver and spleen can be accomplished by classical liposomes. Site-avoidance delivery of potentially toxic agents such as amphotericin B can also be achieved using classical liposomes and is of great value in the treatment of severe fungal infections and the parasitic infection, visceral leishmaniasis.

The therapeutic prospects of classical liposomes in the treatment of infectious diseases caused by extracellular microorganisms or infections by intracellular microorganisms not necessarily located in MPS tissues, have been strongly limited by the short blood residence time due to their rapid uptake by MPS cells after intravenous administration. Striking observations illustrating the limitations of classical liposomes as antibiotic carriers were obtained from studies in mice infected with *Mycobacterium tuberculosis* in the spleen and lungs.[9] Intravenous injection of liposome-entrapped streptomycin led to a significant decrease of the spleen mycobacterium number, but antibacterial effects were not seen in the lungs, as compared to treatment with free streptomycin. Comparative pharmacokinetic studies revealed that streptomycin concentrations in spleen and liver were higher after administration in liposomal form, whereas streptomycin concentrations in the lungs and kidneys were independent from the drug form (liposomal or free) of streptomycin.[10]

Other illustrative observations in this respect are the data from experimental models of *Mycobacterium avium* complex infection in beige mice. In this model of chronic disseminated infection, liver, spleen, lungs, kidneys and lymph nodes become involved in the infection over a period of weeks. The studies clearly demonstrate that liposome-based delivery of antibiotic enhances its antimycobacterial effect. In some studies the superior therapeutic effect of liposomal agent is only seen in the spleen, and not in the lungs.[11-14] However, in other studies the increased therapeutic effect of liposomal agent over nonencapsulated agent was also observed in the lungs.[15-17] The discrepancies between these studies may be explained by differences in the lipid composition, and the total lipid doses used in the studies. The use of liposomes for delivery of antibiotic to infections in non-MPS tissues requires a reduced uptake by the MPS, enabling the liposomes to circulate long enough to reach the infected tissues or cells. In addition, the liposomes should be rather stable against drug leakage during the long blood residence time period, whereas at the same time they should also be able to release their encapsulated agent at the site of infection to a sufficient degree.

Sterically Stabilized Liposomes in the Treatment of Infections

The new generation of liposomes characterized by long circulation half-lives and reduced hepatosplenic uptake show great promise for achieving improved delivery of antibiotics to infections localized in non-MPS tissues. These infections are of high clinical relevance.

The development of the sterically stabilized liposomes is important in this respect—they remain in the vascular circulation for extended periods of time without requiring high lipid dose or rigid bilayers. It is speculated that the long circulating liposomes leave the blood in areas of inflammation as a result of locally increased vascular permeability and endothelial leakage, both developing at the infected site during the progression of the infection. The mechanisms of liposome extravasation and their interstitial movement have not yet been elucidated. The dynamics of extravasation of sterically stabilized liposomes at the microcirculatory level has been examined in various experimental models. Strong evidence has been presented showing that pathological body sites such as tumors and sites of infection and inflammation, are much more accessible to sterically stabilized liposomes than to conventional liposomes.[18-25]

The use of sterically stabilized liposomes for site-specific delivery of antibiotic in infections outside the MPS of liver and spleen was investigated in two experimental animal infection models, described below.

(1) In our laboratory we used an experimental model of unilateral pneumonia caused by *Klebsiella pneumoniae* in rats.[26] In this model a progressive infection in the left lung was created, whereas in the right lung of the same animal signs of an infection were not observed. If untreated, the course of this acute lung infection is fatal within 5 days. Prolonged blood residence time of sterically stabilized liposomes composed of PEG-DSPE: PHEPC:Chol (molar ratio, 0.15:1.85:1), with a mean size of about 80 nm was observed. Circulation half-life for these liposomes was 20 h in uninfected rats. The prolonged residence time in blood was accompanied by a relatively low hepatosplenic uptake. The bacterial lung infection did not affect the circulation time and tissue distribution to a substantial extent. Compared to the sterically stabilized liposomes, 110 nm classical liposomes composed of EPG: EPC:Chol (molar ratio, 1:10:5) exhibited short circulation time in the blood of rats. The degree of lung localization of sterically stabilized liposomes was determined using the high-affinity [^{67}Ga]DF complex as aqueous liposomal marker, at 40 h after administration, when about 10% of liposomes were still circulating in the blood. It was observed that in rats with severe left lung infection the degree of localization of sterically stabilized liposomes in the left lung tissue was up to 10-fold higher compared to liposome localization in the left lung of uninfected rats. In severely infected rats about 9% of the injected dose of liposomes were recovered from the infected lung tissue. The increased localization of liposomes in infected left lung tissue strongly correlated with the severity of infection. In the right lung of the same infected rat (in which lung infection was not developed) the degree of localization of liposomes was not increased compared with the localization observed in uninfected rats. The localization of the classical relatively short circulating EPG: EPC: Chol liposomes in infected left lung tissue at 16 h after administration was significantly lower compared to the sterically stabilized liposomes.

The usefulness of the sterically stabilized liposomes for delivery of antibiotic to the infected lung tissue was studied in the same model.[27] The aminoglycoside gentamicin and the cefalosporine ceftazidime were encapsulated, and administered intravenously. At a single dose treatment schedule started at 24 h after bacterial infection, an increased therapeutic effect of the antibiotic was obtained after administration in the liposome-encapsulated form: increased survival of the infected animals as well as increased killing of bacteria in the infected lung at day 7 after bacterial infection as compared to the effects of free drug. The antibiotic-containing sterically stabilized liposomes appeared to be rather stable during circulation. Relatively high and sustained concentrations of liposome-associated gentamicin or ceftazidine in blood were obtained. Interestingly, the antibiotic-containing liposomes did not show bactericidal activity in vitro.

The data obtained from our studies in the *Klebsiella pneumoniae* lung infection model in rats indicate that sterically stabilized liposomes are "passively" targeted towards the infected tissue. This passive targeting aspect of sterically stabilized liposomes is of great importance with respect to possible clinical application, as in immunocompromised patients which are often leukopenic, the localization as well as the etiology of the infection is often unknown. Substantial leakage of antibiotic during the relatively long circulation in blood was not observed. It is concluded that the superior therapeutic effect is the result of liposome localization and local release of entrapped antibiotic within the site of infection.

(2) The second experimental animal infection model in which sterically stabilized liposomes were used for delivery of antibiotic is the model of *Mycobacterium avium* complex infection in beige mice by Gangadharam et al.[28] In this model the infection is disseminated to various organs such as liver, spleen, lungs and lymph nodes. The aminoglycoside streptomycin was encapsulated in sterically stabilized liposomes composed of PEG-DSPE:DSPC:Chol (molar ratio, 1:9:6.7), with a mean size of about 170 nm. Also classical liposomes composed of EPG:EPC:Chol (molar ratio, 1:9:6.7) containing streptomycin were studied. Antibiotic-containing liposomes were administered intravenously for 2 weeks by twice-weekly injections at doses of 15 mg/kg, starting at 24 h after bacterial infection. Streptomycin-containing sterically stabilized liposomes were bactericidal in the spleen, whereas in the liver bacterial multiplication was prevented. Streptomycin-containing classical liposomes were bactericidal in the spleen as well as in the liver. In the lung both liposome types were not bactericidal, but were able to fully prevent substantial bacterial multiplication which, in control animals, was at least 10^3-fold within 4 weeks.

The investigators conclude that PEG-liposomes containing solid phase lipids such as DSPC and Cholesterol are appropriate in the treatment of the *Mycobacterium avium* complex infections, as encapsulated streptomycin is able to kill or inhibit the growth of the mycobacteria. The study does not show a particular advantage of the sterically stabilized liposomes over the classical liposomes in terms of mycobacterium killing capacity. However, in view of the capacity of sterically stabilized liposomes to localize in deep-seated tissue macrophages, superior efficacy of antibiotics when encapsulated in these liposomes may be expected in mycobacterium infections disseminated in these tissues.

In addition to the use of sterically stabilized liposomes for site-specific delivery of antibiotic used to target infected tissues outside the MPS, these liposomes may be used as microreservoirs of antibiotic during circulation, or to reduce antibiotic toxicity. For the antiviral agent didanosine which is an anti-HIV agent, it was demonstrated in rats that the encapsulation of this agent into sterically stabilized liposomes greatly reduced the agent systemic clearance to 180 times lower than that of the free agent.[29] As a result the dose and frequency of administration of this potentially toxic agent could be reduced. It is expected that when encapsulated in sterically stabilized liposomes, the delivery of antiviral agent to tissues susceptible to HIV infection is increased. Together, this may lead to a potential therapeutic advantage of the liposomal formulation in the treatment of HIV disease.

The microreservoir concept, now focused on certain antibacterial agents that show extremely low half-life in blood, also deserves attention.

Concluding Remarks

The clinical application of sterically stabilized liposomes as carriers of antibiotics is primarily of importance in the treatment of serious bacterial infections which constitute a major cause of morbidity and mortality particularly in immunocompromised patients.

One factor contributing to the lack of success of antibiotic treatment in these patients is a moderate antibiotic susceptibility of the microorganism. In such a case relatively high antibiotic concentrations at the site of infection at the earliest time possible are needed to prevent dissemination of the infection and life-threatening septicemia. In cases when treatment with antibiotic in the free form fails, site-specific targeting of antibiotic by means of sterically stabilized liposomes should be considered.

Another factor contributing to antibiotic treatment failure is that successful antibiotic treatment is hampered by the intracellular location of the microorganism. Disseminated intracellular mycobacterial infections are relevant in this respect. Moreover, mycobacteria show low multiplication rates intracellularly. Eradication of the intracellular mycobacteria in the "dormant state" requires treatment for prolonged periods of time with high doses of antibiotic, to obtain a bactericidal effect. Such treatment schedules facilitate the development of antibiotic resistance. In this respect site-specific antibiotic delivery systems such as sterically stabilized liposomes may be of great value should be thoroughly investigated.

Finally, two characteristics of sterically stabilized liposomes as carriers of antimicrobial agents need to be emphasized. A valuable asset of the sterically stabilized liposomes is that they show dose-independent pharmacokinetics. High lipid doses are not needed to obtain prolonged circulation as was the case with the earlier long circulating liposome types. Therefore MPS saturation resulting in impaired ability to clear bacteria from the blood can be avoided. This is important from a clinical point of view, as for immunocompromised patients in which generalization of the infection frequently occurs, maximal blood clearance capacity of the MPS is needed and should be maintained.

A second characteristic of high importance is that the prolonged blood residence time of sterically stabilized liposomes is independent of the lipid composition (phase transition temperature, charge). In view of the difference in pharmacodynamics of different classes of antibiotics, the rate of release of encapsulated antibiotic from the liposomes is important. Manipulation of the release rate can be achieved by variation of the lipid composition, which in the case of the sterically stabilized liposomes, can be done without compromising the prolonged circulation properties. Therefore these sterically stabilized liposomes show great promise for increased delivery of antibiotics of various classes to sites of bacterial infection.

References

1. Woodle MC. Surface-modified liposomes: assessment and characterization for increased stability and prolonged blood circulation. Chem Phys Lipids 1993; 64:249-262.
2. Woodle MC, Newman MS, Cohen JA. Sterically stabilized liposomes: physical and biological properties. J Drug Targeting 1994; 2:397-403.

3. Allen TM. Long-circulating (sterically stabilized) liposomes for targeted drug delivery. Trends Pharmacol Sci 1994; 15:215-220.
4. Marjan JMJ, Allen TM. Long-circulating liposomes: past, present and future. Biotechnol Adv 1996; 14:151-175.
5. Bakker-Woudenberg IAJM. Liposomes in the treatment of parasitic, viral, fungal and bacterial infections. J Liposome Res 1995; 5:169-191.
6. Bakker-Woudenberg IAJM. Delivery of antimicrobials to infected tissue macrophages. Adv Drug Deliv Rev 1995; 17:5-20.
7. Leenders ACAP, de Marie S. The use of lipid formulations of amphotericin B for systemic fungal infections. Leukemia 1996; 10:1570-1575.
8. Hiemenz JW, Walsh TJ. Lipid formulations of amphotericin B: recent progress and future directions. Clin Infect Dis 1996; 22(Suppl. 2):S133-144.
9. Vladimirsky MA, Ladigina GA. Antibacterial activity of liposome-encapsulated streptomycin in mice infected with *Mycobacterium tuberculosis*. Biomedicine 1982; 36:375-377.
10. Ladigina GA, Vladimirsky MA. The comparative pharmacokinetics of ^{3}H-dihydrostreptomycin in solution and liposomal form in normal and *Mycobacterium tuberculosis* infected mice. Biomed Pharmacother 1986; 40:416-420.
11. Düzgünes N, Perumal VK, Kesavalu L, Goldstein JA, Debs RJ, Gangadharam PRJ. Enhanced effect of liposome-encapsulated amikacin on *Mycobacterium avium-M. intracellulare* complex infection in beige mice. Antimicrob Agents Chemother 1988; 32:1404-1411.
12. Cynamon MH, Swenson CE, Palmer GS, Ginsberg RS. Liposome-encapsulated-amikacin therapy of *Mycobacterium avium* complex infection in beige mice. Antimicrob Agents Chemother 1989; 33:1179-1183.
13. Klemens SP, Cynamon MH, Swenson CE, Ginsberg RS. Liposome-encapsulated gentamicin therapy of *Mycobacterium avium* complex infection in beige mice. Antimicrob Agents Chemother 1990; 34:967-970.
14. Düzgünes N, Ashtekar DR, Flasher DL et al. Treatment of *Mycobacterium avium-intracellulare* complex infection in beige mice with free and liposome-encapsulated streptomycin: role of liposome type and duration of treatment. J Infect Dis 1991; 164:143-151.
15. Saito H, Tomioka H. Therapeutic efficacy of liposome-entrapped rifampin against *Mycobacterium avium* complex infection induced in mice. Antimicrob Agents Chemother 1989; 33:429-433.
16. Tomioka H, Saito H, Sato K, Yoneyama T. Therapeutic efficacy of liposome-encapsulated kanamycin against *Mycobacterium intracellulare* infection induced in mice. Am Rev Respir Dis 1991; 144:575-579.
17. Le Conte P, Le Gallou F, Potel G, Struillou L, Baron D, Drugeon HB. Pharmacokinetics, toxicity and efficacy of liposomal capreomycin in disseminated *Mycobacterium avium* beige mouse model. Antimicrob Agents Chemother 1994; 38:2695-2701.
18. Dvorak HF, Nagy JA, Dvorak JT, Dvorak AM. Identification and characterization of the blood vessels of solid tumors that are leaky to circulating macromolecules. Amer J Pathol 1988; 133:95-109.
19. Jain RK. Vascular and intestitial barriers to delivery of therapeutic agents in tumors. Cancer Metastasis Rev 1990; 9:253-266.
20. Wu NZ, Dewhirst MW. Measurement of material extravasation in microvascular networks using fluorescence video-microscopy. Microvascular Res 1993; 46:231-253.
21. Huang SK, Lee KD, Hong K et al. Microscopic localization of sterically-stabilized liposomes in colon carcinoma-bearing mice. Cancer Res 1992; 52:5135-5143.
22. Wu NZ, Da D, Rudoll TL et al. Increased microvascular permeability contributes to preferential accumulation of stealth liposomes in tumor tissue. Cancer Res 1993; 53:3765-3770.

23. Yuan F, Leunig M, Huang SK et al. Microvascular permeability and interstitial penetration of sterically stabilized (stealth) liposomes in a human tumor xenograft. Cancer Res 1994; 54:3352-3356.
24. Lasic DD, Martin FJ (eds). Stealth Liposomes. Boca Raton, FL: CRC Press, 1995.
25. Huang SK, Hong K, Lee KD, Papahadjopoulos D, Friend DS. Light microscopic localization of silver-enhanced liposome-entrapped colloidal gold in mouse tissues. Biochim Biophys Acta 1991; 1069:117-121.
26. Bakker-Woudenberg IAJM, Lokerse AF, ten Kate MT, Mouton JW, Woodle MC, Storm G. Liposomes with prolonged blood circulation and selective localization in *Klebsiella pneumoniae* infected lung tissue. J Infect Dis 1993; 168:164-171.
27. Bakker-Woudenberg IAJM, ten Kate MT, Stearne-Cullen LET, Woodle MC. Efficacy of gentamicin or ceftazidime entrapped in liposomes with prolonged blood circulation and enhanced localization in *Klebsiella pneumoniae* infected lung tissue. J Infect Dis 1995; 171:938-947.
28. Gangadharam PRJ, Ashtekar DR, Flasher DL, Düzgünes N. Therapy of *Mycobacterium avium* complex infections in beige mice with streptomycin encapsulated in sterically stabilized liposomes. Antimicrob Agents Chemother 1995; 39:725-730.
29. Harvie P, Désormeaux A, Bergeron MC, Tremblay M, Beauchamp D, Poulin L, Bergeron MG. Comparative pharmacokinetics, distributions in tissue, and interactions with blood proteins of conventional and sterically stabilized liposomes containing 2', 3'-dideoxyinosine. Antimicrob Agents Chemother 1996; 40:225-229.

CHAPTER 13

AmBisome®: Long Circulating Liposomal Formulation of Amphotericin B

Jill P. Adler-Moore, Richard T. Proffitt

Introduction

AmBisome® is a newly commercialized formulation of amphotericin B incorporated into small unilamellar liposomes composed of cohesive bilayer forming lipids (see chapter 1). An important feature of this formulation is that the amphotericin B is stably incorporated into the liposome bilayer.[1] In recent years, this and other lipid formulations have been devised to improve the utility of Fungizone, a micellar formulation of amphotericin B which rapidly disassociates upon iv administration. While the standard formulation is proven clinically, it is limited by severe renal toxicity, and many life-threatening fungal infections cannot successfully be treated.[2,3] For the purposes of this chapter, Fungizone is considered "free" amphotericin B (ampB) administration.

The strong association of amphotericin B with the AmBisome liposome bilayer is thought to significantly reduce direct drug exposure to many tissues, including the kidney, thus reducing toxicity.[1] This has been demonstrated in many studies, including the following: in vitro cytotoxicity with mammalian cells, such as red blood cells, macrophage and kidney cells;[4,5] preclinical acute and chronic LD_{50} studies by iv administration;[6] and clinical safety studies.[8,9]

Previous experience has demonstrated that small cohesive liposomes, like those used in AmBisome, are stable in blood and have extended circulation lifetimes (see chapter 1).[10] Consistent with this, AmBisome is relatively long circulating so that it generates a distinct pharmacokinetic profile for amphotericin B in animals[6,11-13] and humans.[14] As the dose is increased from 1-5 mg/kg, the fraction taken up by the reticuloendothelial system (RES) decreases and further redistribution of the drug occurs into other organs, including the kidneys and lungs. Animal studies have shown that the concentration of active drug is highest in the liver and spleen when administered as AmBisome.

In terms of the amphotericin B activity, antimicrobial susceptibility test results uniformly indicate retention of in vitro activity by AmBisome against a wide variety of yeast and filamentous fungi. In animal models of infection, AmBisome produces

Long Circulating Liposomes: Old Drugs, New Therapeutics, edited by Martin C. Woodle and Gerrit Storm.

varying therapeutic responses depending upon the model. Compared to Fungizone, AmBisome has been found significantly better for treatment of murine leishmaniasis[15,16] and similarly efficacious for candidiasis,[18] aspergillosis,[20,21] cryptococcosis[18,22] and histoplasmosis.[23] In some infection models, e.g., blastomycosis[24] and paracoccidioidomycosis,[25] AmBisome (0.2-1.0 mg/kg/dose) was not as effective as comparable doses of Fungizone. However, because of its lower toxicity, AmBisome could be administered at iv doses up to 30 mg/kg in animals with severe infections, thus achieving greater efficacy than Fungizone. This enhanced activity of AmBisome may be a result of sustained drug levels above the MIC as reported in preclinical prophylactic AmBisome studies.[26]

Comparable clinical efficacy has been obtained for AmBisome and Fungizone at 1 mg/kg in double-blind studies in both adults[27] and children.[28] Additional clinical studies have demonstrated that AmBisome is effective in treating a variety of serious, life-threatening fungal infections with doses ranging from 1-5 mg/kg.[29-36]

This chapter describes in greater detail the preclinical and clinical findings on AmBisome. The described characteristics improve the drug's pharmacokinetics and safety profile and maintain or improve its efficacy relative to the conventional micellar Fungizone, increasing the therapeutic potential of Amphotericin B. Further studies of AmBisome are currently being conducted to evaluate the potential for expansion of clinical indications.

AmBisome

Composition

The composition of lipid formulations of ampB has been found to influence its pharmacological properties, primarily toxicity, in numerous studies. The composition chosen for AmBisome, small liposomes composed of HSPC: Chol: DSPG: ampB in a molar ratio of 2:1:0.8:0.4, was selected from studies of the influence on lipid composition on toxicity (see Table 13.1).[1] The hydrogenated phospholipids and cholesterol provide for a cohesive bilayer and stability of the amphotericin B in the bilayer that together provide good in vivo stability and prolonged circulation.[1] Interestingly, the concentration of DSPG also plays a role in binding with the ampB and enhancing its stability in the bilayer; molar ratios of DSPG: ampB of 2:1 are required to reduce toxicity.[1]

The stability of the chosen composition has been demonstrated both in solution and in vivo. Studies of gel permeation (size) chromatography showed that the ampB is not easily separated from the liposomes as both eluted in the same fractions despite large differences in size. Similar results were obtained even after incubation at 37°C in 50% human plasma for up to 72 h, which simulates physiological conditions.[1] Thus, overall, AmBisome is a liposome formulation of ampB that maintains the drug association with the liposomes and minimizes its toxic effects, as discussed in greater detail in the next section.

Toxicity

The strong association of ampB within the AmBisome liposome formulation is thought to be largely responsible for the markedly reduced toxicity of that formulation compared with the conventional bile salt micellar formulation, Fungizone. This effect is seen in studies of toxicity toward mammalian cells, human red blood cells in particular. Substantial hemolysis of these cells occurs (92%) within 2 h when they

Table 13.1. Characterization of various formulations of liposomal amphotericin B

Preparation*	Components	Molar Ratio of Components	% Incorporation of AmB	Mean Diameter** (nm)	Acute Toxicity LD_{50} (mg/kg)
1	HEPC:DSPG:Ampho B	2:0.8:0.4	66	48	<20
2	HEPC:Chol:Ampho B	2:1:0.4	63	50	<20
3	HEPC:Chol:DSPG:Ampho B	2:1:0.4:0.4	43	80	<25
4	HEPC:Chol:DSPG:Ampho B	2:1:0.8:0.4	76	41	>30
5	HEPC:Chol:DLPG:Ampho B	2:1:0.8:0.4	88	35	<20
6	HSPC:Chol:DSPG:Ampho B	2:1:0.8:0.4	85	47	>35
7	HSPC:Chol:DSPG:Ampho B	2:1:0.8:0.4	87	41	>50

*Preparations 1-6 were made by probe sonication and preparation 7 was made using a microemulsification procedure.
**As determined by volume weighted Gaussian analysis using Nicomp laser light scattering apparatus.
HEPC = hydrogenated egg phosphatidylcholine, DSPG = distearoylphosphatidylglycerol, Ampho B = Amphotericin B
Chol = Cholesterol, DLPG = dilaurylphosphatidylglycerol, HSPC = hydrogenated soy phosphatidylcholine

Table 13.2. Mean single dose pharmacokinetics of AmBisome in animals following intravenous administration

Species	n (group)	Dose (mg/kg)	C_{max} (mg/ml)	$AUC_{0-\infty}$ (mg•h/ml)	$t_{1/2}$ (h)	V_d (l/kg)	Cl (ml/h/kg)
Mouse	3	1	8	36	17	0.68	28
Rat	6	1	7.2	64	9.5	0.21	16
		3	30.3	374	7.9	0.10	8.4
		9	141.3	1136	8.0	0.10	8.0
		20	235.2	1807	13.6	0.20	12.6
Rabbit	3-4	0.5	11	20	17	0.16	35
		1.0	26	60	3.6	0.09	16
		2.5	53	207	5.2	0.09	12
		5	132	838	5.5	0.05	5.3
		10	287	2223	7.7	0.05	4.2
Dog	10	0.25	0.21	2.6	7.0	1.66	112
		1	1.9	11	9.3	0.96	79
		4	18	164	8.4	0.29	26
		8	72	986	11.0	0.14	10
		16	174	2595	11.6	0.07	6.0

are incubated with ampB at concentrations as low as 1 µg/ml, whereas AmBisome at concentrations up to 100 µg/ml is well tolerated (5% hemolysis). These results suggest that less than 1% of the ampB is released by AmBisome during the 2 h incubation period.[1] Similar reductions in toxicity have been found with other cell types, such as canine kidney and murine macrophage cell lines where AmBisome was found to be 5- and 2-fold less toxic than Fungizone, respectively.[5] These cell types are of particular interest due to the severity of nephrotoxicity which is the most important dose-limiting toxicity of Fungizone therapy[3] and the tendency of liposomes to be taken up in vivo by phagocytic cells (see chapter 1). Thus these in vitro studies would suggest that AmBisome is potentially less toxic in vivo.

Reduced toxicity was confirmed in extensive preclinical studies in mice,[6] rats[6] and dogs[1] evaluating exposures of up to 30 d. In one acute single dose study in mice,[6] AmBisome was tolerated at doses greater than 175 mg/kg compared to an LD_{50} of 2.3 mg/kg for Fungizone, making AmBisome about 75 times less toxic in mice. In rats this difference in toxicity was about 30 times lower with AmBisome. Multiple dose studies in rodents also showed similar marked reductions in toxicity. In dogs, daily iv doses up to 4 mg/kg were well tolerated whereas the maximum tolerated dose for Fungizone is known to be 0.6 mg/kg.[7] The blood chemistry of Fungizone-treated dogs showed blood urea nitrogen and creatinine levels about 4 and 2.5 times higher than those of dogs treated with 4 mg/kg/day AmBisome for 29 d. These data emphasize the reduction in nephrotoxicity associated with AmBisome in both acute and chronic dosing studies.

Clinical studies have also demonstrated significant reduction in toxicity by AmBisome. The earliest safety data was reported by Meunier et al.[8] In that multicenter study, 126 patients received 133 episodes of AmBisome treatment. The majority of the patients had previously failed conventional ampB (Fungizone) treatment due to toxicity. The mean duration of AmBisome administration was 21 d at an average dose of 2.1 mg/kg (range = 0.45-5 mg/kg). The most common side effect was hypokalemia which was observed in 24 cases. Nausea, vomiting, chills and rigors were observed in a total of 5 instances. Serum chemistries were monitored throughout the study for indications of organ toxicity. Although many patients entered this study with elevated creatinine levels, in only 11 cases did the creatinine level become elevated during AmBisome treatment. In 17 episodes, creatinine was initially high but returned to normal. Glutamic oxaloacetic transaminase became elevated in 19 instances and elevation of alkaline phosphatase was observed in 22 cases. There was no report of discontinuation due to adverse side effects of AmBisome. More recently, AmBisome safety was assessed in a series of 187 transplant recipients.[9] AmBisome was administered daily at dose levels between 1 and 4 mg/kg for a median of 11 d (range 1-112 d). Side-effects attributed to AmBisome therapy, including low serum potassium, low back pain during infusion, dyspnea, allergic reaction and nausea and vomiting, were observed in 7% of the cases and resulted in discontinuation of therapy in only 6 cases. Serum creatinine increases were not statistically significant. Other side effects possibly related to AmBisome included elevated serum urea and alkaline phosphatase, both of which normalized after cessation of AmBisome therapy. In this study, patients were receiving a variety of potentially toxic drugs so the AmBisome side-effect profile was mild and manageable in the vast majority of patients.

Based on the encouraging evidence for reduction in ampB toxicity when given as AmBisome, evaluations of this product for pediatric applications were carried out.[38,39] These reports indicated that AmBisome could be administered without serious side-effects in neonates, even at daily doses up to 6 mg/kg in immunosuppressed children. Clearly, though, additional controlled trials with AmBisome are warranted.

Overall, the preclinical and clinical results that have been reported support a significant reduction of ampB toxicity when administered as AmBisome. This effect may be attributed to the stability of the ampB within the liposome and the stability of the liposome in vivo, both attributable to the specifics of the lipid composition and formulation.

Pharmacokinetics

Studies with small unilamellar liposomes prepared with cohesive bilayers containing cholesterol have indicated that the liposomes can persist in the blood for prolonged periods (see chapter 1). The lipid bilayer composition of AmBisome is also cohesive and pharmacokinetic studies in animals have shown that it exhibits high peak plasma levels, high plasma $AUC_{0-\infty}$, and prolonged plasma elimination (Table 13.2). Overall, the pharmacokinetics of AmBisome in the species studied demonstrated a saturable, nonlinear disposition from plasma, presumably due to uptake by the reticuloendothelial system (RES). In the mouse, the Cmax and $AUC_{0-\infty}$ of amphotericin B as AmBisome increased in a manner greater than the incremental increase in dose. A mean terminal elimination half-life ($t_{1/2}$) with a range of 17-24 h was estimated. Both the volume of distribution (V_d) and the total clearance (Cl) decreased when dose was increased from 1-5 mg/kg. In the rat, rabbit and dog, a similar nonlinear disposition of amphotericin B administered as AmBisome was found:

$t_{1/2}$ of 8-17 h, disproportional increase in $AUC_{0-\infty}$ relative to dose, decrease in V_d and Cl with increased dose.

Tissue distribution studies have been performed in rats and dogs at 1 and 5 mg/kg. With AmBisome, most of the ampB accumulated in the liver and spleen at levels 2-3 times higher than comparable doses of Fungizone. In contrast, ampB accumulation in the kidneys was 6 times lower with AmBisome than Fungizone at 1 mg/kg. Likewise, lung levels were about 2.5 times lower with AmBisome. With an increase from 1 to 5 mg/kg of AmBisome, the relative uptake by liver and spleen decreased and there was a redistribution into other organs including kidneys and lungs.[1] These results suggest that the redistribution that is provided by the liposome is in part responsible for the reduced toxicity, in particular the reduced dose-limiting nephrotoxicity.

The clinical pharmacokinetics of AmBisome are distinctly different from ampB. After a 4 h intravenous infusion of 50-100 mg ampB, the pharmacokinetic profile of ampB is characterized by early peak plasma concentrations of 1.59-2.4 µg/ml accounting for only less than 10% of the total dose and indicating a relatively large volume of distribution.[40] The initial distribution phase with a plasma half-life of 24-48 h is followed with a slow elimination phase with a half-life of about 15 days.[41] Total $AUC_{0-\infty}$ values are consistently very low. The long elimination half-life for ampB is thought to be due to slow release of drug from tissue or cellular membranes, which then circulates bound to plasma lipoproteins.[41] In rabbits, AmBisome at a dose level of 1.0 mg/kg, produced an approximately 5-fold increase in C_{max} and an $AUC_{0-\infty}$ value that was 2-fold greater than that observed for ampB at a comparable dose (Table 13.2). In humans AmBisome volume of distribution is significantly smaller than that observed with ampB and AmBisome has a plasma elimination half-life of approximately 10 h.[14] A recent report describing the pharmacokinetics of AmBisome in febrile neutropenic patients[14] documented that the mean $AUC_{0-\infty}$ determined on the last day of infusion for doses of 1.0, 2.5 and 5.0 mg/kg/day was 69, 206 and 713 µg·hr/ml, respectively. This reflected a nonlinear increase as dose increased. Terminal elimination half-lives ($t_{1/2}$) in the patients were similar for each of the three dose levels (8.6-10.4 h). Thus, in febrile neutropenic patients AmBisome also followed a nonlinear dosage relationship consistent with RES uptake and redistribution.

Preclinical Efficacy

Incorporation of ampB into the AmBisome formulation does not alter its spectrum of antimicrobial activity, either in vitro or in vivo. The in vitro activity of both ampB and AmBisome is similar against yeast and filamentous fungi using standard 24 or 48 h incubations, regardless of whether tube macro- or plate micro-dilution methods are used (Table 13.3).[17,19-21,24,25,42-44] The MIC of AmBisome has ranged from 0.05-2.5 µg/ml as compared to 0.1-2.5 µg/ml for ampB. Thus the stable incorporation of ampB into the bilayer of AmBisome has no inhibitory effect on its in vitro MIC values.

Studies with preclinical animal models also demonstrate that AmBisome retains the ampB antimicrobial activity towards both extracellular and intracellular infections. Note that since the AmBisome pharmacokinetics, biodistribution, and toxicity are distinct, these parameters must be considered in designing efficacious regimens. A summary of studies conducted by various investigators is given in Table 13.4.

Table 13.3. In vitro susceptibility of fungi to AmBisome

Organism Tested	Amphotericin B Antimicrobial Activity		AmBisome Antimicrobial Activity		Reference
	MIC	MFC	MIC	MFC	
Tube Macrodilution Test Method					
C. albicans 44858	0.1 μg/ml	3.2 μg/ml	0.05 μg/ml	6.4 μg/ml	17
Aspergillus fumigatus 4215	1.0 μg/ml	>4.0 μg/ml	1.0 μg/ml	>4.0 μg/ml	20
Blastomyces dermatitidis ATCC 26199	0.5 μg/ml	1.0 μg/ml	0.125 μg/ml	1.0 μg/ml	24
Paracoccidioides brasiliensis, Gar	0.25 μg/ml	0.5 μg/ml	1.0 μg/ml	2.0 μg/ml	25
Plate Microdilution Test Method					
C. albicans (32)*	1.25 μg/ml[a]	1.25 μg/ml[b]	0.62 μg/ml[a]	1.25 μg/ml[b]	42
C. tropicalis (20)	1.25 μg/ml[a]	2.50 μg/ml[b]	0.62 μg/ml[a]	1.25 μg/ml[b]	
C. parapsilosis (20)	1.25 μg/ml[a]	2.50 μg/ml[b]	0.62 μg/ml[a]	1.25 μg/ml[b]	
C$_r$. neoformans (32)	1.25 μg/ml[a]	2.50 μg/ml[b]	0.62 μg/ml[a]	1.25 μg/ml[b]	
Aspergillus spp. (13)	2.50 μg/ml[a]		1.25 μg/ml[a]		
Fusarium spp. (8)	2.50 μg/ml[a]		2.50 μg/ml[a]		
C. albicans 620	0.62 μg/ml	0.62 μg/ml	0.62 μg/ml	0.62 μg/ml	43
C. krusei 2506	0.125 μg/ml[c]		0.125 μg/ml		44
C. krusei 2506	0.125-0.3125 μg/ml[d]		0.125-0.3125 μg/ml[d]		19
C. lusitaniae 1706	0.125-0.3125 μg/ml[d]		0.125-0.3125 μg/ml[d]		
C. lusitaniae 5W31	1.25 μg/ml		1.25 μg/ml		
C. lusitaniae 2819	2.5 μg/ml		2.5 μg/ml		
Aspergillus fumigatus	0.16 μg/ml		0.16 μg/ml		21

[a]MIC_{90} [b]MFC_{90} [c]MIC_{50}[d]MIC_{range}; ( ) = number of isolates tested; C = Candida; Cr = Cryptococcus

Table 13.4. *Summary of preclinical efficacy studies comparing Amphotericin B with AmBisome*

	Amphotericin B			AmBisome		
Infection Model	**Dose Regimen**	**Dose (mg/kg)**	**Major Findings**	**Dose (mg/kg)**	**Major Findings**	**Ref.**
Candida albicans Strain CSPU 39 Mouse, C57BL/6J	Multi: 5 cycles	0.75 (IV)	80% survival at 21 days. 1.0 X 10^3 CFU/mg kidney tissue ($p<0.05$).	2.5, 5, 10, (IV)	100% survival all doses. Dose related kidney clearance of yeast: 68 CFU/mg @2.5 (p(0.05) 16 CFU/mg @5.0(p(0.05) 9 CFU/mg @10.0 (p(0.05)	[18]
Candida albicans ATCC 44858 Mouse, BALB/C; immunocompetent	Multi: 5 cycles	0.4 (IV)	100% survival at 14 days. Significant reduction in CFU/mg of kidney tissue ($p<0.001$) Prevented relapse	0.4, 7.0 (IV)	90-100% survival at 14 days. Significant reduction in CFU/mg of kidney tissue ($p<0.001$). Prevented relapse	[17]
Candida albicans ATCC 44858 Mouse, BALB/C; leukopenic	Multi: 5 cycles	0.3 (IV)	100% survival; Significant reductions in CFU in liver, spleen and lungs($p<0.001$).	0.3 (IV)	50% survival	
			Growth in kidneys inhibited. Infection relapse.	7.0 (IV)	100% Survival. Reduction in CFU/mg of kidney ($p<0.001$). Prevented relapse.	

Candida lusitaniae Mouse CF1; neutropenic	Multi: 3 cycles					[19]
EXPT. 1: strain 1706		1.0 (IV)	Significantly prolonged survival ($p<0.05$); Significant reduction in CFU/g kidney tissue ($p<0.05$).	10, 30 (IV)	Significantly prolonged survival ($p<0.05$); Significant reduction in CFU/g kidney tissue ($p<0.05$).	
EXPT. 2: strain 524		1.0 (IV)	Significantly prolonged survival ($p<0.05$); Significant reduction in CFU/g kidney tissue ($p<0.05$).	10, 15 (IV)	Significantly prolonged survival (p(0.05); Significant reduction in CFU/g kidney tissue ($p<0.05$).	
EXPT. 3: strain 2819		1.0 (IV)	Survival not significantly prolonged; Kidney CFU not significantly reduced	10, 30 (IV)	Significantly reduced CFU/g kidney tissue ($p<0.05$).	
Candida lusitaniae Mouse CF1; neutropenic	Multi: 3 cycles					[19]
EXPT. 4: strain 5W31		1.0 (IV)	Survival not significantly prolonged; Kidney CFU not significantly reduced.	10 (IV)	Survival not significantly prolonged; Kidney CFU not significantly reduced	
				30 (IV)	Survival not prolonged; Significant reduction in CFU/g kidney tissue ($p<0.05$)	
Aspergillus fumigatus Isolate 4215 Rabbit, NZW; granulocytopenic	Multi 10 cycles	1.0(IV)	30% survival at day 10 ($p=0.1$). Approximately 15-fold reduction in mean number of CFU ($p<0.001$).	1.0, 5.0, 10(IV)	80-100% survival at day 10 ($p<0.01$).	[20]
				1.0(IV)	Approximately 8-fold reduction in mean number of CFU ($p<0.01$).	
				5.0, 10(IV)	Approximately 15-fold reduction in mean number of CFU (p(0.001).	

Table 13.4. Summary of preclinical efficacy studies comparing Amphotericin B with AmBisome

Infection Model	Amphotericin B			AmBisome		Ref.
	Dose Regimen	**Dose (mg/kg)**	**Major Findings**	**Dose (mg/kg)**	**Major Findings**	
Cryptococcus neoformans Isolate 89-98 (intracranial) Mouse, ICR	Multi: 7 cycles	0.3(IV), 1.0(IV)	Survival significantly prolonged (p(0.05).	0.3(IV) 1.0(IV)	Survival was not prolonged. Survival significantly prolonged (p(0.05).	[22]
		3.0,7.0(IP) 3.0(IP)	Survival significantly prolonged (p(0.05). Day 37 brain CFU not significantly different from AmBisome (3.0 mg/kg).	3.0, 7.0 (IV) 3.0 (IV)	Survival significantly prolonged (p(0.05). Day 37 brain CFU not significantly different from Amphotericin B (3.0mg/kg).	
				20, 30 (IV)	44% brain cultures negative (20 mg/kg); 78% brain cultures negative (30 mg/kg) on day 30.	[24]
Blastomyces dermatitidis Strain ATCC 26199 Mouse, CD-1	Multi: 6 cycles	0.2(IV) 1.0(IV)	70% survival at day 49. 100% survival (p(0.001); Significant reduction in CFU/lung (p(0.001).	0.2(IV) 1.0, 3.0, 7.5, 15(IV)	50% survival at day 49. 90-100% survival (p(0.001); Significant reduction in CFU/lung (p(0.001).	
Coccidioides immitis Clinical isolate Mouse, ICR	Multi: 6 cycles	0.3,1.0(IV)	Significantly protective 30 days post challenge (p(0.05). ***Endpoint***: Lung weight was used as protection index.	0.3, 1.0(IV)	Significantly protective 30 days post-challenge (p(0.05). ***Endpoint***: Lung weight was used as protection index.	[46]

Histoplasma capsulatum Isolate G217 Mouse, athymic *nu/nu*	Multi: 6 cycles	0.3(IV) 1.0(IV) 3.0(IP)	Survival not prolonged significantly. Survival not prolonged significantly. Survival prolonged significantly (p(0.001).	0.3(IV) 1.0(IV) 3.0(IV)	Survival not prolonged significantly. Survival not prolonged significantly. Survival prolonged significantly (p(0.001).	[23]
EXPT. 1 EXPT. 2 Isolate 93-255 Mouse, athymic *nu/nu*	6 cycles	0.3,0.6(IV)	Survival prolonged significantly (p<0.005)	0.3,0.6(IV)	Survival prolonged significantly (p<0.005)	
		1.0(IV), 3.0(IP)	Survival prolonged significantly (p(0.001).	1.0, 3.0(IV)	Survival prolonged significantly (p(0.001);	
Paracoccidiodes brasiliensis Isolate Gar Mouse, BALB/C	Multi: 6 cycles	0.6(IV)	47% survival (p(0.05) at day 40. Mean Log CFU/lungs = 7.11	0.6(IV) 5.0, 15, 30(IV)	7% survival (p(0.05) 67 to 80% survival (p(0.05-0.0001); Significant reduction in CFU/mg (p(0.01-0.001) vs. 0.6 mg/kg amphotericin B.	[25]
Trichosporon beigelii, Mouse, CF1; immunosuppressed EXPT. 1 Isolate 009	Multi: 10 cycles	1.0(IP)	No significant reduction of CFU/g kidney tissue.	1.0, 5.0(IV) 10(IV)	No significant reduction of CFU/g kidney tissue. Significant reduction (p(0.05) of CFU/g kidney tissue.	[45]
EXPT. 2 Isolate 008	10cycles	1.0(IP)	Significant reduction (p(0.05) of CFU/ g kidney tissue.	1.0, 5.0, 10(IV)	Significant reduction (p(0.05) of CFU/g kidney tissue Dose related decrease in CFU.	

Table 13.4. Summary of preclinical efficacy studies comparing Amphotericin B with AmBisome

		Amphotericin B		AmBisome		
Infection Model	**Dose Regimen**	**Dose (mg/kg)**	**Major Findings**	**Dose (mg/kg)**	**Major Findings**	**Ref.**
Leishmania donovani MHOM/ET/67/L82 Mouse, BALB/C EXPT. 1	Single	0.05(IV)	5.3% Inhibition of amastigotes in liver (p=0.26).	0.25(IV)	53.8% Inhibition of amastigotes in liver (p=0).	[15]
		0.2(IV)	14.1% Inhibition of amastigotes in liver (p=0.03).	1.0(IV)	85.4% Inhibition of amastigotes in liver (p=0).	
		1.0(IV)	42.2% Inhibition of amastigotes in liver (p=0.0002).	5.0(IV)	99.9% Inhibition of amastigotes in liver (p=0).	
				25(IV)	99.2% Inhibition of amastigotes in liver (p=0).	
			Dose related response. ED_{50}=0.154		Dose related response. ED_{50}= 4.914	
EXPT. 2	Single	0.04(IV)	3.4% Inhibition of amastigotes in liver (p=0.69).	0.04(IV)	15.8% Inhibition of amastigotes in liver (p=0.11).	
		0.2(IV)	22.0% Inhibition of amastigotes in liver (p=0.016).	0.2(IV)	41.2% Inhibition of amastigotes in liver (p=0.001).	
		1.0(IV)	52.7% Inhibition of amastigotes in liver (p=0.0003).	1.0(IV)	84.5% Inhibition of amastigotes in liver (p=0).	
				5.0(IV)	99.8% Inhibition of amastigotes in liver (p=0).	
			Dose related response. ED_{50} = 0.951		Dose related response. ED_{50}=0.256	

In some cases, AmBisome has been shown to more safely and effectively treat fungal and parasitic infections in immunocompetent and immunosuppressed animals than comparable doses of Fungizone. For example, in a murine model of leishmaniasis, AmBisome was as effective at 0.26 mg/kg as Fungizone at 0.95 mg/kg in reducing microbial burden in the liver by 50%.[15] In immunosuppressed rabbits with pulmonary aspergillosis, 1 mg/kg Fungizone increased 10 d survival only 30% whereas an equal dose of AmBisome increased survival by 80%.[20] In some models of infection, equal doses of Fungizone and AmBisome gave comparable efficacy, e.g., treatment of infections due to *Candida albicans*,[18] *Trichosporon beigelii*,[45] *Cryptococcus neoformans*[18,22] *Coccidiodes immitis*[46] and *Histoplasma capsulatum*.[23]

When AmBisome has been tested at higher doses (3-30 mg/kg), which are well tolerated due to its reduced toxicity, it has shown equal efficacy in some studies and better efficacy in other studies compared to Fungizone. The advantage associated with using higher nontoxic doses of AmBisome is that elimination of fungus is often not achieved with nonlethal doses of Fungizone whereas treatment with high doses of AmBisome can eradicate the fungus. For example, animals infected with *Blastomyces dermatitidis*[24] or *Paracoccidioides brasiliensis*[25] were cured at a rate of 100% or 47% by AmBisome treatment at 25 or 30 mg/kg, respectively. In comparison there were 0-10% cures with Fungizone treatment of *B. dermatitidis* infected mice and no cures with Fungizone treatment of *P. brasiliensis* infected mice. Likewise, *Cryptococcus neoformans* could not be recovered from the brains of 44% of mice treated at 20 mg/kg or 78% of mice treated at 30 mg/kg AmBisome. Fungizone did not clear the infection.[22]

Therefore, although AmBisome retains or enhances the activity of ampB for the treatment of fungal infections in animal models, the results will vary depending on several factors, including the infection model, fungal isolate, dosing schedule, time of evaluation of efficacy, and immune status of the animals. These same parameters need to be considered in clinical applications of AmBisome.

Clinical Efficacy

Clinical studies of AmBisome have demonstrated a substantially enhanced therapeutic index as compared to Fungizone. An early multicenter clinical trial examined the use of AmBisome as salvage therapy in immunocompromised patients with suspected or proven fungal infections who had failed previous antifungal therapy, or had renal insufficiency or toxicity.[29] The median daily dose level of AmBisome was 2.2 mg/kg (range 0.5-5 mg/kg). Out of a total of 126 patients, there were 64 cases with proven invasive infections. Of these, 37 (58%) were cured, 12 (19%) improved, and 15 (23%) failed to respond. From these results it was concluded that AmBisome was effective in the majority of patients with invasive fungal infections.

AmBisome has also been studied in organ transplant patients who develop opportunistic fungal infections during the time that they are heavily immunosuppressed. Tollemar et al reported on the use of AmBisome in eight organ or bone marrow transplant patients that developed suspected deep fungal infections.[30] AmBisome was administered at a maximum daily dose between 0.9 and 2.5 mg/kg. Five of these patients had fever and chills and/or elevated creatinine levels while receiving Fungizone. When switched to AmBisome, six of the patients were mycologically cured. Of the two patients that died, one was negative for fungal infection at autopsy and the other was not tested. With the exception of cholestasis and elevated alkaline phosphatase in two patients, there were no adverse effects attributed to the AmBisome therapy in this study.

AmBisome has also been investigated in neutropenic patients with suspected or confirmed fungal infections. In one study report, 116 patients who had failed or could not tolerate Fungizone were treated with AmBisome with a median duration of 12 d (range 2-96 d) and with a median total dose of 1684 mg (range 180-10440 mg).[31] Overall, 61% of the patients treated with AmBisome recovered. In 21 patients with proven aspergillosis, 13 (62%) obtained complete or partial resolution of the clinical and radiological signs of infection. In this study adverse events were rare, even in patients receiving 5 mg/kg/day. No clinically significant deterioration in serum urea or creatinine resulted from the AmBisome therapy but electrolyte replacement was required in some patients. Abnormal hepatic function possibly attributable to AmBisome was noted in 17% of the 133 treatment episodes but was severe enough to warrant discontinuation of therapy in only 2 episodes (< 2%).

The treatment of opportunistic fungal infections in AIDS patients with AmBisome has also provided encouraging results. Schurmann et al described the successful treatment of mycologically proven cryptococcal meningitis in three patients.[32] In this case no drug-related hematological toxicity was observed. A multicenter study used AmBisome in AIDS patients with cryptococcosis.[33] Patients were treated at 3 mg/kg/day for at least 42 d. Of the 23 enrolled patients evaluable for clinical efficacy, 18 (78%) were either cured or improved. Of 21 patients that were mycologically evaluable, 14 (67%) were cured. Again, AmBisome was well tolerated in these patients.

The results of a randomized trial of AmBisome (4.0 mg/kg) versus ampB (0.7 mg/kg) for cryptococcal meningitis in HIV positive patients was recently reported.[34] The median time to achieve sterilization of the CSF was 14 d for AmBisome compared to > 21 d for ampB. However, the treatment failures (23%) in both groups were identical. Serum creatinine increases were significantly greater ($p = 0.02$) in the amp B group and two patients had to discontinue therapy due to toxicity.

The improved safety profile of AmBisome has led to other clinical applications with reported success. Viviani et al reported the successful treatment of a *Fusarium* infection in an acute lymphoblastic leukemia patient.[35] Fisher and co-workers reported the successful treatment of advanced mucormycosis in a diabetic patient.[36] An open trial of AmBisome for the treatment of severe drug-resistant visceral leishmaniasis was conducted in Sudan.[37] In this study, 49 patients were entered, of which 6 died of causes not attributed to AmBisome therapy. The patients were treated with one of three regimens with an overall cure rate of 29/43 (67%). The optimal regimen was found to be a course of 4 mg/kg given six times over a two week period.

There have been several preliminary reports of prospective comparative clinical trials with AmBisome. In one study,[27] 193 febrile patients were randomized to receive ampB (1 mg/kg/day), AmBisome (1 mg/kg/day) or AmBisome (3 mg/ kg/ day). Resolution of fever was not significantly different between the three treatment groups. In the confirmed mycoses, the clinical and mycological cure rates were higher with AmBisome treatment, although statistical significance was not reported. Another empiric therapy trial was done in 203 neutropenic pediatric patients to compare the safety and efficacy of the same three dosing regimens described above in children.[28] Resolution of fever and complete resolution of infection was statistically higher ($p = 0.02$) in both AmBisome groups compared to the Fungizone treatment group. The overall incidence of adverse events was higher in the ampB group ($p = 0.02$), but the prevalence of rash, chills and nephrotoxicity was not significantly different. The incidence of hypokalemia, however, was significantly lower ($p = 0.035$) in both AmBisome groups.

The numerous clinical studies of AmBisome have demonstrated that it is often tolerated when Fungizone is not, while maintaining antifungal efficacy. Additional controlled clinical evaluations of AmBisome are in progress. The existing reports summarized here clearly indicate that AmBisome has a place in the treatment of serious life-threatening fungal infections.

Mechanism of Action

The above and other studies have shown that the AmBisome formulation of ampB retains its ability to inhibit the in vivo replication of a broad range of both intracellular and extracellular microbes. To elucidate the antimicrobial mechanism of action, several studies were done, both in vitro and in vivo.

The tissue distribution has been studied by fluorescent microscopy using a modified form of AmBisome containing an aqueous fluorescent dye entrapped inside the liposomes.[1] Kidney tissue sections were taken from *Candida albicans*-infected animals, treated with this fluorescently modified form of AmBisome; controls consisted of unmodified AmBisome and liposomes containing the fluorescent dye but lacking ampB. The sections from the mice given the fluorescent liposomes showed bright fluorescence, above the autofluorescent background seen with unmodified AmBisome, associated with sites of infection. Thus, the AmBisome localized at fungal infection sites in the kidney. However, this did not provide direct evidence of the interactions between the liposomes and the fungi.

Investigations of potential interactions between AmBisome and fungal cells were done using in vitro incubation for 24 h. Fluorescent dye labeled AmBisome incubated with fungal cells showed the dye distributed throughout the fungal cytoplasm, but fluorescent dye remained on the surface of the cells when liposomes lacking ampB were used. Free dye failed to produce cytoplasmic fluorescence. These results suggest that AmBisome can bind to the surface of fungal cells with subsequent breakdown, release of drug and entry into the fungal cytoplasm.[1,47] Visualization of liposomes binding to fungal cells was demonstrated using freeze-fracture analysis of *Candida glabrata* incubated with AmBisome or liposomes lacking ampB (Fig. 13.1).[1] Subsequent studies using a lipid-based fluorescent label were performed to evaluate lipid localization after AmBisome binding to the fungal cells. The results in Figure 13.2 show fluorescent lipid initially on the cell surface and with continued incubation suggest that lipid penetration into the cytoplasm occurs only when ampB is present in the labeled liposomes.[47] Further studies of lipid distribution were performed at the ultrastructural level by incorporation of a small amount of lipid, bound with gold particles, into the ampB liposomes or the liposomes lacking ampB. The results (Fig. 13.3) show that lipid from the liposomes without ampB did not penetrate into the fungal cytoplasm whereas the lipid from the liposomes with ampB did.[47]

Taken together, these and all the other data described, suggest that AmBisome, a stable liposome form of ampB, is able to effectively redistribute the drug away from sensitive tissues yet still provide activity toward fungal infections in vivo. The in vitro studies suggest that interaction of AmBisome with fungal cells is followed by liposome bilayer disruption and delivery of ampB to the fungal cells. This may explain how AmBisome can maintain ampB efficacy while stably incorporating ampB in its cohesive bilayer.

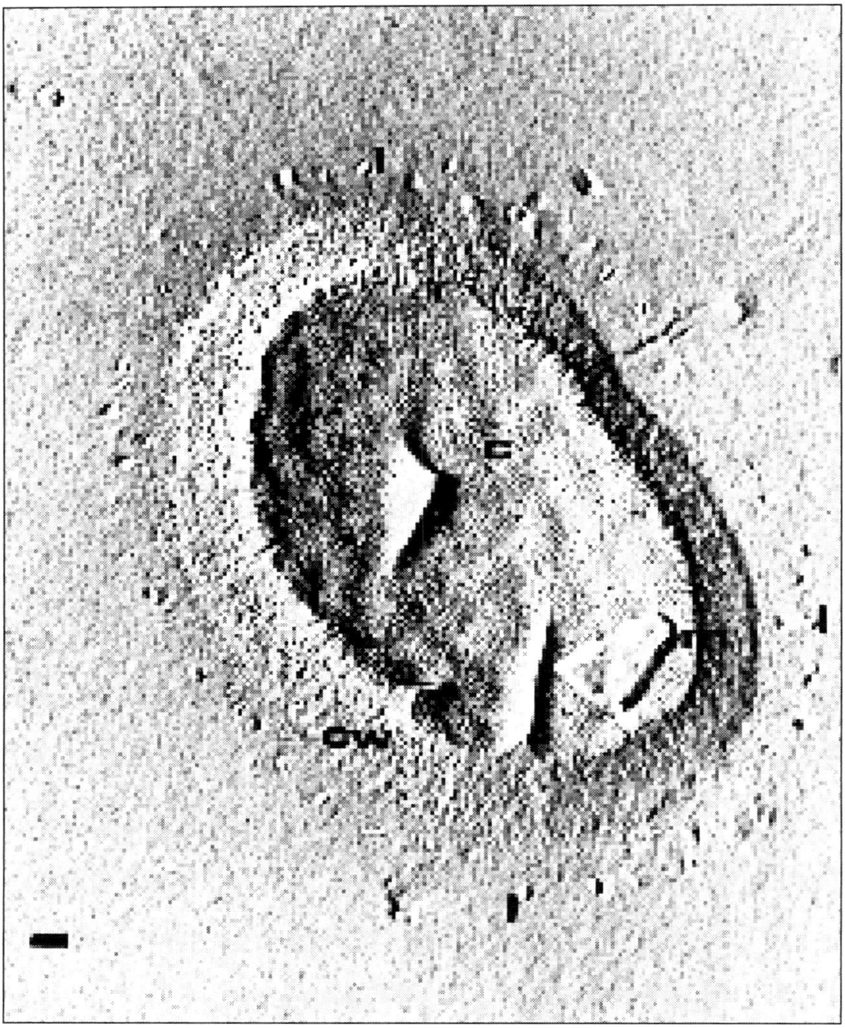

Fig. 13.1. Freeze fracture preparation of the yeast, *Candida glabrata*, incubated for four hours at 37°C with AmBisome. 1 = liposomes; cw = yeast cell wall; m = yeast cell membrane; c = yeast cytoplasm; Bar = 90 nm. (Courtesy of Kevin Franke and Steve Karl, California State University, Fullerton).

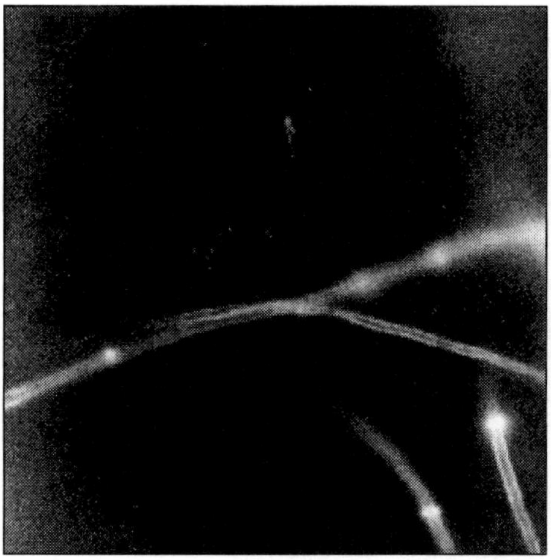

Fig. 13.2. (a) *Aspergillus fumigatus* incubated for 24 h with L-Rho-PE liposomes without drug showing lipid (red fluorescence) from the liposomes on the surface of the hyphae. (b) *Aspergillus fumigatus* incubated for 4 h with L-Rho-PE AmBisome showing lipid (red fluorescence) from the liposomes in the cytoplasm and on the surface of the hyphae. (c) *Aspergillus fumigatus* incubated for 24 h with L-Rho-PE AmBisome showing lipid (red fluorescence) from the liposomes localized in the cytoplasm.

Fig. 13.2. (a)

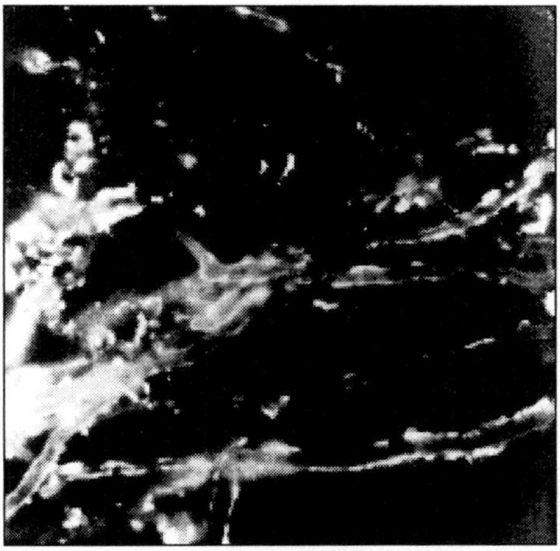

Fig. 13.2. (b)

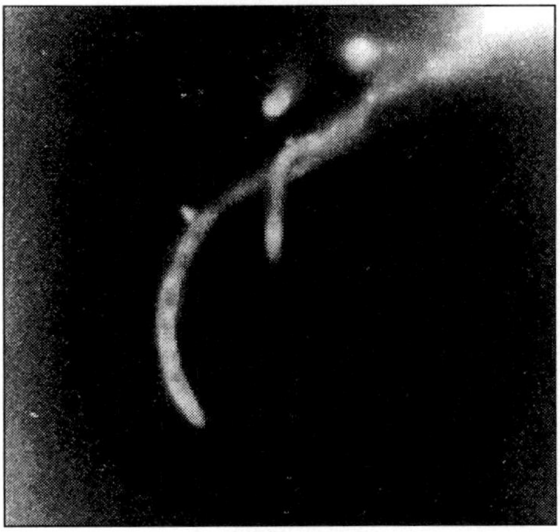

Fig. 13.2. (c)

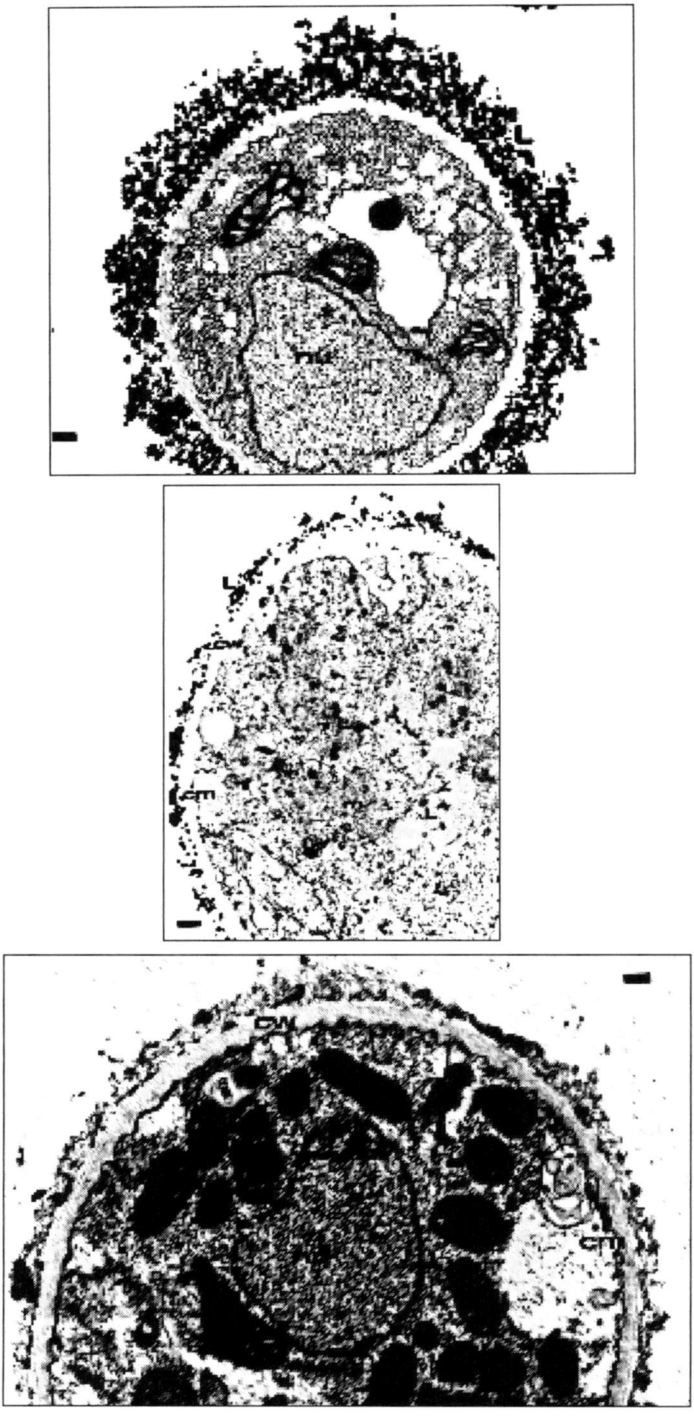
L
CW
CM

Conclusions

The incorporation of amphotericin B into the cohesive bilayer of AmBisome, composed of hydrogenated phospholipids, cholesterol and DSPG, produces a stable, long circulating, small unilamellar liposome. This strong association of ampB within the AmBisome bilayer is largely responsible for markedly reducing both the cellular and in vivo toxicity of this drug in preclinical and clinical studies. As AmBisome, the pharmacokinetics of ampB are altered to produce much higher C_{max} and $AUC_{0-\infty}$ and significantly smaller volume of distribution and total clearance of the drug. Although the toxicity of ampB for animals is significantly decreased when formulated as AmBisome, the drug's efficacy is still retained. AmBisome shows a broad range of antimicrobial in vitro activity and in vivo, it has equivalent or better efficacy than ampB depending upon the preclinical infection model. In comparative clinical trials, at comparable dose levels, AmBisome has been shown to be as efficacious as Fungizone in treating fungal infections in both children and adults.

References

1. Adler-Moore J, Proffitt RT. Development, characterization, efficacy and mode of action of AmBisome,® a unilamellar liposomal formulation of amphotericin B. J Liposome Research 1993; 3(3):429-450.
2. Utz JP. Amphotericin B toxicity: general side effects. Ann Int Med 1964; 61:340-354.
3. McCurdy KK, Frederic M, Elkinton JR. Renal tubular acidosis due to amphotericin B. N Eng J Med 1968; 278:124-131.
4. Proffitt RT, Adler-Moore J, Fujii G et al. Stability and mode of action of AmBisome (liposomal amphotericin B). J Controlled Release 1994; 28:342-343.
5. McAndrews BJ, Lee MJA, Adler-Moore JP. Comparative toxicities of fungizone and AmBisome for cultured kidney cells and macrophages. General Proceeding of the ASM Meeting for American Society for Microbiology. Atlanta, Georgia 1993:11 (Abstract).
6. Proffitt RT, Satorius A, Chiang S et al. Pharmacology and toxicology of a liposomal formulation of amphotericin B (AmBisome) in rodents. J Antimicrobial Chemotherapy. 1991; 28 (Suppl. B):49-61.
7. Fielding RM, Singer AW, Wang LH et al. Relationship of pharmacokinetics and drug distribution in tissue to increased safety of amphotericin B colloidal dispersion in dogs. Antimicrobial Agents Chemotherapy 1992; 36:299-307.
8. Meunier F, Prentice HG, Ringden O. Liposomal amphotericin B (AmBisome): safety data from a phase II/III clinical trial. J Antimicrobial Chemotherapy 1991; 28(suppl B):83-91.

Fig. 13.3. See opposite page. (a) *Aspergillus fumigatus* incubated for 14 h with gold-labeled liposomes without drug showing lipid from the liposomes in association with the surface of the fungal cell wall. Nu = nucleus; L = gold-labeled lipid of liposomes; M = mitochondria. Bar = 200 nm. (b) *Aspergillus fumigatus* incubated for 14 h with gold-labeled AmBisome showing lipid from the liposomes in association with the fungal cell wall, penetrating through the cell wall and lipid accumulating in the cytoplasm. CW = cell wall; CM = cell membrane; L = gold-labeled lipid of liposomes; M = mitochondria. Bar = 200 nm. (c) *Aspergillus fumigatus* incubated for 14 h with phosphate buffer showing characteristic fungal morphology. CW = cell wall; CM = cell membrane; Nu = nucleus. Bar = 200 nm.

9. Ringden O, Andstrom E, Remberger M et al. Safety of liposomal amphotericin B (AmBisome) in 187 transplant recipients treated with cyclosporin. Bone Marrow Transplantation 1994; 14(Suppl.5):S10-S14.
10. Turner AF, Presant CA, Proffitt RT et al. In-111-labeled liposomes: dosimetry and tumor depiction. Radiology 1988; 166(3):761-765.
11. Lee JW, Amantea MA, Francis PA et al. Pharmacokinetics and safety of a unilamellar liposomal formulation of amphotericin B (AmBisome) in rabbits. Antimicrobial Agents Chemotherapy 1994; 38(4):713-718.
12. Hiles R, Bekersky I. Pharmacokinetics (PK) of liposomal amphotericin B (AmBisome) in rats. Pharmaceutical Research 1993; 10(10):S-372 (Abstract #PPDM 8295).
13. Bekersky I, Hiles R. Pharmacokinetics (PK) of liposomal amphotericin B (AmBisome) in dogs. Pharmaceutical Research 1993; 10(10):S-372 (Abstract # PPDM 8296).
14. Walsh TJ, Bekersky I, Yeldandi V et al. Pharmacokinetics of AmBisome in persistently febrile neutropenic patients receiving empirical antifungal therapy. Proceedings of the 35th Interscience Conference on Antimicrobial Agents and Chemotherapy. Abstract A13. 1995:3.
15. Croft SL, Davidson RNN, Thorton EA. Liposomal amphotericin B in the treatment of visceral leishmaniasis. J Antimicrobial Chemotherapy 1991; 28 (Suppl B):111-118.
16. Gangneux JP, Sulahian A, Garin Y et al. Therapy of visceral leishmaniasis due to *Leishmania infantum*: experimental assessment of efficacy of AmBisome. Antimicrobial Agents Chemotherapy 1996; 40(5):1214-1218.
17. Van Etten EWM, Van den Heuvel-de Groot C, Bakker-Woudenberg, IAJM. Efficacies of amphotericin B-desoxycholate (Fungizone), liposomal amphotericin B (AmBisome) and Fluconazole in the treatment of systemic candidosis in immunocompetent and leukopenic mice. J Antimicrobial Chemotherapy 1993; 32:723-739.
18. Adler-Moore JP, Chiang S, Satorius A et al. Treatment of murine candidosis and cryptococcosis with a unilamellar liposomal amphotericin B formulation (AmBisome). J Antimicrobial Chemotherapy 1991; 28(Suppl. B):63-71.
19. Karyotakis NC, Anaissie, E J. Efficacy of escalating doses of liposomal amphotericin B (AmBisome) against hematogenous *Candida lusitaniae* and *Candida krusei* infection in neutropenic mice. Antimicrobial Agents Chemotherapy 1994; 38(11): 2660-2662.
20. Francis P, Lee JW, Hoffman A et al. Efficacy of unilamellar liposomal amphotericin B in treatment of pulmonary aspergillosis in persistently granulocytopenic rabbits: the potential role of D-mannitol and galactomannan as markers of infection and therapeutic response. J Infectious Diseases 1994; 169:356-368.
21. Allen SD, Sorenson KN, Nejdl MJ et al. Prophylactic efficacy of aerosolized liposomal (AmBisome) and non-liposomal (Fungizone) amphotericin B in murine pulmonary aspergillosis. J Antimicrob Chemotherapy 1994; 34:1001-1013.
22. Albert MM, Stahl-Carroll TL, Luther MF et al. Comparison of liposomal amphotericin B to amphotericin B for treatment of murine cryptococcal meningitis. J Mycol Med 1995; 5:1-6.
23. Graybill JR, Boccanegra R. Liposomal amphotericin B therapy of murine histoplasmosis. Antimicrobial Agents Chemotherapy 1995; 39 (8):1885-1887.
24. Clemons KV, Stevens DA. Therapeutic efficacy of a liposomal formulation of amphotericin B (AmBisome) against murine blastomycosis. J Antimicrobial Chemotherapy 1993; 32:465-472.
25. Clemons KV, Stevens DA. Comparison of a liposomal amphotericin B formulation (AmBisome) and desoxycholate amphotericin B (Fungizone) for the treatment of murine paracoccidioidomycosis. J Med Vet Mycol 1993; 31:387-394.

26. Garcia AL. Efficacy of single high dose AmBisome for treatment of murine candidiasis and histoplasmosis. Masters thesis. Pomona CA: California State Polytechnic University, Pomona, 1996;
27. Prentice HG, Catovsky D, Aoun M et al. AmBisome versus amphotericin B in patients with fever unresponsive to antibiotic therapy for 96 h, or with confirmed fungal infection. 2nd International Symposium on Febrile Neutropenia. 1995; Abstract 35.
28. Hann IM, Stevens RF, Pinkerton CR et al. Safety and efficacy of two dose regimes of AmBisome versus amphotericin B as empiric antifungal treatment in neutropenic pediatric patients. 21st Annual Meeting of the European Group for Blood & Bone Marrow Transplantation & 11th Meeting of the Nurses Group, Davos, Switzerland, 1995; Abstract.
29. Ringden O, Meunier F, Tollemar J et al. Efficacy of amphotericin B encapsulated in liposomes (AmBisome) in the treatment of invasive fungal infections in immunocompromised patients. J Antimicrobial Chemotherapy 1991; 28 (suppl B):73-82.
30. Tollemar J, Ringden O, Tyden G. Liposomal amphotericin B (AmBisome) treatment in solid organ and bone marrow transplant recipients. Efficacy and Safety Evaluation. Clinical Transplantation 1990; 4:167-175.
31. Mills W, Chopra R, Linch DC et al. Liposomal amphotericin B in the treatment of fungal infections in neutropenic patients: a single-centre experience of 133 episodes in 116 patients. British Journal Haematology 1994; 86:754-760.
32. Schurmann D, De Matos Marques B, Grunewald T et al. Safety and efficacy of liposomal amphotericin B in treating AIDS-associated disseminated cryptococcosis. J Infectious Diseases 1991; 164:620.
33. Coker RJ, Viviani M, Gazzard BG et al. Treatment of cryptococcosis with liposomal amphotericin B (AmBisome) in 23 patients with AIDS. AIDS 1993; 7:829-835.
34. Leenders ACAP, Reiss P, Portegies P et al. A randomized trial of liposomal amphotericin B (AmBisome) 4 mg/kg versus amphotericin B 0.7 mg/kg for cryptococcal meningitis in HIV infected patients. Proceedings of the 36th Interscience Conference on Antimicrobial Agents Chemotherapy. 1996; Abstract LM35.p 287.
35. Viviani MA, Cofrancesco E, Boschetti C et al. Eradication of *Fusarium* infection in a leukopenic patient treated with liposomal amphotericin B. Mycoses 1991; 34:255.
36. Fisher EW, Toma A, Fisher PH et al. Rhinocerebral mucormycosis: use of liposomal amphotericin B. J Laryngology and Otology 1991;105:575.
37. Seaman J, Boer C, Wilkinson R et al. Liposomal amphotericin B (AmBisome) in the treatment of complicated kala-azar under field conditions. Clin Inf Dis 1995; 21:188-193.
38. Emminger W, Graninger W, Emminger-Schmidmeier W et al. Tolerance of high doses of amphotericin B by infusion of a liposomal formulation in children with cancer. Ann Hematol 1994; 68:27-31.
39. Zoubek A, Emminger W, Emminger-Schmidmeier W et al. Conventional vs. liposomal amphotericin B in immunosuppressed children. Pediatric Hematology Oncology 1992; 9:187-190.
40. Bindschadler DD, Bennett JE. A pharmacologic guide to the clinical use of amphotericin B. J Infectious Diseases 1969; 120(4):427-436.
41. Daneshmend TK, Warnock DW. Clinical pharmacokinetics of systemic antifungal drugs. Clinical Pharmacokinetics 1983; 8:17-42.
42. Anaissie E, Paetznick V, Proffitt R et al. Comparison of the in vitro antifungal activity of free and liposome-encapsulated amphotericin B. Eur J Clin Microbiol Infect Dis 1991; 10:665-668.

43. Garcia AL, Satorius A, Proffitt RT et al. Prophylaxis with single high dose AmBisome (AmBi) minimizes murine histoplasmosis and candidiasis. Abstract, 95th General Meeting of the American Society for Microbiology, Washington DC 1995:164.
44. Karyotakis NC, Anaissie EJ, Hachem R et al. Comparison of the efficacy of polyenes and triazoles against hematogenous *Candida krusei* infection in neutropenic mice. J Infectious Diseases 1993;168:1311-1313.
45. Anaissie EJ, Hachem R, Karyotakis NC et al. Comparative efficacies of amphotericin B, triazoles, and combination of both as experimental therapy for murine trichosporonosis. Antimicrobial Agents Chemotherapy 1994; 38(11):2541-2544.
46. Albert MM, Adams K, Luther MJ et al. Efficacy of AmBisome in murine coccidioidomycosis. J Med Vet Mycol 1994; 32:467-471.
47. Adler-Moore JP. AmBisome targeting to fungal infections. Bone Marrow Transplantation 1994; 14 (Suppl 5):S3-S7.

CHAPTER 14

Steric Stabilization of Liposomes Containing Amphotericin B: A Future Direction in Antifungal Therapy?

Els W.M. van Etten, Irma A.J.M. Bakker-Woudenberg

In the past ten years lipid formulations of amphotericin B (AMB) have been developed by the pharmaceutical industry with the primary aim of reduction of AMB's toxicity. At present, three different industrially produced AMB-lipid formulations are available, with substantial lower toxicity than conventional AMB (Fungizone®). One formulation is AmBisome®, and is discussed in chapter 13 of this book.

Fundamental studies on AmBisome® show that next to substantial uptake of AmBisome® by the macrophages of the mononuclear phagocyte system (MPS) in the liver and spleen, intact AmBisome® can reach sites of infection outside MPS-tissues. At the site of infection direct interaction between AmBisome® and the fungal cell may occur or AMB may be released from AmBisome® in close vicinity of the fungus.[1]

Therefore, the presence of a substantial amount of intact liposomes at the site of infection seems to be highly important for antifungal efficacy. It has been previously shown that a relatively long blood residence time of liposomes is needed for increased localization of intact liposomes at sites of infection outside MPS-tissues.[2] The blood residence time of AmBisome®, that has a small particle size and rigid bilayers, is primarily dependent on the lipid dose administered. AmBisome® shows a prolonged blood circulation time only at relatively high lipid doses.[3] By using sterically stabilized liposomes a prolonged blood circulation time of liposomes can be achieved, without the constraints of high lipid dose, small particle size, or rigid nature of the bilayers.

At our laboratory we prepared sterically stabilized liposomes containing AMB. In these liposomes AMB was complexed to PEG-DSPE.[4] The liposomes consisted of PEG-DSPE:HSPC:Chol:AMB in a molar ratio of 0.21:1.79:1:0.32, further referred to as PEG-AMB. In Figure 14.1 the blood residence of conventional AMB, AmBisome® and PEG-AMB is shown.

For AmBisome® blood residence time is primarily dependent on the lipid dose administered.[5] The elimination half-life of AmBisome® in mice was about 8 h at a lipid dose of 66 μmol lipid/kg.[5] Further increase of the blood residence time, independent of the lipid dose, was achieved with PEG-AMB, the elimination half-life being approximately 20 h for a dose range of 4.7-47 μmol lipid/kg.[6]

Long Circulating Liposomes: Old Drugs, New Therapeutics, edited by Martin C. Woodle and Gerrit Storm.

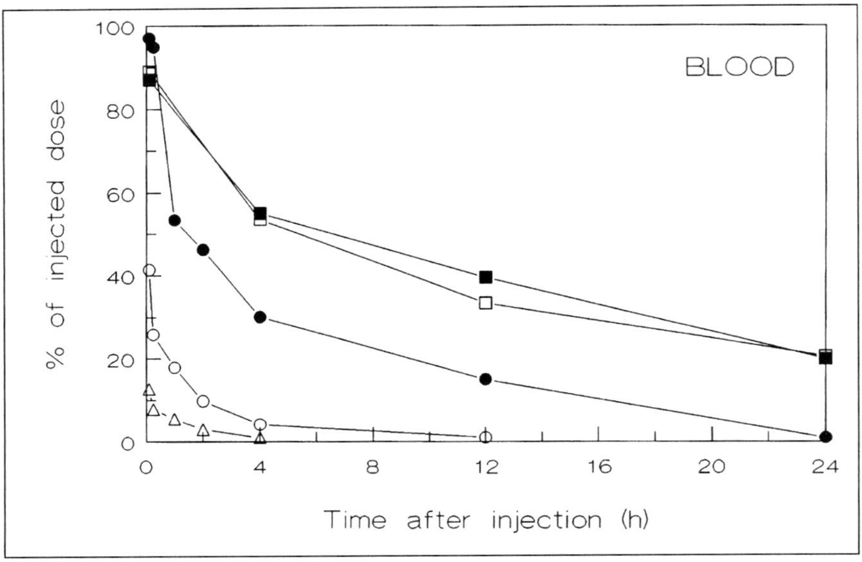

Fig. 14.1. Blood levels of AMB and AMB-liposomes after a single dose iv of conventional AMB, AmBisome® or PEG-AMB in uninfected mice. Δ, conventional AMB at 0.3 mg AMB/kg; O, AmBisome® at 0.3 mg AMB/kg (2.8 µmol lipid/kg); ●, AmBisome® at 7 mg AMB/kg (66 µmol lipid/kg); □, PEG-AMB-LIP at 0.5 mg AMB/kg (4.7 µmol lipid/kg); ■, PEG-AMB-LIP at 5 mg AMB/kg (47 µmol lipid/kg). Each symbol represents the mean of two experiments. Blood samples from five mice were pooled.

Table 14.1. Toxicity versus antifungal activity of conventional AMB, AmBisome®, and PEG-AMB.

In vivo toxicity	AMB formulation	In vitro antifungal activity
MTD[a] mg AMB/kg		AMB conc (mg/l) required for >99.9% killing of *C. albicans* after 6 h exposure
0.8	conventional AMB	0.2
>31	AmBisome®	12.8
9	PEG-AMB	0.4

[a]MTD = maximum tolerated dosage in uninfected mice

Besides the relatively long blood circulation time of PEG-AMB, another characteristic of this formulation, important for therapeutic efficacy was the high antifungal activity. Comparative data on in vivo toxicity and in vitro antifungal activity for conventional AMB, AmBisome®, and PEG-AMB are shown in Table 14.1.

These data clearly show that for AmBisome® the reduction of AMB's toxicity following liposomal encapsulation is associated with a substantial loss of antifungal

Table 14.2. ***Effect[a] of single dose PEG-AMB versus single- or multidose AmBisome® on survival of mice and killing of*** **Candida albicans** ***in the kidney in leukopenic mice.***

		Start of treatment	Day 3[b]		Day 7[b]	
Treatment	No. of doses	Log_{10} no. *Candida* in kidney[c]	Survival (%)	Log_{10} no. *Candida* in kidney of surviving mice[c]	Survival (%)	Log_{10} no. *Candida* in kidney of surviving mice[c]
		4.2 ± 0.2				
PEG-AMB	1		100	3.1 ± 0.7[d]	100	2.0 ± 0.4[d]
AmBisome®	1		70	5.3 ± 0.6	40	3.9 ± 0.5
AmBisome®	5		60	5.4 ± 0.9	30	4.6 ± 1.1
placebo	1		50	6.6 ± 0.2	10	7.4

[a]Leukopenic mice were inoculated iv at zero time with 3×10^4 *C. albicans*. Treatment was started at 20 h after *C.albicans* inoculation. PEG-AMB was administered iv as a single dose at 5 mg AMB/kg. AmBisome® was administered iv as a single dose, or daily for five consecutive days at 5 mg AMB/kg/day.
[b]Effect of treatment was determined at day 3 (n = 10) as well as day 7 (n = 10) after inoculation of *C. albicans*.
[c]Each value represents the geometric mean ± SD.
[d]$p \leq 0.01$ compared with number of viable *C. albicans* at start of treatment.

activity.[6] It should also be emphasized that in a number of animal models of fungal infection high dosages of AmBisome® were needed for treatment to be effective, especially when the animals were severely immunocompromised.[7] However, with PEG-AMB it is now possible to reduce the toxicity of AMB by liposomal encapsulation without reducing its antifungal activity.[6]

Thus, the PEG-AMB formulation shows two characteristics that are expected to be important for improved antifungal efficacy: low toxicity while maintaining high antifungal activity, and prolonged circulation time of intact liposomes in blood, that is independent of the dose administered. Supporting evidence for improved efficacy of PEG-AMB in the treatment of fungal infections was obtained in two different animal models of invasive fungal infections. Clinically relevant issues including persistent leukopenia and dissemination of infection were addressed. It was shown in an animal model of severe invasive *Candida albicans* infection in persistently leukopenic mice (Table 14.2), that only treatment with PEG-AMB, given as a single dose, resulted in survival of all animals and killing of viable *Candida albicans* in the kidney within a short period of time after *Candida* infection.[6] This effect could not be achieved with AmBisome® at an equivalent dose, not even when administered repeatedly.[6,8]

In a model of *Aspergillus fumigatus* one-sided pulmonary infection in persistently leukopenic rats[9] it was demonstrated that survival of the animals was significantly prolonged after only a single dose of PEG-AMB.[10] Similar therapeutic efficacy in this infection model was reported for AmBisome®, when administered repeatedly.[9]

We anticipate that sterically stabilized liposome formulations containing AMB may be of substantial value for the treatment of severe invasive fungal infections in humans. It requires thorough investigation, therefore further development of AMB-liposome-technology remains highly relevant.

References

1. Adler-Moore JP, Proffitt RT. Development, characterization, efficacy and mode of action of AmBisome®, a unilammelar formulation of amphotercin B. J Liposome Res 1993; 3:429-450.
2. Bakker-Woudenberg IAJM, Lokerse AF, Ten Kate MT, Mouton JW, Woodle MC, Storm G: Liposomes with prolonged blood circulation and selective localization in *Klebsiella pneumoniae* infected lung tissue. J Infect Dis 1993; 168:164-171.
3. AmBisome® Liposomal Amphotericin B, Product Monograph. NeXstar Pharmaceuticals, Inc. 1994.
4. Van Etten EWM, Van Vianen W, Tijhuis RHG, Storm G, and Bakker-Woudenberg IAJM. Sterically stabilized amphotericin B-liposomes: toxicity and biodistribution in mice. J Control Release 1995; 37:123-129.
5. Van Etten EWM, Otte-Lambillion M, Van Vianen W, Ten Kate MT, Bakker-Woudenberg IAJM. Biodistribution of liposomal amphotericin B (AmBisome) versus amphotericin B-desoxycholate (Fungizone) in immunocompetent uninfected mice as well as in leucopenic mice infected with *Candida albicans.* J Antimicrob Chemother 1995; 35:509-519.
6. Van Etten EWM, Ten Kate MT, Stearne-Cullen LET, Bakker-Woudenberg IAJM: Amphotericin B liposomes with prolonged circulation in blood: in vitro antifungal activity, toxicity, and efficacy in systemic candidiasis in leukopenic mice. Antimicrob Agents Chemother 1995; 39:1954-1958.
7. Leenders ACAP, De Marie S. The use of lipid formulations of amphotericin B for systemic fungal infections. Leukemia 1996; 10:1570-1575.

8. Van Etten EWM, Snijders SV, Verbrugh HA, Bakker-Woudenberg IAJM. Efficacy of pegylated long circulating amphotericin B-liposomes versus AmBisome® in the treatment of systemic candidiasis in leukopenic mice in relation to the severity of infection. (manuscript in preparation)
9. Leenders ACAP, de Marie S, Ten Kate MT, Bakker-Woudenberg IAJM, and Verbrugh HA. Liposomal amphotericin B (AmBisome®) reduces dissemination of infection as compared to amphotercin B deoxycholate (Fungizone®) in a newly developed animal model of one-sided pulmonary aspergillosis. J Antimicrob Chemother 1996; 38:215-225.
10. Van Etten EWM, Stearne-Cullen LET, Snijders SV, Verbrugh HA, Bakker-Woudenberg IAJM. Efficacy of pegylated long circulating amphotericin B-liposomes in the treatment of pulmonary aspergillosis in leukopenic rats. (Manuscript in preparation).

CHAPTER 15

Clinical and Preclinical Studies with Low-Clearance Liposomal Amikacin (MiKasome®)

Robert M. Fielding, Geoffrey Mukwaya, Robert A. Sandhaus

Introduction

The introduction of aminoglycoside antibiotics in the mid-twentieth century led to dramatic changes in the treatment of tuberculosis followed by improved therapy of the emerging problem of Gram-negative bacterial infections. Clinical experience soon revealed that although the aminoglycosides possess a wide spectrum of antimicrobial activity, they may also produce serious, dose-related damage to the kidneys and the auditory/vestibular apparatus. As a result, aminoglycosides require therapeutic drug monitoring to maintain their plasma concentrations within a narrow therapeutic window, and aminoglycosides are now often reserved for the most severe infections despite their excellent antimicrobial properties.

The encapsulation of aminoglycosides in liposomes has been widely investigated (reviewed in ref. 1, and in chapter 12 of this book). The increased efficacy against intracellular organisms, including mycobacteria observed in these studies is attributed to the rapid uptake of circulating liposomes by the mononuclear phagocyte system (MPS). However, the potential of liposomes to improve the therapeutic profile of aminoglycosides against extracellular infections remains largely unexplored, as does the impact of differences between liposomal aminoglycoside formulations on their therapeutic potential.

Most previous studies with liposomal aminoglycosides used conventional (high-clearance) liposome formulations. Conventional liposomes may be limited in their effectiveness against extracellular infections due to their short half-life and rapid MPS uptake. Small, low-clearance liposomes might improve efficacy against extracellular organisms by achieving a wider tissue distribution, better penetration to sites of inflammation and longer duration of action. MiKasome, a formulation of the aminoglycoside amikacin in small unilamellar liposomes, was developed to evaluate the possibility that substantial improvements in the safety and pharmacology of an aminoglycoside could be obtained using small, low-clearance liposomes.

Long Circulating Liposomes: Old Drugs, New Therapeutics, edited by Martin C. Woodle and Gerrit Storm.

Methods

Materials

MiKasome® (amikacin liposome injection), a sterile intravenous formulation of amikacin in small unilamellar liposomes (median diameter < 100 nm) was supplied by NeXstar Pharmaceuticals, Inc., Boulder, CO. Each vial of MiKasome contained approximately 500 mg of amikacin base, 2000 mg of hydrogenated soy phosphatidylcholine (HSPC), 100 mg of distearoylphosphatidyl glycerol (DSPG), and 500 mg of cholesterol in a 9% sucrose/succinate buffer (pH 6). The concentration of amikacin in the formulation was approximately 10 mg/ml.

Preclinical Studies

MiKasome was administered by intravenous injection or infusion to rats, rabbits, dogs and rhesus monkeys at doses between 2.5 mg/kg and 360 mg/kg. In some studies, conventional amikacin sulfate solution was administered intravenously as a control. For determination of pharmacokinetic parameters, samples of plasma, urine and tissues were obtained from animals at various timepoints and frozen. Total amikacin was measured in plasma samples by adding detergent to disrupt liposomes prior to the determination of amikacin concentration. In some studies, concentrations of free amikacin in plasma were estimated by measuring amikacin in ultrafiltrates of plasma (Centrifree®, Amicon, Inc.). Amikacin concentrations in all samples were measured by a fluorescence polarization immunoassay with a limit of quantitation of ≤ 1 µg/ml. Plasma pharmacokinetic parameters were obtained from nonlinear, least-squares fits to plasma concentration vs. time data (RSTRIP, MicroMath), using model-independent calculation methods. For characterization of MiKasome safety, groups of 3 male and 3 female dogs received daily or every-third-day injections of MiKasome, empty liposomes or conventional amikacin over a one-month period. Safety parameters monitored included weekly clinical pathology (clinical chemistry, hematology, urinalysis and coagulation), body weights, clinical observations, vital signs, with gross necropsy and histopathologic examination of tissues at the end of the study.

Clinical Studies

The pharmacokinetics and safety of intravenous MiKasome were studied in asymptomatic, HIV-seropositive volunteers. Subjects were enrolled in one of three dose regimens: a single 1 h infusion of 2.5 mg/kg (n = 5), 7 consecutive daily infusions of 2.5 mg/kg (n = 5), or 7 consecutive daily infusions of 5 mg/kg (n = 5). All subjects were males, with a mean age of 36.8, 40.2 and 42.2 years for the three dose groups, respectively. Plasma and urine samples were obtained during the dosing phase, and at timepoints up to 2 weeks after the last dose was administered. Total amikacin was measured in plasma samples by adding detergent to disrupt liposomes prior to determination of amikacin. Free amikacin was estimated by measuring amikacin in samples of plasma ultrafiltrates (Centrifree, Amicon, Inc.) obtained at each timepoint. Amikacin was measured in plasma, plasma ultrafiltrates and urine samples by a fluorescence polarization immunoassay (Abbott TDx/FLx) with a limit of quantitation of 3 µg/ml. Safety parameters in this Phase 1 study included renal function (monitored by measuring creatinine clearance), and VIIIth cranial nerve function (monitored by pure tone audiometry and vestibular autorotation testing).

Results

Preclinical Pharmacokinetics of MiKasome

The pharmacokinetics of intravenous MiKasome were characterized in several species after both single and repeated dose administration.[2,3] These studies demonstrated that Mikasome has a pharmacokinetic profile that differs substantially from both conventional amikacin and from other liposomal aminoglycosides, but which is relatively similar between species.

After equivalent doses in rabbits (15 or 40 mg/kg), MiKasome increased the area under the plasma concentration curve by approx. 150-fold, and the half-life by approx. 25-fold compared to conventional amikacin (Fig. 15.1). Pharmacokinetic parameters calculated after single doses of similar magnitude (10-50 mg/kg) in four species are summarized in Table 15.1. Conventional amikacin, which is rapidly excreted in the urine as unchanged drug, has a clearance close to the glomerular filtration rate (> 3 ml/min/kg in animals).[3,4] MiKasome clearances were between 0.02-0.04 ml/min/kg in all four species, indicating that sequestration of the amikacin payload within circulating liposomes greatly reduced the renal filtration of amikacin in this formulation. As a consequence, Mikasome half-lives (14-40 h) were substantially greater than the half-life of conventional amikacin (0.35 h in rats,[3] 1 h in dogs).[4] As observed with other liposomal formulations, these low-clearance liposomes had a low volume of distribution, approaching the volume of the plasma compartment (≤100 ml/kg).

The levels of free (ultrafiltrable) amikacin in plasma samples were < 10% of the total plasma amikacin concentration at all timepoints, and often fell to below 1 µg/ml within a few hours after MiKasome injection. The majority of the free amikacin in the plasma may result from release of a small fraction (< 15%) of the amikacin in the formulation externally associated with the liposomes.

Both the urinary clearance and recovery of amikacin after MiKasome administration were lower than observed for conventional amikacin. These data are again consistent with the hypothesis that amikacin remains largely sequestered within slowly-cleared, circulating liposomes. Total urinary recoveries over a two-week period after a single dose in rhesus monkeys, and during steady-state conditions in dogs were less than one-third of the total dose, indicating that a major part of the administered antibiotic remained in the tissues, or was eliminated by an alternate mechanism.

MiKasome clearances as a function of species and dose are shown in Figure 15.2. MiKasome clearance remained below 0.05 ml/min/kg in all species and doses, but appeared to decrease slightly with dose (from 0.05 ml/min/kg in rabbits at 2.5 mg/kg, to 0.01 ml/min/kg in dogs at 360 mg/kg). There was no apparent species difference in the clearance of MiKasome. Steady-state volumes of distribution observed after MiKasome (53-120 ml/kg) approached the volume of the plasma compartment, and did not appear to vary with either species or dose. As a result, observed half-lives for MiKasome also exhibited dose or concentration dependence, ranging from > 200 h at plasma concentrations > 4000 µg/ml (after 30 d exposures at 120 mg/kg/day in dogs), to less than 20 h at plasma concentrations ≤ 50 µg/ml (data from several species).

Despite its apparently nonlinear disposition, MiKasome reached steady state within 2 weeks of repeated dosing in dogs (Fig. 15.3), with steady state plasma concentrations close to those predicted from single dose plasma data. During repeated

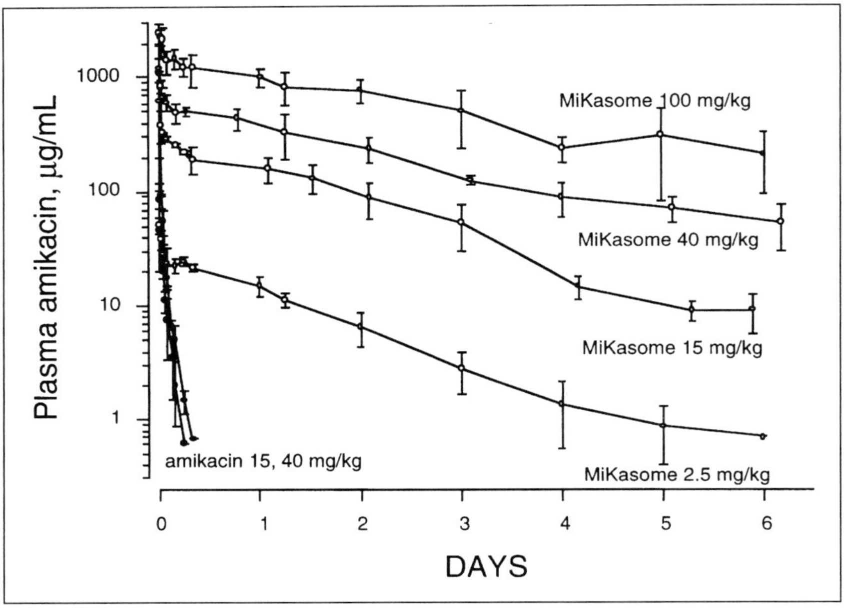

Fig. 15.1. Pharmacokinetic profiles of intravenous MiKasome and conventional amikacin in rabbits. Values are group mean (± SD, n = 3) total amikacin concentrations in plasma.

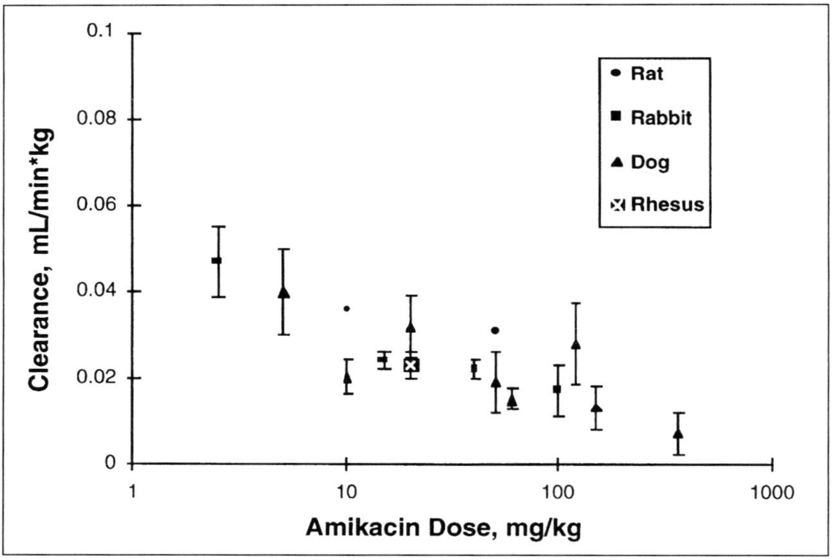

Fig. 15.2. Interspecies comparison of intravenous MiKasome clearance as a function of amikacin dose. The doses shown correspond to total lipid doses ranging from approx. 20 µmol/kg to 2800 µmol/kg.

Table 15.1. Interspecies summary of MiKasome pharmacokinetic parameters

Species **Dose**	**—** **mg/kg**	**Monkey** **20**	**Dog** **10, 20**	**Rabbit** **15, 40**	**Rat** **10, 50**
Plasma Clearance	ml/min/kg	0.023	0.02-0.03	0.02	0.03-0.04
Terminal Half-life	h	36	40	28-40	14-36
C_{max}	μg/ml	447	318-518	472-1184	300-1500
V_c	ml/kg	45	20-55	35	25-36
V_{ss}	ml/kg	71	66-100	57-75	65-94
Urinary Clearance	ml/min/kg	0.006-0.013	0.0015	—	—
Urinary Recovery	% Dose	29.8 (in 14 days)	4 (in 2 days) 23 (steady state)	—	12 (in 7 days)
"Free" amikacin in plasma	% Total	< 5%	0-10%	—	—

Pharmacokinetic parameters were determined after a single intravenous dose of MiKasome in each species. "Free" amikacin in plasma was determined by measuring amikacin concentration in plasma ultrafiltrates.
Abbreviations: C_{max}, Maximum plasma drug concentration; Vc, Volume of the central compartment; Vss, Volume of distribution at steady state.

administration of MiKasome at doses ranging from 20-120 mg/kg/day, trough plasma levels between 500-5000 μg/ml amikacin were maintained in dogs.

Preclinical Efficacy and Safety of MiKasome

The antimicrobial activity of MiKasome was measured in in vitro microtiter broth assays[5] and in animal models of infection.[6-9] In vitro, MiKasome was active against a range of organisms, including *Staphylococcus* spp. and *Pseudomonas* spp., although in vitro activity measurements appear to underestimate in vivo activity, as reported for other liposomal aminoglycosides.[10] MiKasome demonstrated in vivo antimicrobial activity in models of extracellular infection, including *Klebsiella* pneumonia[8] and *Pseudomonas* endocarditis.[9]

Given the efficacy of MiKasome by postinfection treatment summarized above, in order to demonstrate the duration of therapeutic activity after MiKasome injection, and to eliminate the possibility that the observed activity of MiKasome was due to rapidly released, externally associated amikacin in the liposomal formulation, a prophylactic study was performed in immunosuppressed mice exposed to a lethal intraperitoneal challenge with *Klebsiella pneumoniae*.[6,7] In this study, MiKasome (70 mg/kg) administered 1, 2 and 3 d prior to the challenge was effective in sterilizing the blood, liver and spleen, and in prolonging the survival of all treated mice (Fig. 15.4). Conventional amikacin (70 mg/kg) and ampicillin/sulbactam

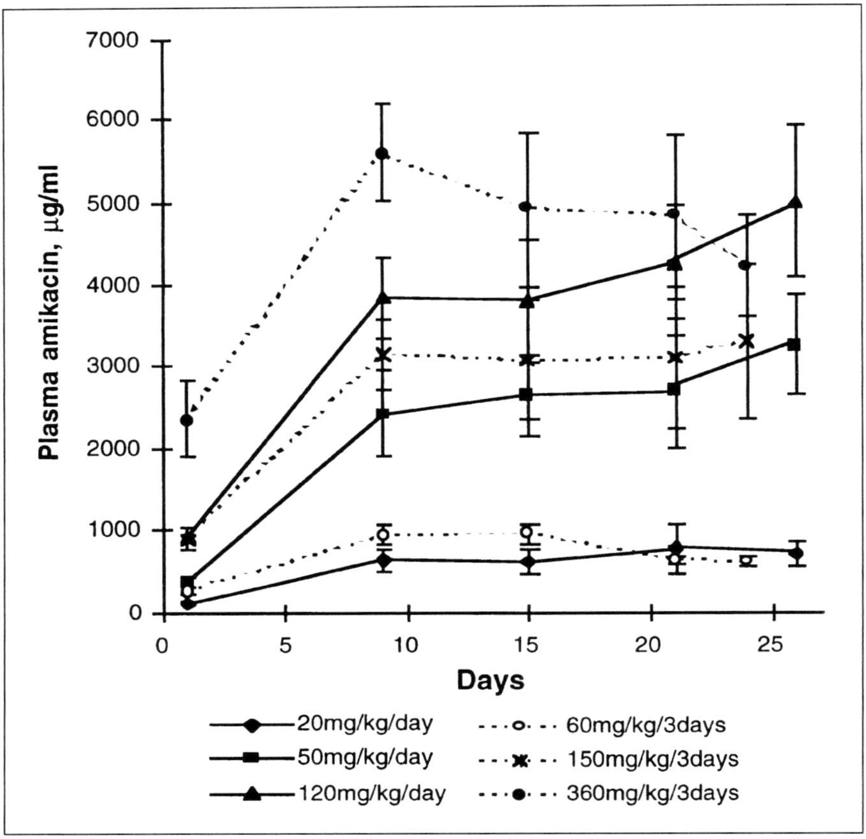

Fig. 15.3. Group mean (± SD, n = 6 per group) trough levels of plasma amikacin observed after repeated intravenous MiKasome dosing in Beagle dogs. Trough plasma samples were obtained 24 h after dosing (daily dosing groups) or 72 h after dosing (every-third-day dosing groups).

(Unasyn®, Pfizer, Inc., 1050 mg/kg) given on the same schedule were ineffective in prolonging survival or in reducing tissue burdens compared to buffer controls ($p \le 0.05$). In this study, the amikacin levels in plasma at the time of bacterial challenge were significantly higher in the MiKasome group (115 ± 46 μg/ml) vs. the conventional amikacin group (0.7 ± 0.1 μg/ml).

In a safety study of intravenous MiKasome, male and female Beagle dogs were exposed to daily or every-third-day injections of MiKasome for one month.[11] No significant toxic changes were observed in animals that received 20 mg/kg/day (approx. 160 μmol total lipid/kg/day) or 60 mg/kg/3d (approx. 470 μmol total lipid/kg/3d) in this study. These doses resulted in steady-state plasma concentrations ≥ 750 μg/ml, with trough levels > 600 μg/ml throughout the one-month study. In contrast, dogs who received daily injections of conventional amikacin (50 mg/kg/day) had steady state concentrations of 8 μg/ml and trough levels < 1 μg/ml. Thus, the No-

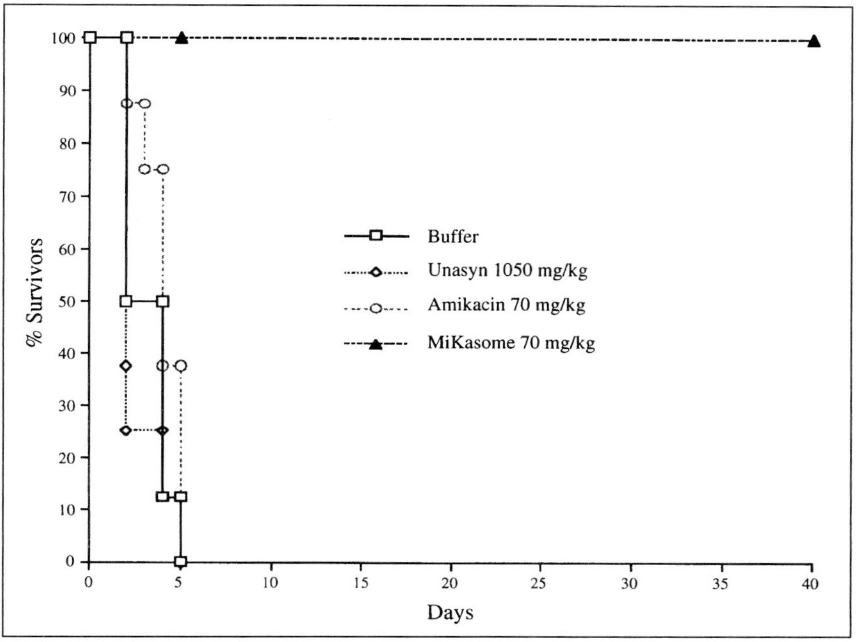

Fig. 15.4. Survival of mice (n = 8 per treatment) challenged with *Klebsiella pneumoniae* (IP). Mice received intravenous prophylactic treatment 3 d, 2 d and 1 d prior to the challenge, with buffer, conventional amikacin, MiKasome or Unasyn at the doses indicated.

Adverse-Effect dose of MiKasome represented an exposure at least 100-fold larger than observed for a similar dose of conventional amikacin.

Analysis of amikacin in samples of kidney, liver, spleen, lung, adrenals, bone marrow, brain, gut, lymph node, muscle and skin obtained at the end of the one-month study showed that MiKasome increased amikacin levels in all tissues measured compared to conventional amikacin. However, MiKasome reduced the ratio of cortical to medullary amikacin in the kidneys, suggesting it alters the site or mechanism of renal drug accumulation. Kidney amikacin concentrations > 1000 µg/g were not associated with elevation of creatinine or urea nitrogen levels in MiKasome-treated animals. Microvacuolated macrophages containing unesterified cholesterol and amikacin were observed in most tissues after the administration of MiKasome, suggesting that MPS-associated cells form an important reservoir of antibiotic in many tissues.

Clinical Pharmacokinetics of MiKasome

As in animals, the pharmacokinetics of MiKasome in man differed markedly from those of conventional amikacin (Fig. 15.5, Table 15.2).[12] The observed clearances of MiKasome (0.015-0.017 ml/min/kg) were more than 70-fold lower than the clearance of conventional amikacin in man (1.3 ml/min/kg),[13] while the terminal half-lives of MiKasome (83-114 h) were more than 35-fold longer than the principal plasma half-life of conventional amikacin (2.3 h).[13] In addition, steady-state volumes of distribution for MiKasome (0.10-0.17 l/kg) were lower than the volume of distribution of conventional amikacin (0.27 l/kg)[13] Peak plasma levels after MiKasome (44.5 μg/ml after a single dose, 242 μg/ml after multiple dosing at 5 mg/kg) were markedly higher than those expected after similar doses of conventional amikacin.

Although plasma levels increased during repeated MiKasome administration, MiKasome clearance after the last dose was similar to the clearance after a single dose. No indications of nonlinear pharmacokinetics were observed at these doses.

Due to the long half-life of MiKasome, the fluctuation in plasma amikacin levels during one dosing interval was small (< 1.4-fold). Plasma levels measured two weeks after completion of the 7-day multiple-dose exposure still exceeded 10 μg/ml.

The levels of free amikacin after MiKasome administration were much lower than total amikacin levels in plasma. Concentrations of free amikacin measured in plasma ultrafiltrates were < 4.1 μg/ml in all patients at all timepoints, and the total AUC of free amikacin was < 1% of the total amikacin AUC. Free amikacin levels appeared to decline rapidly after dosing, and were often undetectable at later timepoints.

After a single dose of MiKasome, approx. 30% of the administered amikacin was recovered in the urine, mostly within the first 24 h. After repeated dosing, 33-45% of the dose was recovered in the urine during the last dosing interval. This suggests that at steady-state between one-third and one-half of the amikacin in MiKasome is excreted unchanged in the urine.

No renal or auditory/vestibular toxicity was observed in the subjects dosed with MiKasome for one week at 2.5 or 5 mg/kg/day.

Discussion

Preclinical Studies of MiKasome

MiKasome, a formulation of amikacin in small, low-clearance liposomes, presents a pharmacokinetic profile substantially different from conventional amikacin in all species studied. MiKasome has a low clearance and low volume of distribution over a wide range of doses, indicating it sequesters its antibiotic payload within slowly-cleared, circulating liposomes. Although MiKasome pharmacokinetics were nonlinear in animals (half-life increased with dose or concentration), plasma levels reached steady-state after multiple dosing at levels close to those predicted from single dose data. The long half-lives observed for MiKasome suggest that dosing 1-2 times per week may be sufficient to maintain effective levels of MiKasome.

At therapeutic doses, the amount of lipid administered in MiKasome may be in the range associated with MPS saturation after administration of conventional multilamellar vesicle (MLV) liposomes. However, the clearance of MiKasome remains low over a wide range of doses (Fig. 15.2), suggesting that saturation of MPS clearance is not the major factor in determining its clearance. The clearance of MiKasome is substantially lower, and less dose-dependent than that of conventional MLVs or aminoglycoside MLVs, and is similar to that for sterically stabilized aminoglycoside liposomes (Fig. 15.6).

MiKasome demonstrated in vitro activity and in vivo activity with postinfection treatments against a range of organisms. The prophylactic activity of MiKasome against *Klebsiella pneumoniae* demonstrated the extended duration of antimicro-

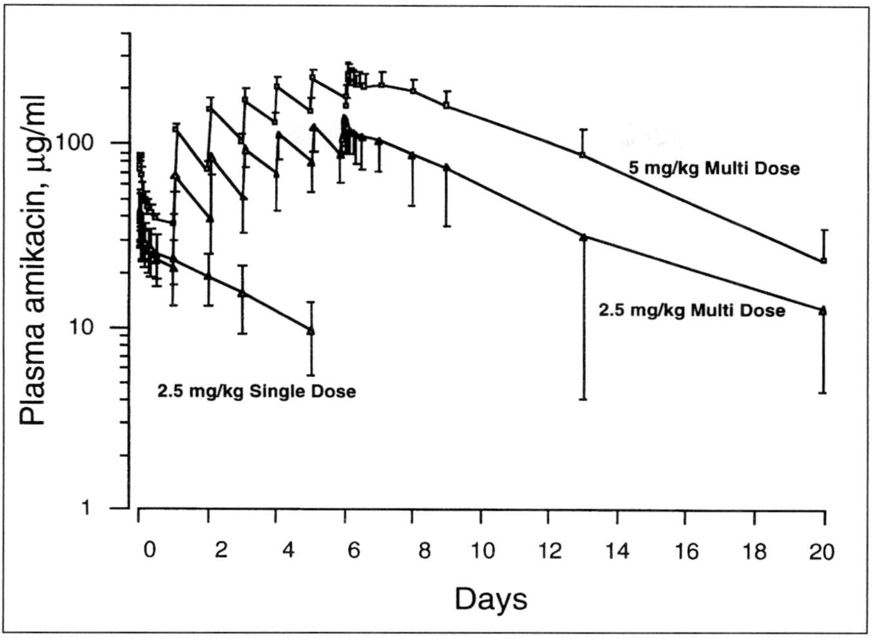

Fig. 15.5. Mean (± SD, n = 5 per group) plasma total amikacin concentrations after a single dose, or seven consecutive daily doses of intravenous MiKasome in HIV-seropositive subjects.

bial activity after MiKasome administration compared to conventional treatments.

In safety studies, continuous exposure to MiKasome at plasma levels > 600 µg/ml in a one-month study produced no indications of toxicity or significant physiologic changes. The demonstration of safety after repeated MiKasome doses of 60 mg amikacin/kg and 470 µmol total lipid/kg is in contrast to a report that PEG-PE stabilized liposomes caused acute mortality in rats after a single dose containing 450 µmol phospholipid/kg.[10]

Phase 1 Study of Mikasome

The pharmacokinetics of MiKasome in man were similar to those observed in animals: greatly decreased clearance, decreased volume of distribution, longer half-life and higher plasma concentrations for MiKasome compared to conventional amikacin. The observed pharmacokinetics of MiKasome support the hypothesis that amikacin sequestered within circulating liposomes in the plasma is not subject to rapid renal filtration clearance.

Although nonlinear pharmacokinetics were observed in animals, there was no indication of nonlinear kinetics in the dose regimens studied in man. Plasma levels increased during repeated dosing of MiKasome, but significant changes in terminal half-lives or clearance were not observed after multiple dosing. Although steady-state concentrations were not achieved after 1 week of dosing, the data suggest that steady-state might be reached within 2 weeks of repeat dosing in man, as observed

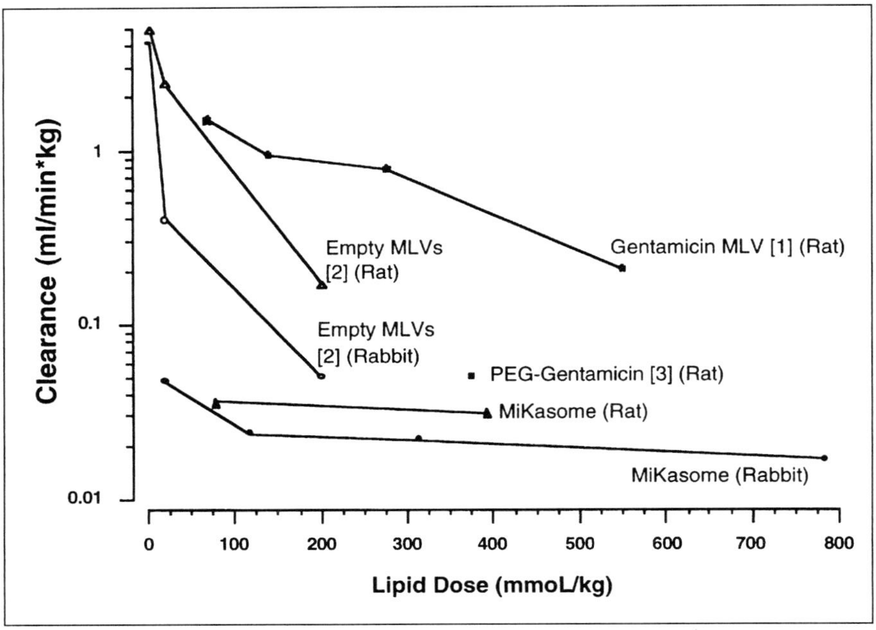

Fig. 15.6. Intravenous MiKasome clearance compared to reported clearances of other intravenous liposomes, as a function of total lipid dose. [1], conventional multilamellar gentamicin liposomes, reported to be > 1 µm dia., data from ref. 15. [2], conventional multilamellar liposomes without drug, reported to be 780 nm dia., data from ref. 16. [3], sterically stabilized (PEG-PE) gentamicin liposomes, reported to be approx. 100 nm dia., clearance calculated from data in ref. 10. For comparison, MiKasome has a dia. < 100 nm.

in animals.

Due to the long plasma half-life of MiKasome, plasma levels were maintained between 104-135 µg/ml (2.5 mg/kg) or 198-229 µg/ml (5 mg/kg) after 1 week of daily dosing. Plasma levels fluctuated less than 1.4-fold during the 24 h dosing interval. Plasma levels declined slowly after the last dose (half-life: 114 h), so that plasma amikacin levels of 10-20 µg/ml persisted 2 weeks after administration of the final dose. These data suggest that therapeutic levels of MiKasome could be maintained with less frequent dosing, perhaps as infrequently as once weekly.

Only a small fraction of the amikacin present in plasma after MiKasome administration was "free" (ultrafiltrable) amikacin. Free amikacin levels in plasma never exceeded 4 µg/ml in any patient during the course of the study, and the AUC of free amikacin was <1% of the total amikacin AUC. The rapid initial decline in free amikacin concentrations observed after MiKasome administration may correlate with the fact that a small fraction (< 15%) of the amikacin in MiKasome is associated with the exterior surface of the liposomes, and can be rapidly released from the liposomes after injection. The majority of the amikacin in MiKasome appears to remain encapsulated within the slowly-cleared liposomes in the plasma compartment.

Table 15.2. Clinical pharmacokinetics of Mikasome in HIV-seropositive male subjects

Parameter	**Units**	**2.5 mg/kg Single Dose**	**2.5 mg/kg Seventh Dose**	**5 mg/kg Seventh Dose**
		TOTAL AMIKACIN		
AUC	µg*h/ml	3553	2625	4969
		1516	*793*	*850*
Clearance	ml/min*kg	0.015	0.017	0.016
		0.01	*0.01*	*0.003*
Half-life	h	83	114	114
		28	*57*	*32*
Vss	l/kg	0.10	0.16	0.17
		0.01	*0.08*	*0.06*
C_{max}	µg/ml	44.5	137	242
		14	*34*	*40*
Css	µg/ml	148	109	207
		63	*33*	*35*
		FREE (ULTRAFILTRABLE) AMIKACIN		
AUC	µg*h/ml	14.2	14.3	16.5
		5	*6*	*2*
Half-life	h	4.6	7.8	3.33
		1.6	*11.2*	*0.67*
C_{max}	µg/ml	BLOQ	BLOQ	3.06
		—	—	*0.7*

Group Means (± *SD*, n = 5 per group). AUC and clearance determined from t = 0 to infinity for single dose. Steady-state AUC and clearance determined over one dosing interval after the seventh dose. Abbreviations: AUC, Area under the plasma concentration vs. time curve; C_{max}, Maximum plasma drug concentration; Vc, Volume of the central compartment; Vss, Volume of distribution at steady state; BLOQ, below validated assay limit of quantitation (3 µg/ml).

The low urinary recovery of amikacin after MiKasome administration is also consistent with the hypothesis that MiKasome sequesters its amikacin payload in the plasma within slowly-cleared liposomes. Again, the presence of a small fraction of the amikacin on the exterior of the liposomes may account for a substantial fraction of the observed urinary clearance and recovery after MiKasome administration. The fact that most of the urinary recovery occurred soon after administration, when most of the administered MiKasome was still within the body, suggests that the small pool of unencapsulated amikacin is rapidly released and excreted unchanged in the urine, while the majority of the drug in the body remains encapsulated.

A previous clinical study reported the plasma levels of a conventional liposomal aminoglycoside formulation administered to HIV patients infected with *Mycobacterium avium* complex.[14] The pharmacokinetics of gentamicin in these large (> 1 µm) multilamellar liposomes differed substantially from those of MiKasome. Mean peak levels, obtained 10 min. after a 2-hour infusion of gentamicin (3.4 mg/kg) were 8 µg/ml, while peak amikacin levels after a 1 h infusion of MiKasome (2.5 mg/kg) exceeded 40 µg/ml. The levels of free gentamicin (measured by ultrafiltration) were 33% of total plasma levels. Free amikacin levels after MiKasome did not exceed 6% of total levels, and were frequently undetectable. The half-life of the gentamicin MLV formulation was less than 12 h, and plasma levels 24 h after a 5 mg/kg dose were ≤ 2 µg/ml. Urinary recovery of gentamicin was 75% of the dose following the last of eight infusions, while urinary recovery of MiKasome was < 50%.

The marked pharmacokinetic differences between these two liposomal aminoglycoside formulations suggest that substantial differences may exist in their safety and efficacy profiles as well.

Conclusion

The pharmacokinetics of MiKasome were similar in man and animals, and differed markedly from those of conventional amikacin and other liposomes. MiKasome markedly increased amikacin plasma concentrations and half-life, while reducing clearance, volume of distribution and urinary recovery of amikacin.

A comparison of MiKasome pharmacokinetics and safety with those of other aminoglycoside liposomes suggests that differences in liposomal composition have a marked effect on their therapeutic properties. MiKasome, a small, low-clearance liposomal aminoglycoside formulation, demonstrated a longer half-life than conventional liposomes and an increased safety margin compared to sterically stabilized liposomes. MiKasome was found to be active against a range of organisms, including in vivo models of extracellular infection.

The pharmacokinetics of MiKasome appeared to be linear at daily doses up to at least 5 mg/kg in man. Plasma levels were maintained at > 100 µg/ml during the entire dosing interval after repeated daily dosing. The long terminal half-life of MiKasome (114 h in man) suggests that relatively constant antibiotic levels could be maintained in plasma and tissues with a MiKasome dosing interval as long as one week. The prophylactic efficacy of MiKasome against *Klebsiella pneumoniae* suggests that the duration of MiKasome activity against extracellular infections in vivo may parallel its long pharmacokinetic half-life.

Acknowledgments

For their contributions to this work, the authors gratefully acknowledge Ramilla Lewis, Lotus Moon-McDermott, Philip M. Zack, Jill Adler-Moore, Stanley C. Gill (NeXstar Pharmaceuticals, Inc.), Robert E. Kates (Analytical Solutions, Inc.), Stanley Deresinski, Carol Kemper, Carla Mansfield (Santa Clara Valley Medical Center), and Terrence Blaschke, Carla Washington (Stanford University Medical Center).

References

1. Karlowsky JA, Zhanel GG. Concepts on the use of liposomal antimicrobial agents: applications for aminoglycosides. Clin Infect Dis 1992; 15:654-67.
2. Fielding RM, Feistner B, Moon-McDermott L et al. Liposomal amikacin (MiKasome) prolongs plasma and tissue residence and reduces urinary excretion of amikacin in

rats. American Society for Microbiology, 97th General Meeting, Miami Beach, May 4-8, 1997.
3. Fielding RM, Feistner B, Moon-McDermott L et al. Liposomal amikacin (MiKasome®): reduced clearance and volume of distribution in rats. Pharm Res 1996; 13:S-479.
4. Baggot JD, Ling GV, Chatfield RC. Clinical pharmacokinetics of amikcain in dogs. Am J Vet Res 1985; 46:1793-6.
5. Fabrey R, Olson J, Adler-Moore J. Differences in in vitro antimicrobial activity of amikacin (Ami) following its encapsulation in stable liposomes (MiKasome). American Society for Microbiology, 97th General Meeting, Miami Beach, May 4-8, 1997.
6. Eng E, Satorius A, Proffitt RT, Adler-Moore JP. Prophylaxis of Klebsiella pneumonia sepsis by MiKasome (MK), a liposomal formulation of amikacin (AK). American Society for Microbiology, 96th General Meeting, New Orleans, May 19-23, 1996.
7. Eng ET. Prophylactic and therapeutic treatment of gram negative septicemia with liposomal and non-liposomal encapsulated amikacin in immunocompromized mice. Thesis presented to California State Polytechnic University, Pomona, CA, 1996.
8. Eng E, McAndrews B, Satorius A et al. Comparative efficacy of amikacin (Ami) and liposomal amikacin (MiKasome) in the treatment of Klebsiella pneumoniae infection in mice. American Society for Microbiology, 97th General Meeting, Miami Beach, May 4-8, 1997.
9. Xiong Y-Q, Adler-Moore J, Zack P, Bayer AS. Efficacy of MiKasome (a liposomal amikacin formulation) vs free amikacin in experimental endocarditis due to Pseudomonas aeruginosa. American Society for Microbiology, 97th General Meeting, Miami Beach, May 4-8, 1997.
10. Bakker-Woudenberg IAJM, ten Kate MT, Stearne-Cullen LET, Woodle MC. Efficacy of gentamicin or ceftazidime entrapped in liposomes with prolonged blood circulation and enhanced localization in Klebsiella pneumoniae infected lung tissue. J Infect Dis 1995; 171:938-947.
11. Zack PM, Fielding RM. Unpublished data.
12. Fielding RM, Mukwaya GM, Washington C et al. A multiple dose phase 1 safety and pharmacokinetic study of low-clearance liposomal Amikacin (MiKasome) in HIV serpositive patients. Infectious Diseases Society of America, 35th Annual Meeting, San Francisco, September 15, 1997.
13. Benet LZ et al. Design and optimization of dosage regimens; pharmacokinetic data. In: Hardman JG et al, eds. The Pharmacological Basis of Therapeutics. 9th ed. New York: McGraw-Hill, 1996:1714.
14. Nightingale SD, Saletan SL, Swenson CE et al. Liposome-encapsulated gentamicin treatment of mycobacterium avium—mycobacterium intracellulare complex bacteremia in AIDS patients. Antimicrob Agents Chemother 1993; 37:1869-72.
15. Swenson CE, Stewart KA, Hammett JL et al. Pharmacokinetics and in vivo activity of liposome-encapsulated gentamicin. Antimicrob Agents Chemother 1990; 34:235-40.
16. Harashima H, Homatsu S, Kojima S et al. Species difference in the disposition of liposomes among mice, rats, and rabbits: allometric relationship and species dependent hepatic uptake mechanism. Pharm Res 1996; 13:1049-1054.

SECTION II

Current Medical Applications

C. Diagnostics

CHAPTER 16

Sterically Stabilized Liposomes to Image Infection and Inflammation

Otto C. Boerman, Wim J.G. Oyen, Frans H.M. Corstens, Gerrit Storm

Introduction

Determination of the localization of acute, subacute and chronic infectious foci may have important implications for the management of patients. Imaging techniques like ultrasound, CT-scanning or MRI are in most cases unable to discriminate active processes from scar tissue due to a cured infection or after surgery. In contrast, scintigraphic images are not determined morphologically, they are based on physiochemical processes in tissues. In clinical practice not only the presence of an active inflammatory process, but also the localization and the number of foci needs to be determined; scintigraphic imaging is an excellent noninvasive method of whole-body scanning. In clinical practice several radiopharmaceuticals are used to image foci of infection and inflammation. In this chapter the development and the potential of a liposome-based radiopharmaceutical for infection and inflammation imaging is described and compared with conventional radiopharmaceuticals.

Conventional Imaging Agents

During the past two decades several radiopharmaceuticals have been developed for infection and inflammation imaging (Gallium-67, radiolabeled leukocytes, Indium-111-IgG). As summarized in Table 16.1, each of the currently available scintigraphic agents for infection and inflammation imaging has its particular disadvantages, stimulating the search for new radiopharmaceuticals. Gallium-67-citrate has been used ever since its discovery in 1971, especially for evaluation of abnormalities of the lungs and the motoric system.[1,2] Scanning of the abdomen is difficult due to the high physiological uptake in the GI and urinary tract, specificity is low and its radiation burden is relatively high.[3,4] In addition, due to its slow background clearance in most cases a final diagnosis can only be made based on the images made 48 h after injection of [^{67}Ga]citrate. The use of radiolabeled autologous leukocytes for imaging infection and inflammation has been a major step forward.[5,6] This approach exploits the ability of leukocytes to migrate towards the site of infection and/or inflammation (chemotaxis). In acute infection, high quality images of infectious foci can be obtained.[7,8] In chronic infection, the sensitivity of labeled leukocyte scintigraphy appears to be limited probably due to the relatively low influx of

Long Circulating Liposomes: Old Drugs, New Therapeutics, edited by Martin C. Woodle and Gerrit Storm.

granulocytes in this type of infection.[9] In addition, this technique is laborious, expensive and requires manipulation of potentially (HBV- or HIV-) contaminated blood.[10]

Recently, ^{111}In-labeled human polyclonal immunoglobulin G [^{111}In]IgG) has been introduced as an infection and inflammation imaging agent. [^{111}In]IgG accumulates nonspecifically in infectious and inflammatory foci and can be administered safely without any side-effects.[11] As reviewed earlier, good results have been obtained in several groups of patients.[12] In patients suspected of infections in the locomotor system, the sensitivity of [^{111}In]IgG for visualizing infectious foci is extremely high (sensitivity: 97%),[13] while a negative scan in most cases rules out the absence of infection (specificity: 85%). In 38 patients suspected of abdominal infections, 25 of 28 foci were detected with [^{111}In]IgG scintigraphy.[14] In febrile granulocytopenic patients (n = 20)—where a labeled leukocyte scan is not feasible—pulmonary, abdominal, joint and soft-tissue infections of both bacterial and fungal origin were detected adequately.[15] Satisfying results have also been obtained in patients with diabetic foot infections, patients with inflammatory bowel disease, and in patients with fever of unknown origin.[16,17] Limitations of [^{111}In]IgG are:

1) The inability to image vascular lesions due to its slow blood clearance.
2) Moderate sensitivity in particular patient groups.
3) Relatively long time span between injection and final diagnosis (± 48 h).
4) The use of Indium-111 as a gamma emitter with its relatively high radiation burden.

In nuclear medicine the most ideal radionuclide for diagnostic use would be ^{99m}Tc, because it is cheap, widely available and its 140 keV gamma is ideal for scintigraphy. Therefore, human polyclonal IgG labeled with ^{99m}Tchas been developed for infection and inflammation imaging.[18,19] However, the 6 h half-life of the label limits the time period to obtain adequate scintigraphic imaging to maximally 24 h postinjection. In addition, stable labeling of IgG with ^{99m}Tc is difficult to achieve, resulting in loss of the radiolabel in vivo and subsequent accumulation of the released label in nontarget organs, especially the kidneys.[20]

Short Circulating Liposomes for Infection and Inflammation Imaging

Liposomes, microscopic lipid vesicles consisting of one or more concentric phospholipid bilayers surrounding aquaeous compartments, can be labeled with gamma emitters such as ^{67}Ga, ^{111}In and ^{99m}Tc. Liposomes have been investigated extensively as carriers for drugs in attempts to achieve selective deposition and/or reduced toxicity.[21,22] At sites of inflammation the vascular permeability is known to be increased.[23-25] Liposomes may extravasate at sites of inflammation and thus liposomes have been proposed as vehicles to target infectious and inflammatory foci.

In the early 1980s, Morgan and co-workers were among the first to study liposomes labeled with ^{99m}Tc for imaging infection.[26] Negatively charged large multilamellar liposomes were labeled with ^{99m}Tc according to the stannous chloride technique (reducing $^{99m}TcO_4^-$ with $SnCl_2$ in the presence of liposomes), and tested in rats with focal *S. aureus* infection.[26] The infectious foci were clearly visualized. Uptake in the kidneys and the bladder was rather high, suggesting extensive release of the radiolabel from the liposomes in vivo. This liposomal formulation has been used successfully in patients with deep-seated infections.[27] Infected prosthetic joints, tubo-

Table 16.1. Summary of the pros and cons of conventional agents for imaging infection and inflammation

Imaging Agent	Advantages	Disadvantages
[^{67}Ga]-citrate	– inexpensive – good imaging of chronic infection/inflammation	– GI and urinary tract activity – high radiation burden – modest target/background ratio – long time span between injection and diagnosis
labeled leukocytes	– good imaging of acute infection/inflammation – rapid background clearance	– laborious preparation – handling of potentially contaminated blood – modest imaging of chronic infection
[^{111}In]IgG	– good imaging of chronic infection/inflammation – one step preparation	– high radiation burden – long time span between injection and diagnosis – inability to image vascular lesions
[^{99m}Tc]IgG	– good imaging of chronic injection and infection/ inflammation – low radiation burden – one step preparation	– long time span between injection and diagnosis – inability to image vascular lesions – high physiologic uptake in kidneys

ovarian abscess and osteomyelitis were visualized on the scan 24 h postinjection. A study in patients with rheumatoid arthritis indicated that mainly clinically active disease was visualized with the ^{99m}Tc labeled liposomes.[28] The authors showed that these liposome scans could discriminate between different grades of inflammation. In another study in arthritis patients, the scans obtained with these [^{99m}Tc]liposomes were compared with the [^{99m}Tc]MDP bone scans and radiographical images of the joints of 13 patients. Again, the liposome scans were positive only in clinically active inflammatory disease.[29] The liposomes used in these early studies were rapidly taken up by cells of the mononuclear phagocyte system (MPS), primarily those located in liver and spleen. In rats more than 50% of the injected dose of this formulation was found in the liver and spleen within hours, while the circulatory half-life ($t_{1/2}\beta$) was less than 2 h.[30] This rapid removal by MPS cells, probably limited their ability to reach infectious and inflammatory foci in vivo. In addition, the stannous chloride labeling method used in these studies has been shown to result in an unstable association of the radiolabel with the liposomes, as excessive loss of the label from the liposome in vivo has been reported.[31]

Long Circulating Liposomes

In recent years, the new development of formulations of liposomes with long circulating properties has greatly broadened the potential applications of liposomes for targeted delivery. It has been demonstrated that small, neutral, cholesterol-rich liposomes composed of "rigid" phospholipids of high-phase-transition temperature can attain prolonged circulation times at relatively high lipid doses.[32] More recently, specific glycolipids, the most well-known examples being monosialoganglioside (G_{M1}) and hydrogenated phosphatidylinositol (HPI), have been shown to prolong the circulation time, without the constraint of high lipid doses.[33-36] Subsequently, it was demonstrated that inclusion of polyethylene glycol (PEG), conjugated to phosphatidylethanolamine, in the bilayer increased the blood circulation time.[37-40] The prolonged circulation half-life of PEG-liposomes, also referred to as 'sterically stabilized' or 'Stealth' liposomes, is caused by their reduced recognition by the MPS, as reflected by their delayed and diminished hepatic and splenic accumulation.[40,41] (Stealth liposomes is a registered trademark of SEQUUS Pharm., Inc. Menlo Park, CA).

Since liposomal accretion in infectious and inflammatory foci is driven by high blood levels, long circulating liposomes may be more suitable for imaging infections and inflammations than short-circulating liposomes.

In a rat model liposomes sterically stabilized with PEG showed preferential uptake in *Klebsiella pneumoniae*-infected lung tissue.[35] Uptake of the liposomes in the infected left lung appeared to be correlated with the intensity of the infection and was not dependent on the lipid dose administered. Up to 9% of the iv injected liposomal dose localized in the infected lung.[36] These studies confirmed the improved outlook for long circulating liposomes for infection and inflammation imaging.

Goins et al[42] proposed that prolonged circulation would improve the application of ^{99m}Tc-labeled long circulating liposomes for imaging focal infection. Unilamellar cholesterol-rich liposomes (185 nm) were labeled with ^{99m}Tc with D,L-hexamethylpropylene amine oxime (HMPAO). These long circulating ^{99m}Tc-labeled liposomes were tested in rats with focal *S. aureus* infection. Twenty-four hours postinjection 13% of the injected dose was still found in the blood compartment, indicating the enhanced circulatory half-life of this liposomal formulation ($t_{1/2}$ = 10 h). Abscess-to-muscle ratios—calculated from 24 h biodistribution data—as high as 35 were obtained, while these ratios did not exceed 10 with conventional imaging agents ([^{67}Ga] or [^{99m}Tc]HSA) in this model.[42] However, these liposomes still showed considerable splenic uptake: 27% of the injected dose was found in the spleen at 24 h postinjection.

^{111}In-Labeled PEG-Liposomes for Imaging Infection and Inflammation

Our group evaluated radiolabeled PEG-liposomes for infection imaging. By encapsulating desferal in PEG-liposomes, these liposomes could be remotely labeled with ^{111}In-oxine with high efficiency (>90%). The biodistribution of ^{111}In-labeled PEG-liposomes (mean diameter approximately 100 nm) was studied in rats with focal *S. aureus* infection using [^{111}In]IgG as a reference agent.[43] The circulatory half-life of the ^{111}In-labeled PEG-liposomes in this model ($t_{1/2}$ = 20 h) was similar to that of [^{111}In]IgG. However, the uptake of the ^{111}In-labeled liposomes in the abscess was more than twice as high as the abscess uptake obtained with [^{111}In]IgG (2.7% ID/g vs. 1.1% ID/g at 48 h pi). Besides high uptake in the abscess, marked localization was also observed in the spleen (16% ID/g at 48 h pi). Uptake in the other nontarget organs was relatively low and decreased with time. Consequently, target-to-background ratios steadily increased during the 48 h following iv administration. Based on tissue counting, abscess-to-muscle ratios were approximately four times as high as those obtained with [^{111}In]IgG in this model (34 versus 8 at 48 h pi) (Fig. 16.1). The distribution of ^{111}In-labeled PEG-liposomes was also monitored by gamma camera imaging. With [^{111}In]liposomes, the abscess was well delineated as early as 1 h postinjection. In terms of contrast between the infect and the background, the images improved with time. From the biodistribution data based on gamma counting of dissected tissues as well as from imaging the rats on the gamma camera, it was clear that the [^{111}In]PEG-liposomes had favorable imaging characteristics as compared to [^{111}In]IgG. One of the most striking beneficial characteristics of the ^{111}In-labeled PEG-liposomes in this model was the almost complete absence of kidney uptake (< 1.0% ID/g). With [^{111}In]IgG, kidney activity exceeds 4.0% ID/g throughout the 48 h time interval. In patients, abdominal abscess localization with [^{111}In]IgG is hampered by the relatively high uptake in the kidneys.[14] The abscess-to-blood ratios obtained with [^{111}In]liposomes were also much higher than those obtained with [^{111}In]IgG in this model (3.3 vs. 1.0 at 48 h). Based on this favorable characteristic one might anticipate that PEG-liposomes—like radiolabeled autologous leukocytes—may be used to image vascular lesions.

^{99m}Tc-Labeled PEG-Liposomes for Imaging Infection and Inflammation

Because of its low costs, good availability, superb physical characteristics and low radiation burden, technetium-99m is by far the most ideal radionuclide for scintigraphy. Several methods have been described to label liposomes with ^{99m}Tc. Philips et al[44] have labeled liposomes with encapsulated glutathione analogously to the widely applied leukocyte labeling method, using the lipophilic chelator HMPAO for transport of the ^{99m}Tc over the lipid bilayer. The [^{99m}Tc]HMPAO complex is reduced by glutathione to its more hydrophilic form and thus trapped within the liposome. We have studied the imaging potential of ^{99m}Tc-labeled PEG-liposomes (100 nm) in several models of infection and inflammation.[45] In rats with *S. aureus* infection, high absolute abscess uptake (up to 1.0% ID/g) and abscess-to-muscle ratios (up to 42, at 24 h pi) were observed. Uptake in background tissues was relatively low, except for the spleen (approximately 12.5% ID/g). Excretion of radioactivity from the body was much higher than observed with the ^{111}In-labeled PEG-liposomes (<5% ID for [^{111}In]liposomes versus 23% ID for [^{99m}Tc]liposomes over 24 h). The enhanced whole-body clearance of the ^{99m}Tc label may be either due to the reduced stability of the ^{99m}Tc-labeled liposomes in vivo, or due to association of the ^{111}In-label with metal binding proteins following the release from the liposomes.

The ^{99m}Tc-labeled PEG-liposomes were also tested in a rat model of sterile inflammation. An inflammatory focus in the left calf muscle was induced by intramuscular injection of 0.1 ml turpentine. This type of inflammatory focus was also clearly visualized, indicating that the presence of microorganisms is not required for accumulation of the liposomes in the inflammatory focus.

Target uptake and retention of the ^{99m}Tc-labeled PEG-liposomes appeared to correlate with the intensity of inflammatory response. *E. coli* infection—with a milder inflammatory response at visual inspection and palpation compared to *S. aureus* infection—showed abscess uptake of 0.6% ID/g and abscess-to-muscle ratios of 25

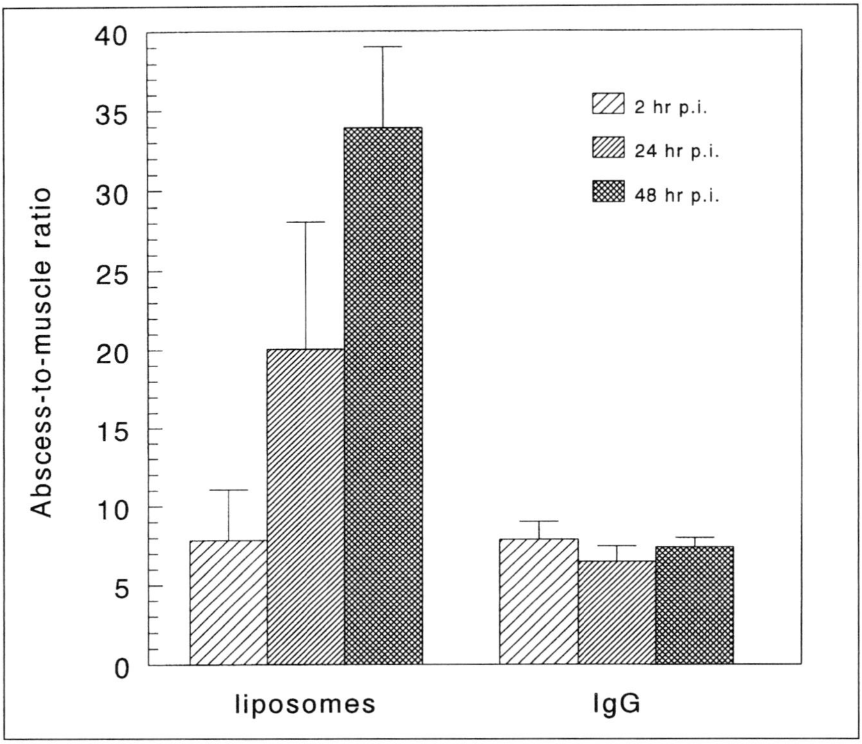

Fig. 16.1. Abscess-to-muscle ratios at 2 h, 24 h and 48 h postinjection for ^{111}In-labeled PEG-liposomes and [^{111}In]IgG in rats with *S. aureus* infection. The biodistribution data of five rats per time point were used. Error bars represent standard deviations.

at 24 h postinjection. In the case of the more severe *S. aureus* infection abscess uptake amounted to 1.0% ID/g and abscess-to-muscle ratios exceeded 40 (24 h pi). Uptake of PEG-liposomes in turpentine-induced inflammation (1.2% ID/g) was even higher than the abscess uptake observed in the *S. aureus* infection (Fig. 16.2).

[^{99m}Tc]liposomes compared favorably with [^{99m}Tc]IgG, a commercially available radiopharmaceutical for imaging infection and inflammation in humans.[18,19] Both abscess uptake and abscess-to-muscle ratios were significantly higher for the labeled PEG-liposomes in a variety of animal models.

The success of scintigraphic imaging of infectious or inflammatory foci, apart from considerations such as availability, costs toxicity and ease of preparation, largely depends upon two factors. First, there needs to be sufficient absolute uptake in the target (the signal) and second, there should be a good target to background ratio (contrast). In a subsequent study we investigated whether the formulation of the PEG-liposomes could be tailored for this application.[46] The mean diameter of the liposomes had a marked effect on their in vivo distribution upon iv administration. PEG-liposomes of 90 nm in diameter combined the two favorable characteristics for imaging: optimal uptake in the infectious focus and minimal splenic accumulation. This preparation not only revealed high absolute uptake of the radiolabel in the tar-

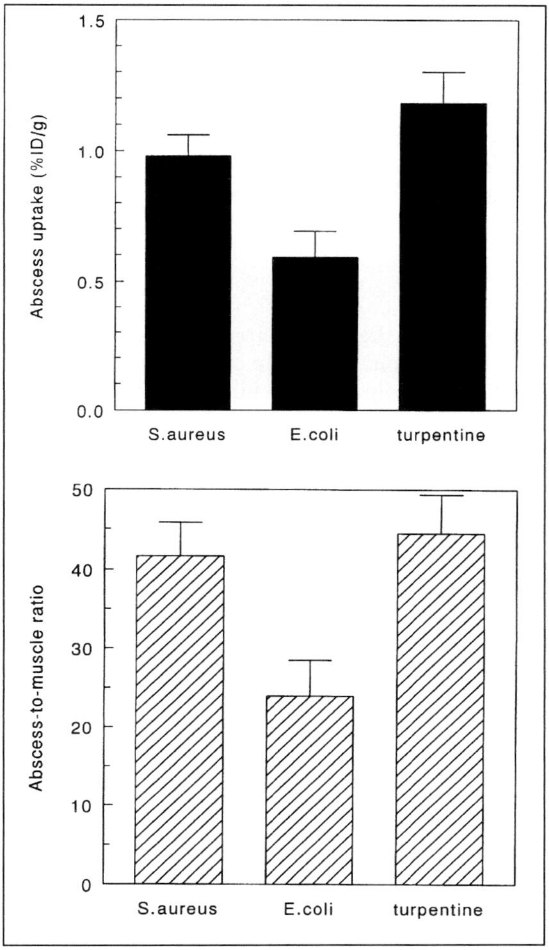

Fig. 16.2. Abscess uptake (top panel) and abscess-to-muscle ratios (bottom panel) at 24 h postinjection for ^{99m}Tc-labeled PEG-liposomes. The biodistribution data of five rats per time point were used. Error bars represent standard errors of the mean.

get (1.6 ± 0.4% ID/g, 24 h pi), but also high abscess-to-contralateral muscle ratios (34.8 ± 8.4). Most importantly, the 90 nm liposomes showed marked reduced splenic uptake (6.9 ± 0.7% ID/g) as compared to the larger liposomes (Fig. 16.3). It has been described that somewhat larger long circulating liposomes tend to accumulate in the spleen.[36,43,45] The relatively high and rapid splenic uptake of larger PEG-liposomes is most likely due to physical filtration rather than phagocytosis by spleen macrophages. In addition, in this study the effect of reduced circulation time of PEG-liposomes was investigated by including increasing amounts of phosphatidylserine (PS) in the lipid bilayer. In the rat model, liposomes with 1.0 mol% PS in the bilayer showed a three-fold higher abscess-to-blood ratio. However, while improved abscess-to-blood ratios were obtained with these PS-containing liposomes, abscess-to-background ratios for liver and spleen decreased.[46] Overall, these studies showed that the in vivo behavior of PEG-liposomes can be modulated to optimize their characteristics for infection imaging. Especially, by carefully tuning their size the localization of long circulating liposomes in the spleen could be reduced.

In terms of diagnostic accuracy, radiolabeled leukocytes are currently the best radiopharmaceutical available for imaging infection and inflammation. However, as indicated above their preparation bears some important disadvantages (blood handling, lack of quality control method, laborious and time-consuming preparation). A liposome-based radiopharmaceutical would have several advantages over radiolabeled leukocytes:

1) Continuous availability.
2) Easy labeling procedure.
3) No need to handle blood.

In patient care, labeled liposomes would then be an attractive agent to replace labeled leukocytes, especially in acute infectious disease. In order to compare the in vivo imaging characteristics of radiolabeled liposomes with those of radiolabeled leukocytes, a comparative study was carried out in a model of infection. Because radiolabeled leukocytes cannot be used in rats, we compared the imaging potential of radiolabeled leukocytes and liposomes in rabbits with focal *S. aureus* infection.[47] The images obtained with ^{111}In and ^{99m}Tc-labeled liposomes (120 nm) and leukocytes are shown in Figure 16.4. Accumulation of the ^{111}In-labeled leukocytes was most rapid as compared to the other imaging agents used in this study. However, from 24 h onwards the images obtained with the ^{111}In-labeled liposomes were as good as those

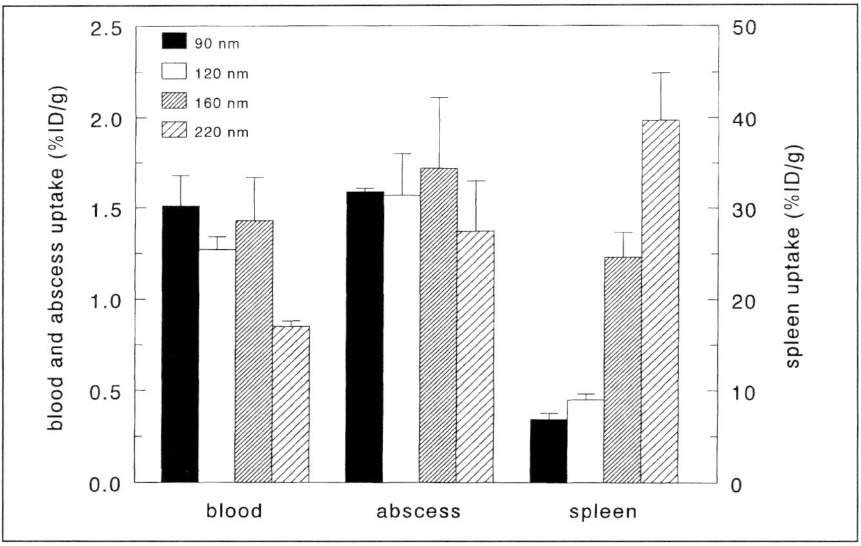

Fig. 16.3. Effect of liposome diameter on the blood level, abscess uptake and spleen uptake of the ^{99m}Tc-labeled PEG-liposomes in rats with *S.aureus* infection of the left calf muscle 24 h postinjection. Five rats per group were used. Error bars represent standard deviations.

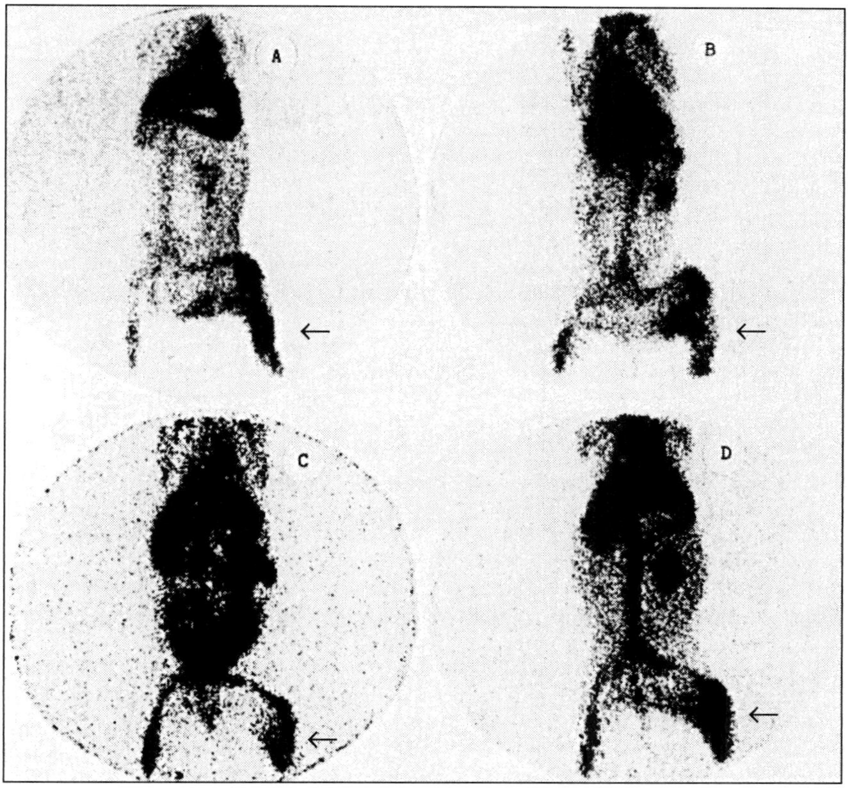

Fig. 16.4. [^{111}In]leukocytes (A) and [^{111}In]liposomes (B), 48 h postinjection; [^{99m}Tc]leukocytes (C) and [^{99m}Tc]liposomes (D), 20 h postinjection in rabbits with *S.aureus* infection. In all images, the *S. aureus* abscess (arrow) in the left hindleg is clearly visualized.

obtained with the [^{111}In]leukocytes. Absolute abscess uptake of the liposomes and the leukocytes (0.16 ± 0.02% ID/g vs. 0.14 ± 0.04% ID/g) as well as the abscess-to-muscle ratios (30 ± 10 vs. 21 ± 4) were similar. When comparing the ^{99m}Tc-labeled preparations in this model, the ^{99m}Tc-labeled liposomes were superior to the [^{99m}Tc]leukocytes, both in terms of abscess uptake as well as abscess-to-muscle ratio.

Summary and Conclusion

Experiments in various experimental models have shown that radiolabeled PEG-liposomes are superior to currently commercially available radiopharmaceuticals for imaging infectious and inflammatory foci. The promising preclinical work along with other attractive features of the formulation of PEG-liposomes (in terms of availability and ease of preparation) represent a significant advance in the field of infection and inflammation imaging. Clinical studies with radiolabeled PEG-liposomes to determine their potential in clinical practice are warranted.

References

1. Lavender JP, Lowe J, Barker JR, Burn JI, Chaudri MA. Gallium-67 citrate scanning in neoplastic and inflammatory lesions. Brit J Radiol 1971; 44:361-366.
2. Ito Y, Okuyama S, Awano T, Takahashi K, Sato T, Kanno I. Diagnostic evaluation of Ga-67 scanning of lung cancer and other diseases. Radiology 1971; 101:355-362.
3. Staab EV, McCartney WH. Role of gallium-67 in inflammatory disease. Semin Nucl Med 1978; 8:219-234.
4. Siemsen JK, Gebe SF, Waxman AD. The use of gallium-67 in pulmonary disorders. Semin Nucl Med 1978; 8:235-249.
5. McAfee JG, Thakur ML. Survey of radioactive agents for the in vitro labeling of phagocytic leucocytes. I Soluble agents. II Particles. J Nucl Med 1976; 17:480-492.
6. Peters AM, Danpure HJ, Osman S, Hawker RJ, Henderson BL, Hodgson HJ, Kelly JD, Neirinckx RD, Lanvender JP. Clinical experience with Tc-99m-HMPAO for labelling leucocytes and imaging infection. Lancet 1986; 2:946-949.
7. Froelich JW, Swanson D. Imaging of inflammatory processes with labeled cells. Semin Nucl Med 1984; 14:128-140.
8. Baba AA, McKillop JH, Cuthbert GF, Neilson W, Gray HW, Anderson JR. Indium-111 leucocyte scintigraphy in abdominal sepsis: do the results affect management? Eur J Nucl Med 1990; 16:307-309.
9. Coleman RE. Radiolabeled leukocytes. In: Freeman LM, Weissmann HS, eds. Nuclear Medicine Annual 1982. New York: Raven Press 1982:119-141.
10. Lange JMA, Boucher CAB, Hollak CEM, Wiltink EH, Reiss P, van Royen EA, Roos M, Danner SA, Goudsmit J. Failure of zidovudine prophylaxis after accidental exposure to HIV-1. N Eng J Med 1990; 323:915-916.
11. Rubin RH, Fischman AJ, Callahan RJ, Khaw BA, Keech F, Ahmad M, Wilkinson R, Strauss HW. ^{111}In-labeled nonspecific immunoglobulin scanning in the detection of focal infection. N Eng J Med 1989; 321:935-940.
12. Oyen WJG, Claessens RAMJ, van der Meer JWM, Rubin RH, Strauss HW, Corstens FHM. Indium-111-labeled human nonspecific immunoglobulin G: A new radiopharmaceutical for imaging infectious and inflammatory foci. Clin Infect Dis 1992; 14:1110-1119.
13. Oyen WJG, van Horn JR, Claessens RAMJ, Slooff TJJH, van der Meer JWM, Corstens FHM. Diagnosis of bone, joint and joint prothesis infections with indium-111 labeled nonspecific human immunoglobulin G scintigraphy. Radiology 1992; 182:195-199.
14. Oyen WJG, Claessens RAMJ, van der Meer JWM, Corstens FHM. Detection of subacute infectious foci with indium-111 labeled autologous leukocytes and indium-111 labeled human nonspecific immunoglobulin G: a prospective comparative study. J Nucl Med 1991; 32:1854-1860.
15. Oyen WJG, Claessens RAMJ, Raemaekers JMM, de Pauw BE, van der Meer JWM, Corstens FHM. Diagnosing infection in febrile granulocytopenic patients with indium-111 labeled human IgG. J Clin Oncol 1992; 10:61-68.
16. Oyen WJG, Netten PM, Lemmens JAM, Claessens RAMJ, Lutterman JA, van der Vliet JA, Goris JA, van der Meer JWM, Corstens FHM. Evaluation of infectious diabetic foot complications with Indium-111-labeled human nonspecific immunoglobulin G. J Nucl Med 1992; 33:1330-1336.
17. De Klein EMHA, Oyen WJG, Claessens RAMJ, Corstens FHM, van der Meer JWM. Arch Intern Med 1995; 155:1989-1994.
18. Blok D, van Ogtrop M, Arndt JW, Camps JAJ, Feitsma RIJ, Goedemans W, Pauwels EKJ. Detection of inflammatory lesions with radiolabelled immunoglobulins. Eur J Nucl Med 1990; 16:303-305.

19. Buscombe JR, Lui D, Ensing G, de Jong R, Ell PJ. [^{99m}Tc]human immunoglobulin (HIG)—first results of a new agent for the localization of infection and inflammation. Eur J Nucl Med 1990; 16:649-655.
20. Claessens RAMJ, Boerman OC, Koenders EB, Oyen WJG, van der Meer JWM, Corstens FHM. Technetium-99m labelled hydrazinonicotinamido human non-specific polyclonal immunoglobulin G for detection of infectious foci: a comparison with two other technetium-labelled immunoglobulin preparations. Eur J Nucl Med 1996; 23:414-421.
21. Nässander UK, Storm G, Peeters PAM, Crommelin DJA. Liposomes. In: Biodegradable Polymers as Drug Delivery Systems (eds. M. Chasin, R. Langer), New York: Marcel Dekker, 1990; 261-338.
22. Crommelin DJA, Schreier H. Liposomes. In: Colloidal Drug Delivery Systems (J. Kreuter, ed.). New York: Marcel Dekker Inc, 1994; 73-89.
23. Jain RK. Determinants of tumor blood flow: A review. Cancer Res 1988; 48:2641-2658.
24. Morrel EM, Tompkins RG, Fischman AJ, Wilkinson RA, Burke JF, Rubin RH, Strauss HW, Yarmush ML. Autoradiographic method for quantitation of radiolabeled proteins in tissues using Indium-111. J Nucl Med 1989; 30:1538-1545.
25. Juweid M, Strauss HW, Yaoita H, Rubin RH, Fischman AJ. Accumulation of immunoglobulin G at focal sites of inflammation. Eur J Nucl Med 1992; 19:159-165.
26. Morgan JR, Williams KE, Davies RL, Leach K, Thomson M, Williams LAP. Localisation of experimental staphylococcal abscesses by [^{99m}Tc]technetium-labelled liposomes. J Med Microbiol 1981; 14:213-217.
27. Morgan JR, Williams LA, Howard CB. Technetium-labelled liposome imaging for deep-seated infection. Br J Radiol 1985; 58:35-39.
28. Williams BD, O'Sullivan MM, Saggu GS, Williams KE, Williams LA, Morgan JR. Synovial accumulation of technetium labelled liposomes in rheumatoid arthritis. Ann Rheum Dis 1987; 46:314-318.
29. O'Sullivan MM, Powell N, French AP, Williams KE, Morgan JR, Williams BD. Inflammatory joint disease: a comparison of liposome scanning, bone scanning, and radiography. Ann Rheum Dis 1988; 47:485-491.
30. Love WG, Amos N, Kellaway IW, Williams BD. Specific accumulation of cholesterol-rich liposomes in the inflammatory tissue of rats with adjuvant arthritis. Ann Rheum Dis 1990; 49:611-614.
31. Barratt GM, Tuzel NS, Tyman BE. The labeling of liposomal membranes with radioactive technetium. In: Gregoriadis G, ed. Liposome technology, volume II. Boca Raton, FL: CRC Press 1984; 94-105.
32. Senior JH. Fate and behavior of liposomes in vivo: a review of controlling factors. Crit Rev Ther Drug Carrier Syst 1987; 3:123-193.
33. Liu D, Mori A, Huang L. Large liposomes containing ganglioside G_{M1} accumulate effectively in spleen. Biochim Biophys Acta 1991; 1066:159-165.
34. Allen TM, Chonn A. Large unilamellar liposomes with low uptake into the reticuloendothial system. FEBS Lett 1987; 223:42-46.
35. Bakker-Woudenberg IAJM, Lokerse AF, ten Kate MT, Storm G. Enhanced localization of liposomes with prolonged blood circulation time in infected lung tissue. Biochim Biophys Acta 1992; 1138:318-326.
36. Bakker Woudenberg IA, Lokerse AF, ten Kate MT, Mouton JW, Woodle MC, Storm G. Liposomes with prolonged blood circulation and selective localization in *Klebsiella pneumoniae*-infected lung tissue. J Infect Dis 1993; 168:164-171.
37. Gabizon A, Papahadjopoulos D. Liposome formulation with prolonged circulation time in blood and enhanced uptake by tumors. Proc Natl Acad Sci USA 1988; 85:6949-6953.

38. Allen TM, Hansen C, Martin F, Redemann C, Yan-Young A. Liposomes containing synthetic lipid derivatives of polyethylene glycol show prolonged circulation half-lives in vivo. Biochim Biophys Acta 1991; 1066:29-36.
39. Blume G, Cevc G. Liposomes for the sustained drug release in vivo. Biochim Biophys Acta 1990; 1029:91-97.
40. Klibanov AL, Maruyama K, Torchilin VP et al. Amphipathic polyethyleneglycols effectively prolong the circulation time of liposomes FEBS Lett 1990; 268:235-237.
41. Storm G, Belliot SO, Daemen T, Lasic DD. Surface modification of nanoparticles to oppose uptake by the mononuclear phagocyte system. Adv Drug Delivery Rev 1995; 17:31-48.
42. Goins B, Klipper R, Rudolph AS, Cliff RO, Blumhardt R, Phillips WT. Biodistribution and imaging studies of technetium-99m-labeled liposomes in rats with focal infection. J Nucl Med 1993; 34:2160-2168.
43. Boerman OC, Storm G, Oyen WJG, van Bloois L, van der Meer JWM, Claessens RAMJ, Crommelin DJA, Corstens FHM. Sterically stabilized liposomes labeled with indium-111 to image focal infection in rats. J Nucl Med 1995; 36:1639-1644.
44. Phillips WT, Rudolf AS, Goins B, Timmons JH, Klipper R, Blumhardt R. A simple method for producing a technetium-99m-labeled liposome which is stable in vivo. Nucl Med Biol 1992; 19:539-547.
45. Oyen WJG, Boerman OC, Storm G, van Bloois L, Koenders EB, Claessens RAMJ, Perenboom RM, Crommelin DJA, van der Meer JWM, Corstens FHM. Detecting infection and inflammation with ^{99m}Tc labeled Stealth® liposomes. J Nucl Med 1996; 37:1392-1397.
46. Boerman OC, Oyen WJG, van Bloois L, van der Meer JWM, Koenders EB, Claessens RAMJ, Crommelin DJA, Corstens FHM, Storm G. Optimization of technetium-qqm-labeled PEG-liposomes to image focal infection: effects of particle size and circulation time. J Nucl Med 1997; 38:489-493.
47. Oyen WJG, Boerman OC, Storm G, van Bloois L, Koenders EB, Crommelin DJA, van der Meer JWM, Corstens FHM. Labeled Stealth® liposomes in experimental infection: an alternative to leukocyte scintigraphy? Nuclear Med Commun 17; 742-748, 1996.

CHAPTER 17

Long Circulating Liposomes for Diagnostic Imaging

Vladimir S. Trubetskoy, Vladimir P. Torchilin

Introduction

The basic property of liposomes to entrap water-soluble substances into their aqueous interior and lipophilic ones in the lipid bilayer makes them potentially attractive carriers for a variety of diagnostic reporter moieties. Since the conception of using liposomes for this purpose,[1] all four major fields of diagnostic imaging, gamma-radioscintigraphy, magnetic resonance imaging (MRI), radiography and ultrasonography, have been studied with liposomal-bound reporters.[2] In fact, the use of liposomal imaging agents advanced to a stage of industrial production and Phase II/III clinical testing of ^{111}In-radiolabeled liposomes.[3] This has been followed by clinical testing with long circulating liposomes, initially for blood pool and tumor imaging[4] (see chapter 18). The particular value of long circulating liposomes for diagnostic imaging is considered here with special focus on cardiovascular and lymph applications.

For liposomal preparations to be useful for virtually any imaging application, the chosen diagnostic moieties or reporter groups should remain tightly associated with them in vivo during the relevant imaging time interval. For computed tomography (CT) imaging, for example, water soluble X-ray contrast agents (primarily, iodine-containing organic compounds) have been entrapped in the inner aqueous compartment of liposomes as a means to obtain stability.[2,5] Interestingly, even liposomes for sonography were prepared incorporating a gas bubble inside the liposome.[6,7] Alternatively, chelated radioactive or paramagnetic heavy metal ions for radioscintigraphy and MRI, respectively, have been covalently bound to phospholipids incorporated into the liposome bilayer,[8,9] but this approach can be more problematic since the moieties in the liposome external surface are subject to disassociation, especially in vivo. Thus approaches have been found to keep a variety of reporter moieties with the liposome in vivo.

Apart from their ability to retain different reporter moieties, other properties have led to the emergence of liposomes as an indispensable tool in diagnostic imaging. As became clear over time, liposomes possess a property which makes them an extremely versatile instrument for different fields of biomedical engineering: preparation by modular assembly of different membrane components. Thus merely by adding or substitution of lipidic modifiers, one can easily control important properties

Long Circulating Liposomes: Old Drugs, New Therapeutics, edited by Martin C. Woodle and Gerrit Storm.

of the liposome, including surface charge, membrane fluidity, ligand-directed specific interactions, etc. Also, due to their particulate nature liposomes are recognized in vivo and taken up by macrophages. The idea to exploit this general nature of particles for imaging macrophage-rich tissues first appeared approximately forty years ago using a radioactive ^{198}Au colloid for imaging lymph nodes.[10] Subsequently, liposome-aided imaging of liver and spleen followed this strategy.[5,8,11]

On the other hand, there are many imaging applications where macrophage uptake is undesirable. In this case, only the liposomes that escape rapid phagocytosis can help to visualize body compartments other than organs and cells of the mononuclear phagocytic system (MPS), primarily the liver and spleen and, to a lesser extent, the bone marrow. Clear examples can be found in cardiac and vascular imaging (visualization of blood pool). For these purposes, the liposomes should stay in circulation for prolonged periods. To this end, the acidic glycolipid ganglioside G_{M1} emerged as the first bilayer constituent able to enhance circulation of liposomes bearing surface charge,[12] generally required for obtaining surface bound metal chelates. Later, amphiphilic PEG incorporated into liposomes modifies their surface characteristics and also was found to convey prolonged circulation[13] (see chapter 1). The ability of polymer-modified liposomes to avoid MPS uptake, was initially studied using in vivo therapeutic models[14,15] but later brought to life the variety of diagnostic applications discussed here. The development of long circulating liposomes with a focus on imaging is summarized in the next section of this chapter.

As one might assume, the achievement of prolonged blood persistence of liposomes loaded with various diagnostic reporter moieties benefits diagnostic blood-pool imaging. An obvious unmet need is better alternatives for ^{99m}Tc-labeled autologous red blood cells in such scintigraphic applications as determination of myocardial ejection fraction and detection of gastrointestinal tract bleedings. In addition to blood pool imaging, prolonged circulation also favors liposome accumulation in tumors or other sites of pathology where enhanced vascular permeability retains macromolecular agents[16] (see chapter 1) and thus can serve as a basis for corresponding imaging agents. Such potential diagnostic applications of long circulating liposomes for infarcted myocardium and percutaneous lymphography will be discussed below with examples taken from our data following a description of the general development of liposomal diagnostic imaging.

Generally speaking, all imaging modalities such as gamma-imaging, MRI, CT and sonography, might benefit from using long circulating contrast liposomes. In addition to imaging blood pool and areas with leaky vasculature where increased circulation poses a clear advantage to obtain sufficient levels of the contrast agent signal, many other targets of small size or limited blood flow might be imaged better when high concentrations of contrast agents are achieved. Long circulating liposomes loaded with contrast agents can provide this high concentration in many cases.

Development of Liposomes for Diagnostic Imaging

The restricted value of imaging agents limited to MPS-rich tissues led to early efforts to obtain liposomal imaging agents that overcome this barrier to reach other tissues. One of the early approaches to change liposome blood clearance time was to presaturate an animal's MPS system with high lipid concentrations (often referred to as RES blockade) before the injection of radiolabeled liposomes.[17] For example, preinjection of "cold" liposomes (DSPC:Chol:DMPG at 5:4:1) at a phospholipid dose of 200 mg/kg resulted in circulation half-life of radiolabeled liposomes of approximately 10 h in rabbits without any manipulations of the liposome properties.[18] It was demonstrated that this approach can improve tumor localization of small liposomes in EMT6 tumor-bearing mice by 50% when compared with nonpreinjected animals.[19]

As mentioned above, one of the factors which can be used to govern behavior of particulates in vivo is their surface modification. For example, extended blood half-lives of liposomes are obtained by ganglioside G_{M1} incorporation. This composition was tested for visualization of implanted subcutaneous tumors in mice using [^{111}In]bleomycin-containing liposomes.[20] Intratumor radioactivity levels up to 10% of ID/g of tissue were achieved 24 h after intravenous administration of the G_{M1}-containing liposomes, an increase by about a factor of 3 to 5. It should be noted, however, that ganglioside G_{M1} does not prolong liposome circulation at increased lipid doses.[21] Other carbohydrate-based amphiphilic bilayer modifiers have been found to prolong liposome circulation and subsequently used for diagnostic imaging. For example, liposomes composed of DPPC:Chol:palmitoyl-D-glucuronide (40:40:20) with encapsulated [^{99m}Tc]DTPA demonstrated prolonged circulation in S180 tumor-bearing mice.[22] A tumor-to-blood ratio of 19.4 was observed compared with 13.8 for unmodified DPPC:Chol at a 24 h time point. Tumor imaging with long circulating liposomes is described in more detail in chapter 18.

Some hydrophilic flexible polymeric chains (like polyethylene glycol, PEG) have been found to be attractive alternatives to the carbohydrates for prolonged circulation of liposomes (see chapter 1). The effect of surface grafted polymers is thought to function via a mechanism involving steric protection of the liposome surface from opsonizing proteins in body fluids including plasma, lymph etc. The mechanism of this phenomenon appears to depend on a flexible nature of the surface-grafted polymer chains.[23,24] In fact, the surface modification with polymers has been found to provide several advantages, including fewer complications of interactions with the reporter moiety and, importantly, demonstrated applicability in humans with a low dose dependence on blood clearance (see chapters 8 and 11).

Regardless of formulation specifics, many imaging applications can benefit from prolonged liposome circulation. Imaging agents associated with long circulating liposomes have clear application in blood-pool imaging, including the measurements of vital cardiac parameters (such as cardiac ejection fraction and wall movement) or gastrointestinal tract bleeding sites. Eventually, such preparations might replace ^{99m}Tc-labeled autologous red blood cells currently used for such procedures in clinical practice. Importantly, liposomes with longer times in the blood have been shown to exhibit passive accumulation in the sites with leaky endothelial lining including tumors[25] and infection sites (see also chapters 1, 5, 6, 12 and 16.[26] Thus long circulating liposomes, and in particular those formed with simple inclusion of PEG-PE, are of considerable interest as carriers for the full range of reporter moieties used in diagnostic imaging not only of the blood pool but also using their passive localization.

The evaluation of PEG-PE containing long circulating liposomes has advanced quite rapidly since the initial suggestion of their value to carry radioactive metals.[27] Early on, Woodle suggested using the [^{67}Ga]desferroxamine complex with PEG-modified liposomes.[27] In this procedure the chelator is entrapped in the aqueous interior during liposome formation, and ^{67}Ga added as an oxine complex to preformed vesicles. With concentrations of desferroxamine high enough (> 10 mM), essentially 100% labeling efficiency was achieved. The preparation assembled demonstrated a blood circulation half-life in rats of approximately 14-15 h but actual imaging tests were not performed.

Another radioactive preparation of PEG-PE-containing liposomes was developed by Tilcock et al for gamma imaging.[28] In their work, ^{99m}Tc was used to label the liposome surface using DPPE-DTTA incorporation. In this case, the radioisotope was bound to the exterior chelator present on preformed liposomes. Generally, this method suffers from nonspecific ^{99m}Tc interactions with the liposomes leading to rapid in vivo leakage with subsequent bladder accumulation. However, the labeling procedure was optimized to minimize this adverse effect. Addition of MPEG-PE to the liposomes increased Tc circulation half-life to 12 h from 30 min with unmodified liposomes. The preparation was even stable to lyophilization in the presence of 150 mM sucrose. Interestingly, the incorporation of ganglioside G_{M1} failed to produce any noticeably prolongation in circulation time for these liposomes. The authors speculate that this might be explained by competition of DTTA and N-acetylneuraminic acid on the G_{M1} moieties for Tc binding on the liposome surface. Unfortunately, bilayer-associated but non-DTTA chelated technetium seems to leak from the liposome in vivo which decreases the apparent circulation time.[28]

To overcome the problems with surface chelators, PEG-PE-containing liposomes with the ^{99m}Tc encapsulated in the internal aqueous compartment were prepared using hexamethylpropyleneamine oxime (HMPAO).[4] The results demonstrated substantial circulation time prolongation in humans of PEG-liposomes compared to nonmodified neutral liposomes or autologous red blood cells with half-lives of 35, 20.5, and 8.5 h, respectively. The sharply increased blood circulation of 99m-Tc-labeled PEG liposomes makes them a good potential contrast agent for the gamma-imaging of the blood pool.

PEG-modified liposomes carrying ^{99m}Tc have been also used for visualization of focal infection sites in rats.[29] The study reported infection site accumulation with PEG-PE-containing liposomes giving an abscess to muscle ratio of 42 (at 24 h after iv administration). This represented 2.7% ID/g which was about three times that of radiolabeled polyclonal IgG, 1.1% ID/g. For imaging, the abscess in the calf muscle of *Staphylococcus aureus*-infected rats was clearly visible 1 h after liposome administration. Thus these long circulating liposomes appear superior to the current standard polyclonal IgG method for infection imaging, discussed in more detail in chapter 16.

Long circulating liposomes have also been considered for use in MRI. Schwendener developed PEG surface modified liposomes with entrapped water-soluble paramagnetic label, gadolinium benzoyloxypropionic tetraacetate (Gd-BOPTA) for MR imaging of vascular system and recently reported preliminary in vivo results.[30] For biodistribution studies, trace amounts of $^{157}GdCl_3$ were added to the BOPTA chelating mixture. The PEG-modified liposomes demonstrated half-life in mice of approx. 90 min, less than generally found using radioactive labels or encapsulated drugs, but substantially more than nonmodified liposomes (15 min). Whether loss of the label, its adverse effect on blood circulation, or another mechanism is responsible for the reduced blood circulation in this study is not clear. Regardless, long circulating liposomes are potentially important for improved MRI contrast agents.

The preclinical and clinical studies summarized above have demonstrated that long circulating liposomes can provide important advantages for blood pool, tumor, and infection imaging using the different imaging techniques and the relevant reporter moieties. Of course, each technique and reporter moiety requires development of the appropriate long circulating liposome preparation that fulfills the many requirements: i.e., signal to noise, target to normal tissue ratio, low label leakage, etc. Nonetheless, the fact that the Tc-labeled liposomes have reached clinical studies for the blood pool demonstrates that clinical studies of other potential roles can be expected to follow quickly. The remaining discussion in this chapter is devoted to potential application to cardiovascular and lymph imaging based on our own work and expanding the technology from simple "passive" targeting to enhancements obtained by use of antibodies to provide greater selectivity, an "active" targeting approach.

Infarcted Myocardium Imaging with Radioscintigraphy

An interesting extension of the results obtained with long circulating liposomes is to use antibody targeting combined with PEG-modification for visualization of lesions readily accessible from the blood. The use of antimyosin antibodies coimmobilized on the liposome surface with PEG chains, can serve as an example of the preparation capable of prolonged circulation and specific binding to the infarcted area of the rabbit myocardium. The specific antibody used (mAb R11D10) effectively binds myosin inside ischemic and/or necrotic cardiomyocytes with affected or destroyed cellular membranes, but does not interact with normal cells with an intact plasma membrane.[31,32] A Fab fragment of this antimyosin-Ab was used to construct a targeted form of radiolabeled long circulating liposomes for imaging the region of ischemically compromised myocardium.[33,34] The liposomes in these studies were prepared with or without the Fab fragment to identify the advance conveyed by the targeting ligand. They also were prepared with 1 mol% of ^{111}In-labeled diethylenetriamine pentaacetic acid conjugated to stearylamine and with and without PEG-PE, giving a total of four test conditions. The following liposome preparations were tested: antimyosin-liposomes with no PEG, liposomes with 4 mol% PEG and without antimyosin, antimyosin-liposomes with 4 mol% PEG, and antimyosin-liposomes with 10 mol% PEG. The half-life of PEG-free immunoliposomes in rabbits was 40 min, and increased to about 17 h for liposomes with 10 mol% PEG-PE. We performed experimental visualization of the myocardial infarction in dogs using such PEG-antimyosin-^{111}In-labeled liposomes. The imaging data are presented on Figure 17.1. The scintigraphic image taken immediately after injection is labeled 0H, that at 8 h later is labeled 8H, showing the infarction site pointed out with the arrow (Fig. 17.1). An image of postmortem isolated heart with myocardium infarction marked with arrows is labeled EX and a series of heart transverse slices is labeled SL. The gamma-scintigraphic image taken immediately after injection, labeled as 0H, does not show any radioactivity accumulation in experimental infarct. However, after 8 h, in the image labeled 8H, one can easily see the accumulation of radioactivity as indicated by the arrow in a certain zone identified as an area of myocardial infarction. The same area is clearly seen on the expanded images of a postmortem heart (image labeled EX with arrows) and the series of individual heart transverse slices, (images labeled SL). Importantly, the radioactive areas on individual slices exactly coincide with the histochemical staining for the necrotic tissues with triphenyl tetrasodium chloride.

For further biodistribution experiments, we used only antimyosin-liposomes with 4 mol% PEG and the control preparation of nonmodified liposomes. The biodistribution of the four types of liposomes in selected tissues are summarized in Figure 17.2. The data obtained show the typical organ distribution found for liposomes with a tendency to accumulate in the liver and spleen. Addition of PEG-PE,

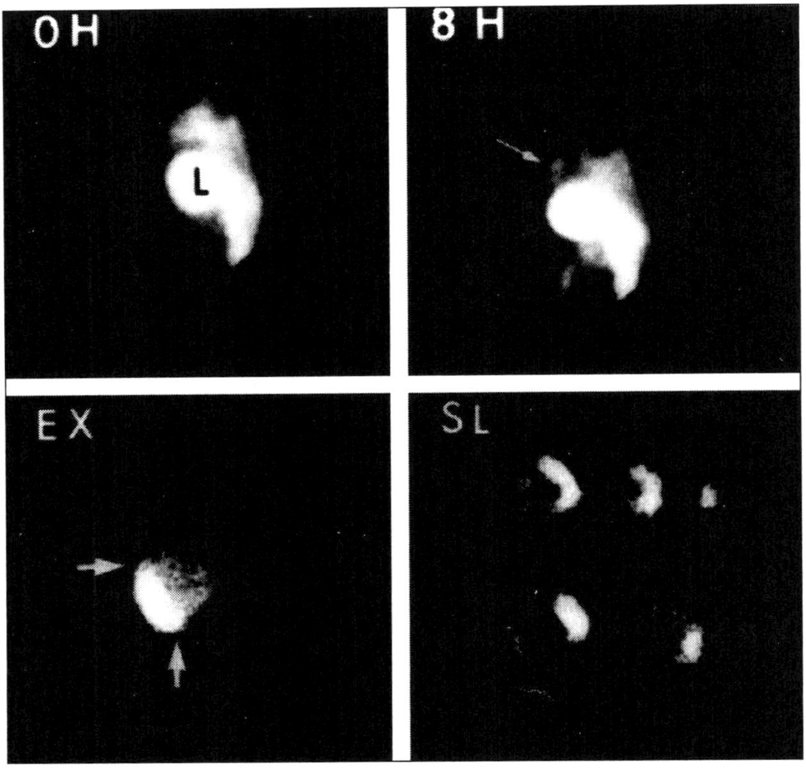

Fig. 17.1. Gamma-scintigraphic images of experimental myocardial infarction in rabbit after injection of ^{111}In-labeled PEG-modified liposomes with antimyosin antibody (see explanations in the text).

even with the Fab, sharply decreases the liposome accumulation in the liver and spleen and increases blood levels. Moreover, the combination of Fab and PEG on the liposome surface gives the greatest accumulation in the infarcted myocardium—almost 0.25% ID/g—about twice that of the Fab-liposomes without PEG-PE. It appears that two factors contribute to target binding: the specific one provided by the antibodies on the surface of the liposomes permitting binding to the target even with a relatively short circulation time; and the nonspecific one due to prolonged circulation that may depend upon filtration in affected tissues apparently requiring many passages through the region. Thus, the maximal absolute delivery can be achieved in the case of liposomes with coimmobilized Fab and PEG. These results suggest that such ligand-targeted long circulating liposome preparations are good candidates for delivery of both imaging and pharmaceutical agents to infarcted myocardium.

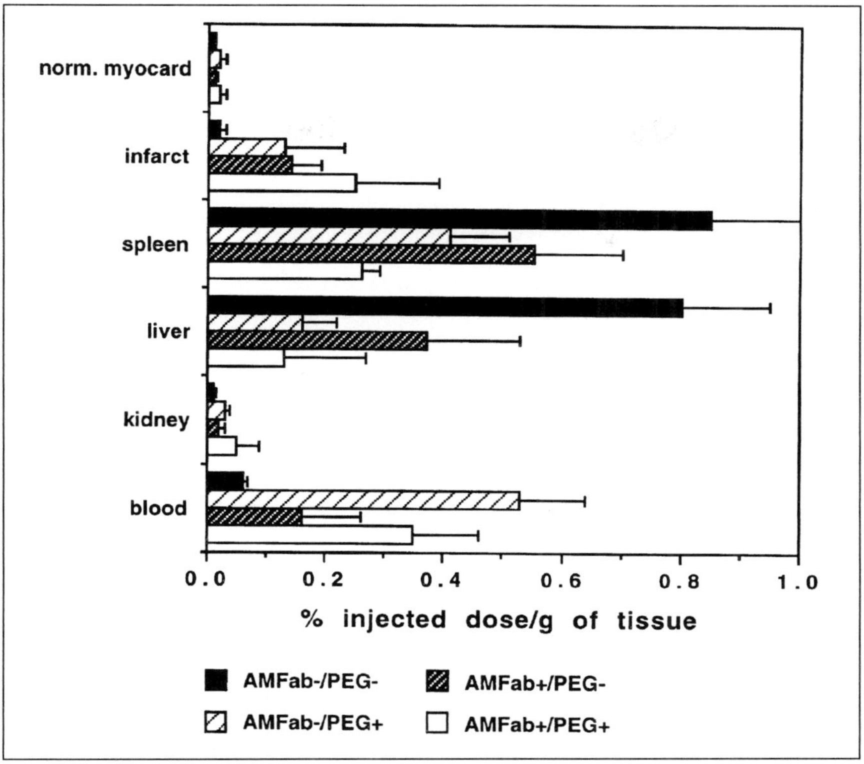

Fig. 17.2. Effects of surface-bound PEG and antimyosin antibody Fab on liposome biodistribution in rabbits 5 h postinjection.

Indirect Percutaneous Lymphography with Magnetic Resonance

Long circulating liposomes can also be used for diagnostic imaging applications beyond intravenous administration of contrast agents. Percutaneous lymphography using subcutaneous administration is another diagnostic use of liposomes. Lymphographic agents which are currently in use or under development generally fall into two major categories: soluble agents (like [^{99m}Tc]human serum albumin) which give measurements of lymph fluid and particulate agents (like radiolabeled colloids or iron oxide-based MR agents) which have the tendency to be accumulated within nodal macrophages. Soluble agents are helpful in depicting lymph vessels and generally depend upon avoidance of macrophage uptake. On the other hand, the action of particulate agents is based on the ability of nodal macrophages to recognize and, in effect, concentrate particulates.

Liposomes, as typical colloids, are phagocytosed by the lymph node macrophages following subcutaneous, intraperitoneal and intramuscular administration and can be used for lymph node visualization.[35] Thus the presence of polymers on the surface of liposomes reducing macrophage uptake can change their behavior following subcutaneous injection, making them more similar to the soluble media. Recent work of Moghimi et al[36] represents a good example of this effect. They reported interesting results in their examination of the influence of the surface PEG chain length absorbed onto polystyrene nanoparticles on the sequestration of these particles by lymph node macrophages following SC administration. It was shown that such surface modification of polystyrene nanospheres can substantially increase their drainage into the lymphatics from the injection site with the best lymph node uptake obtained when using relatively short PEG chains of less than 1000 molecular weight. In our own experiments, we demonstrated that PEG-PE-containing liposomes with Gd-DTPA-PE rapidly and substantially increase MR images of lymph node working with the T_1-weighed signal after subcutaneous injection into rabbit's front paw.[37] The node-to-muscle intensity ratio reached about 2.5 within the first 4 min after administration to the axillary node. The unmodified liposomes only slightly enhanced the lymph nodes and exhibited a substantially slower accumulation rate, as shown in Figure 17.3. Fig. 17.4 shows pre- and 4 min transverse images for unmodified (A), dextran-modified (B) and PEG-modified (C) Gd-liposomes. Note that conclusive visualization of subscapular node can be seen only with PEG-Gd-liposomes.

Lymph node imaging can also be obtained using radioactive liposomes. Interestingly, gentle massage of injection site further increases both the rapidity of visu-

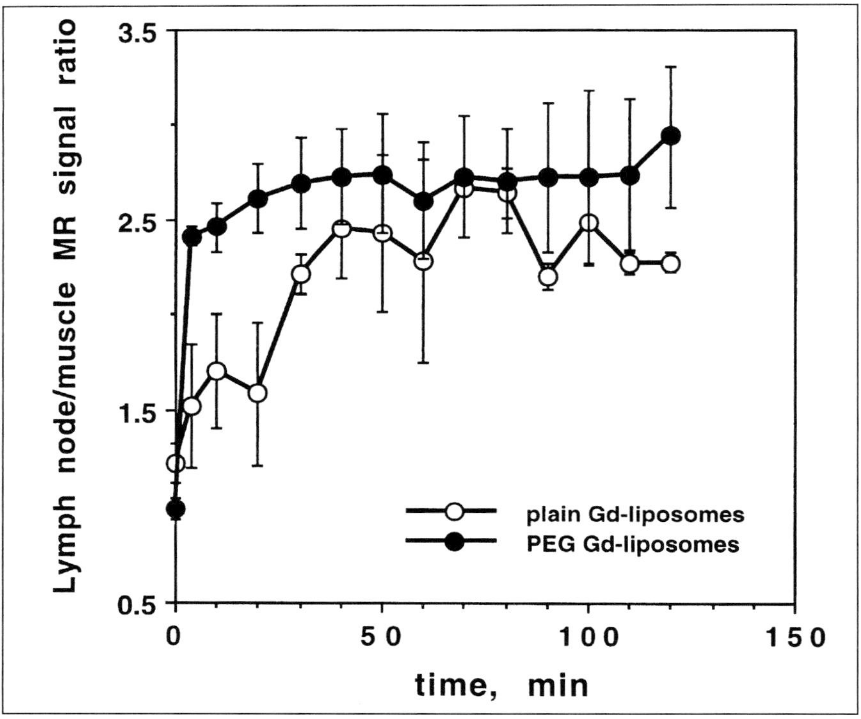

Fig. 17.3. Rabbit subscapular lymph node T_1 signal kinetics after subcutaneous administration of PEG-modified and nonmodified Gd-containing liposomes (1 µmol Gd per injection site in 0.5 ml of saline).

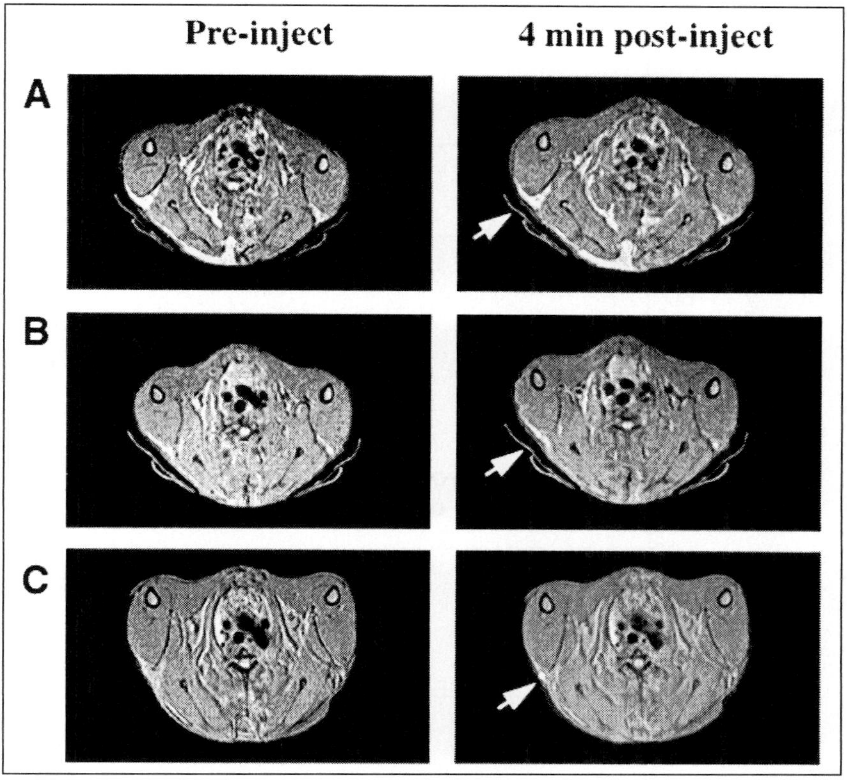

Fig. 17.4. Transverse T_1 images of axillary/subscapular lymph node area in rabbit before and 4 min after subcutaneous administration of plain (A), dextran-modified (B) and PEG-modified (C) Gd-containing liposomes. Subscapular nodes are marked by arrows.

alization and the delivery of the contrast agent to the primary lymph node after subcutaneous administration. Figure 17.5 represents the influence of massage on the popliteal lymph node localization of ^{111}In-labeled PEG-modified liposomes. One can see that massage of the injection site can increase the delivery of liposomes to the primary lymph node by a factor of 5 to 6. Apparently the PEG-modified liposomes can be "squeezed" further down the lymphatic chain by direct massage. On the scintigraphic images of the rabbit lower body (Fig. 17.6) taken immediately after subcutaneous injection (A), before injection site massage (B), after injection site massage (C), and after popliteal node massage (D), one can clearly see the increase in the intranodal radiolabel localization as a result of the massage, and the visualization of the thoracic duct after direct massage of the popliteal node.

It is also possible to use the polymer-modified liposomes containing Gd to visualize intranodal tumors by MRI. Studies were performed with the VX2 carcinoma in rabbit for this application as an imaging contrast agent. To initiate malignant growth, tumor cells were injected directly into the one of the popliteal lymph nodes under

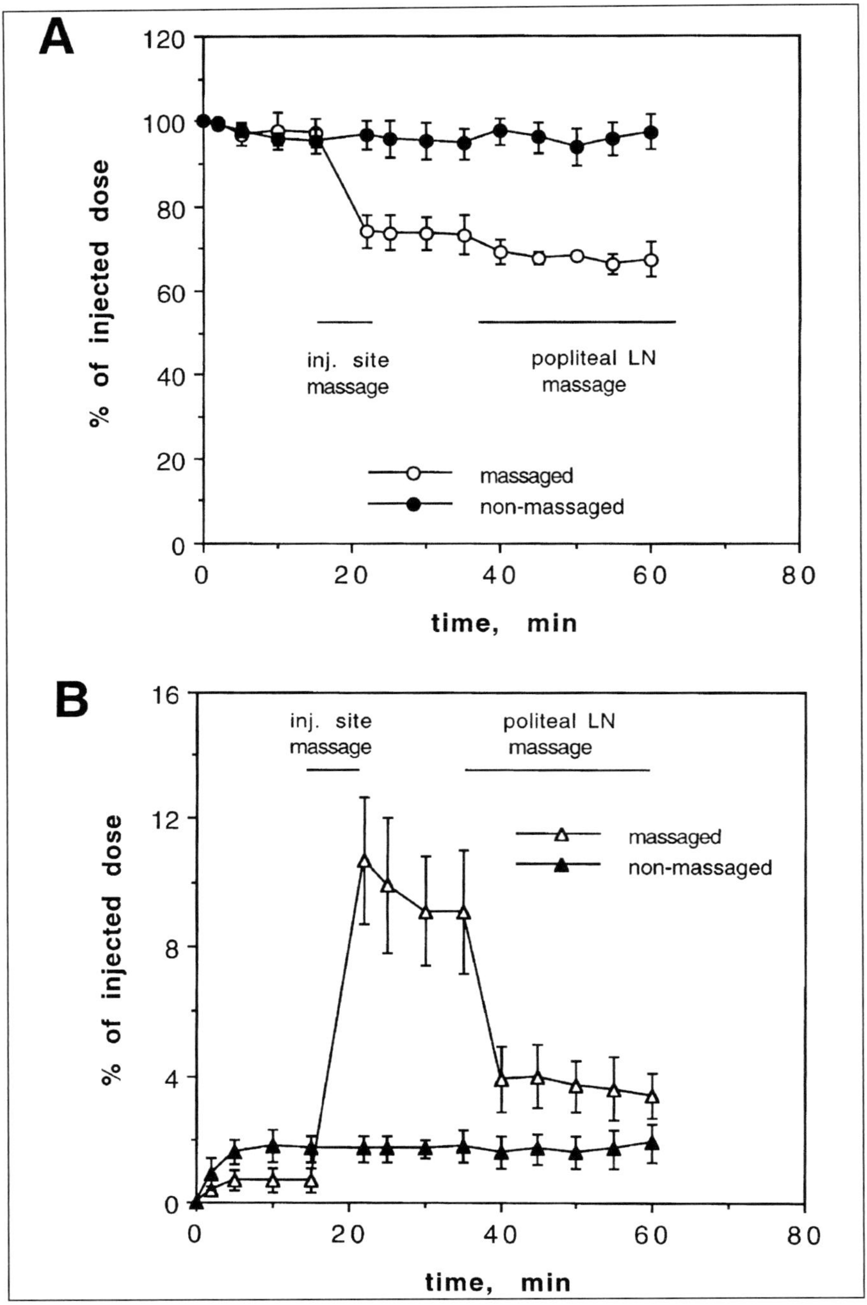

Fig. 17.5. Localization of ^{111}In-labeled PEG-liposomes at the injection site (A) and in the popliteal lymph node (B) after subcutaneous injection into dorsum of rabbit's hind paw.

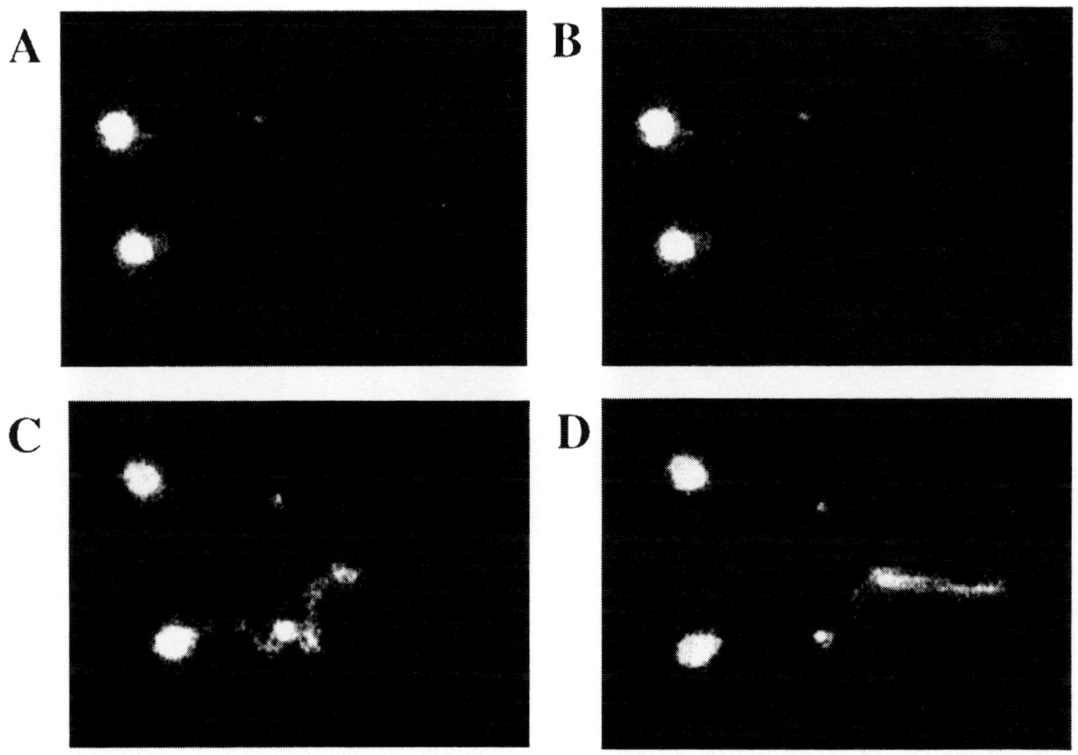

Fig. 17.6. Rabbit lower body scintigraphy after subcutaneous injection of ^{111}In-labeled PEG-liposomes (20 μCi) into dorsum of both hind paws. A, 2 min post-injection image; B, before massage of the injection site; C, after massage of the injection site; D, after massage of the popliteal lymph node.

Fig. 17.7. Transverse T_1 images of popliteal lymph node area in rabbit with intranodal VX2 tumor before and after subcutaneous injection of PEG-modified Gd-containing liposomes (0.5 μmol Gd per injection site). Water- and fat-containing phantoms were placed on top of the rabbit for achieving proper fat suppression.

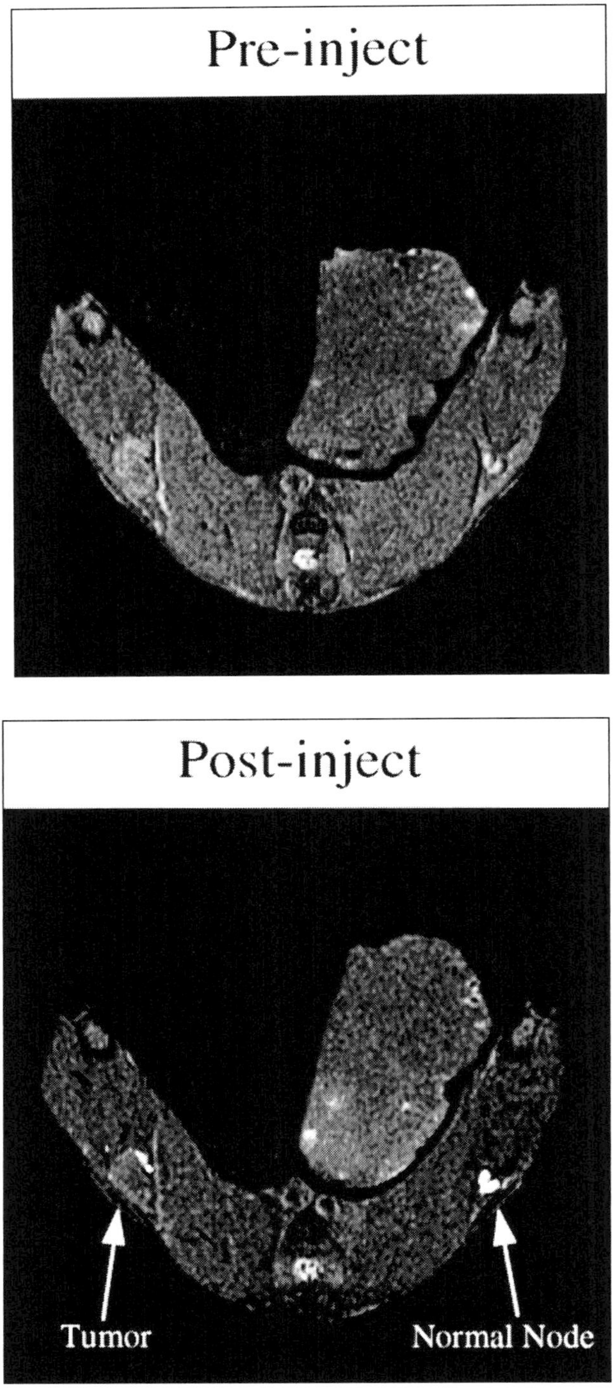

X-ray guidance. Contralateral uninjected node served as a control. Injections of liposomes containing MPEG-PE and Gd-DTPA-PE were performed 10 days after cell inoculation. Tumor growth displaces both sinusoids (lymph fluid pool) and macrophage populations inside the node thus causing filling defects in intranodal contrast agent distribution. This uneven partitioning of the contrast material in the tumor-containing node (as opposed to complete filling for nontumorous nodes) is clearly seen on the T_1-weighted postcontrast image (shown in Fig. 17.7). The above data suggest that subcutaneous injection of polymer-modified diagnostic liposomes results in their association with the lymph fluid and a greater content in normal nodes. Whether their uptake by nodal macrophages occurs or not is not clear from these results but might be determined from kinetic studies. If macrophage uptake is limited, then, unlike other colloidal lymphotropic contrast agents whose actions are based on entrapment by nodal macrophages, these liposomes would be acting like lymphangiographic scintigraphy or MR agents depicting lymph fluid flow through the node. Such behavior would be useful in the visualization of lymphatic system components other than those in the lymph node (e.g., vessels, trunks).

Conclusions

With the emergence of long circulating liposomes, the use of liposomes for contrast agent-aided imaging has received a new impetus. The advantages of these novel carriers over previously used preparations are obvious, especially in such areas as diagnosis of heart disease (determination of vital cardiac parameters, myocardial infarction imaging), cancer (detection of solid tumors and lymph node metastases), and other critical pathologies. The preclinical experimental results obtained in vivo are very promising and support further evaluation through clinical studies. The determination and optimization of the many parameters required to obtain adequate signals and specificity for accurate images can be made via preclinical studies, but the final demonstration of value to the patients and physicians will be made in the clinic.

References

1. McDougal IR, Dunnick JK, Goris ML. In vivo distribution of vesicles loaded with radiopharmaceuticals: A study of different routes of administration. J Nucl Med 1975; 16:488-491.
2. Seltzer SE. The role of liposomes in diagnostic imaging. Radiology 1989; 171:19-21.
3. Presant CA, Turner AF, Proffitt RT. Potential for improvement in clinical decision-making: tumor imaging with In-111 labeled liposomes. Results of a Phase II-III study. J Liposome Res 1994; 4:985-1009.
4. Goins B, Klipper R, Blumhardt, R, Phillips WT. Development of PEG-coated liposomes with extended circulation time for blood pool imaging. Clin Nucl Med 1994; 19:262.
5. Rozenberg OA, Hanson KP. Radiopaque liposomes for imaging of the spleen and liver. Radiology 1983; 149:877-878.
6. Unger E, Fritz T, Shen D-K, Lund P, Sahn D, Ramaswami R, Matsunaga T, Yellowhair D, Kulik B. Gas filled lipid bilayers as imaging contrast agents. J Liposome Res 1994; 4:861-874.
7. Unger E, Shen DK, Fritz T, Lund P, Wu GL, Kulik B, DeYoung D, Standen J, Ovott T, Matsunaga T. Gas-filled liposomes as echocardiographic contrast agents in rabbits with myocardial infarction. Invest Radiol 1993; 28:1155-1159.

8. Kabalka GW, Davis MA, Holmberg E, Maruyama K, Huang L. Gadolinium-labeled liposomes containing amphiphilic Gd-DTPA derivatives of varying chain length: targeted MRI contrast enhancement agents for the liver. Magn Res Imaging 1991; 9:373-377.
9. Schwendener RA, Wuthrich R, Duewell S, Wehrli E, von Schulthess GK. A pharmacokinetic and MRI study of unilamellar gadolinium-, manganese-, and iron-DTPA-stearate liposomes as organ-specific contrast agents. Invest Radiol 1990; 25:922-932.
10. Hultborn KA, Larsson LG, Ragnhult I. The lymph drainage from the breast to the axillary and parasternal lymph nodes, studied with the aid of colloidal Au-198. Acta Radiol 1955; 43:52-64.
11. Unger E, Cardenas D, Zerella A, Fajardo LL, Tilcock C. Biodistribution and clearance of liposomal gadolinium-DTPA. Invest Radiol 1990; 25:638-644.
12. Allen TM, Chonn A. Large unilamellar liposomes with low uptake into the reticuloendothelial system. FEBS Lett 1987; 223:42-46.
13. Klibanov AL, Maruyama K, Torchilin VP, Huang L. Amphipatic polyethyleneglycols effectively prolong the circulation time of liposomes. FEBS Lett 1990; 268:235-238.
14. Papahadjopoulos D, Allen TM, Gabizon A, Mayhew E, Matthay K, Huang SK, Lee K.-D, Woodle MC, Lasic DD, Redemann C, Martin FJ. Sterically stabilized liposomes: improvements in pharmacokinetics and antitumor therapeutic efficacy. Proc Natl Acad Sci USA 1991; 88:11460-11464.
15. Vaage J, Mayhew E, Lasic D, Martin F. Therapy of primary and metastatic mouse mammary carcinomas with doxorubucin encapsulated in long circulating liposomes. Int J Cancer 1992; 51:942-948.
16. Maeda H, Matsumura Y. Tumoritropic and lymphotropic principles of macromolecular drugs. Crit Rev Ther Drug Carrier Systems 1989; 6:193-210.
17. Senior JH. Fate and behavior of liposomes in vivo: a review of controlling factors. CRC Crit Rev Ther Drug Carrier Syst 1987; 3:123-193.
18. Rudolph AS, Klipper RW, Goins B, Phillips WT. In vivo biodistribution of radiolabeled blood substitute: ^{99m}Tc-labeled liposome-encapsulated hemoglobin in anesthetized rabbit. Proc Natl Acad Sci USA 1991; 88:10976-10980.
19. Proffitt RT, Willimas LE, Presant CA, Tin GW, Uliana JA, Gamble RC, Baldeschweiler JD. Liposomal blockade of the reticuloendothelial system: improved tumor imaging with small unilamellar vesicles. Science 1983; 220:502-505.
20. Gabizon A, Price DC, Huberty J, Bresalier RS, Papahadjopoulos D. Effect of liposome composition and other factors on the targeting of liposomes to experimental tumors: biodistribution and imaging studies. Cancer Res 1990; 50:6371-6378.
21. Goins B, Ligler FS, Rudolph AS. Inclusion of ganglioside G_{M_1} into liposome-encapsulated hemoglobin does not extend circulation persistence at clinically relevant doses. Artificial Cells, Blood Substitutes and Immobilization Biotechnology 1994; 22:9-25.
22. Oku N, Namba Y, Takeda A, Okada S. Tumor Imaging with technetium-99m-DTPA encapsulated in RES-avoiding liposomes. Nucl Med Biol 1993; 20:407-412.
23. Torchilin VP, Papisov MI. Why do polyethyleneglycol-coated liposomes circulate so long? J Liposome Res 1994; 4:725-739.
24. Torchilin VP, Omelyanenko VG, Papisov MI, Bogdanov, Jr. AA, Trubetskoy VS, Herron JN, Gentry CA. Poly(ethyleneglycol) on the liposome surface: on the mechanism of polymer-coated liposome longevity. Biochim Biophys Acta 1994; 1195:11-20.
25. Peterson HI. Tumor Blood Circulation. Boca Raton: CRC Press, 1979.
26. Arfors K-E, Rutili G, Sevensjo E. Microvascular transport of macromolecules in normal and inflammatory conditions. Acta Physiol Scand 1979; 463:93.
27. Woodle MC, 67Gallium-labeled liposomes with prolonged circulation:preparation and potential as nuclear imaging agents. Nucl Med Biol 1993; 20:149-155.

28. Tilcock C, Ahkong QF, Fisher D. ^{99m}Tc-labeling of lipid vesicles containing the lipophilic chelator PE-DTTA: effect of tin-to-chelate ratio, chelate content and surface polymer on labelin efficiency and biodistribution behavior. Nucl Med Biol 1994; 21:89-96.
29. Boerman OC, Storm G, Oyen WJG, van Bloois L, van der Meer JWM, Crommelin DJA, Claessens RAMJ, Corstens FHM. Sterically stabilized liposomes labeled with indium-111 to image focal infection in rats. J Nucl Med 1995; 36:1639-1644.
30. Schwendener RA. Liposomes as carriers for paramagnetic gadolinium chelates as organ specific contrast agents for magnetic resonance imaging (MRI). J Liposome Res 1994; 4:837-857.
31. Khaw BA, Mattis JA, Melnicoff G, Strauss HW, Gold HK, Haber E, Monoclonal antibody to cardiac myosin: imaging of experimental myocardial infarction. Hybridoma 1984; 3:11-23.
32. Khaw BA, Yasuda T, Gold HK, Strauss HW, Haber E, Acute myocardial infarct imaging with indium-111-labeled monoclonal antimyosin Fab J Nucl Med 1987; 28:1671-1678.
33. Torchilin VP, Klibanov AL, Huang L, O'Donnel S, Nossiff N, Khaw BA. Targeted accumulation of polyethylene-coated immunoliposomes in infarcted rabbit myocardium. FASEB J 1992; 6:2716-2719.
34. Torchilin VP, Narula J, Halpern E, Khaw BA. Poly(ethylene glycol)-coated anti-cardiac myosin immunoliposomes: factors influencing targeted accumulation in the infarcted myocardium. Biochim Biophys Acta 1996; 1279:75-83.
35. Patel HM, Boodle KM, Vaughan-Jones R. Assessment of potential use of liposomes for lymphoscintigraphy and lymphatic drug delivery. Biochim Biophys Acta 1984; 801:76-86.
36. Moghimi SM, Hawley AE, Christy NM, Gray T, Illum L, Davis SS, Surface engineered nanospheres with enhanced drainage into lymphatics and uptake by macrophages of the regional lymph node. FEBS Lett 1994; 344:25-30.
37. Trubetskoy VS, Cannillo JA, Milstein A, Wolf GL, Torchilin VP. Controlled delivery of Gd-containing liposomes to lymph nodes: surface modification may enhance MRI contrast properties. Magn Res Imaging 1995; 13:31-37.

CHAPTER 18

Tumor Imaging with Long Circulating Liposomes

Naoto Oku

Introduction

Liposomes can be used as ideal drug carriers, and many previous studies have demonstrated the enhanced efficacy and reduced toxicity of encapsulated drugs.[1-6] In the field of nuclear medicine, liposomes have been used as carriers of radionuclides,[7,8] especially in the field of cancer diagnosis and imaging of infection and inflammation (see chapter 16). Effective tumor imaging has been successful using liposomes, since they can encapsulate and deliver a large amount of imaging agents to the target organ, and can reduce the rapid excretion of these agents from the body. Conventional liposomes, however, have a limitation for tumor imaging especially when the tumor reside near the reticuloendothelial system (RES) such as in the liver, since the vesicles tend to be trapped by the RES.

The clearance and RES-trapping of liposomes from the circulation are known to be greatly affected by liposomal size and lipid composition. Previous results indicate that the rigid small-sized (100-200 nm) liposomes tend to be retained in the bloodstream without disruption for a prolonged period of time.[9] Small-sized liposomes composed of distearoylphosphatidylcholine (DSPC) and cholesterol are one such example, as described below.

In view of the lifespan of erythrocytes, which is about 120 d, vesicles composed of biomembrane components might have potential for long circulation times. Thus, attempts have been made to prepare liposomes with longer circulation times by the modification of the liposomal surface to mimic biomembranes. To avoid the RES-trapping of liposomes and to obtain longer half-lives of liposomes in the bloodstream, various investigators have modified the liposomal surface with gangliosides or hydrophilic bulky molecules; successful results were obtained by the modification of liposomes with monosialoganglioside G_{M1},[10-12] glucuronide derivative,[13,14] and polyethylene glycol (PEG).[15-19] Interestingly, G_{M1} modification was shown to be effective only in mice and not in rats[20] nor rabbits.[21]

These long circulating liposomes show a further advantage for tumor imaging since they accumulate effectively in tumor tissues, as the vasculature in the tumor tissues is leaky enough to allow extravasation of small-sized liposomes. Although the systemic background is increased by use of long circulating liposomes, effective tumor accumulation of liposomes gives clear images of tumors, as shown below.

Long Circulating Liposomes: Old Drugs, New Therapeutics, edited by Martin C. Woodle and Gerrit Storm.

Possible solutions to the problem of high systemic background are being studied. Reduction of the systemic background of liposomal agents in the bloodstream was also attempted by inducing RES-mediated removal of the residual circulating liposomes from the blood after appropriate amounts of liposomes had accumulated in tumor tissue (see below).[22] In this review, I will focus on radionuclide imaging only. However, liposomes have been applied as carriers of magnetic resonance imaging agents,[23,24] with good results. Also, owing to the air solubility in hydrophobic regions, liposomes have been recently applied for ultrasound echo imaging.[25]

Tumor Imaging Using Rigid Small-Sized Liposomes

Based on the previous studies, rigid small-sized liposomes have been shown to have rather long circulating capacity among the 'conventional' liposome types. One of such type of liposomes composed of DSPC and cholesterol (2:1), is currently commercialized. These liposomes contain ionophore A23187, which enables the incorporation of Indium-111 in the aqueous phase of preformed liposomes. Liposomal ^{111}In imaging in nude mice bearing human tumor xenografts indicated that these [^{111}In]liposomes were useful for tumor imaging, although the RES was also imaged clearly.[26] Clinical studies of these liposomes have been performed. Such liposomes also accumulated preferentially in RES after systemic imaging, although spot imaging (as shown in Fig. 8.1) with [^{111}In]liposomes accurately defined a metastatic site, that could not be well visualized by conventional scanning.[27] Through a Phase III trial with 38 patients, the sensitivity of ^{111}In-liposomes for detecting tumors in primary sites was 82% (10/12) and at metastatic sites was 65% (17/26).[28] Tumors imaged through Phase II and Phase III studies were breast, lung, head-neck, prostate, colon, ovary, cervix, thyroid, kidney, testes, melanoma, sarcoma and lymphoma. The frequency of false-positive scans was also quite low (2%).[28]

To enhance the trapping efficiency of liposomes for the radionuclide Technetium-99m, Goins and co-workers used a lipophilic chelator, hexamethylprophleneamine oxide (HMPAO), which carried ^{99m}Tc into preformed DSPC-based liposomes containing reduced glutathione. The [^{99m}Tc]HMPAO molecule in liposomes is thought to be reduced by glutathione, becoming more hydrophilic and trapped within the liposomes. Actually, the labeling efficiencies were 91% for negative liposomes and 72% for neutral liposomes.[29] They also showed the effective imaging of tumors by neutral [^{99m}Tc]liposomes in nude mice which were intramuscularly implanted with Chinese hamster ovary cells.

Tumor Imaging Using Long Circulating Liposomes

As described above, liposomes are recognized and trapped by RES, and so suppression of the RES-recognition should result in an increased half-life of liposomes in bloodstream. Long circulating liposomes tend to accumulate in tumor tissues by passive targeting. Therefore, they are useful tools for tumor imaging.

Gabizon and co-workers examined the biodistribution and tumor-imaging ability of radionuclide-encapsulated long circulating liposomes composed of hydrogenated phosphatidylcholine (HPC), cholesterol and a small molar fraction of G_{M1} or hydrogenated phosphatidylinositol (HPI).[30] Phosphatidylinositol (PI) is a common component of biomembranes, and its incorporation into liposomal membrane is known to prolong liposomal circulation.[11] Gallium-67-desferoxamine or [^{111}In]bleomycin was incorporated into liposomes containing G_{M1} or HPI, and these liposomes were injected intravenously into mice bearing B16 melanoma, J6456

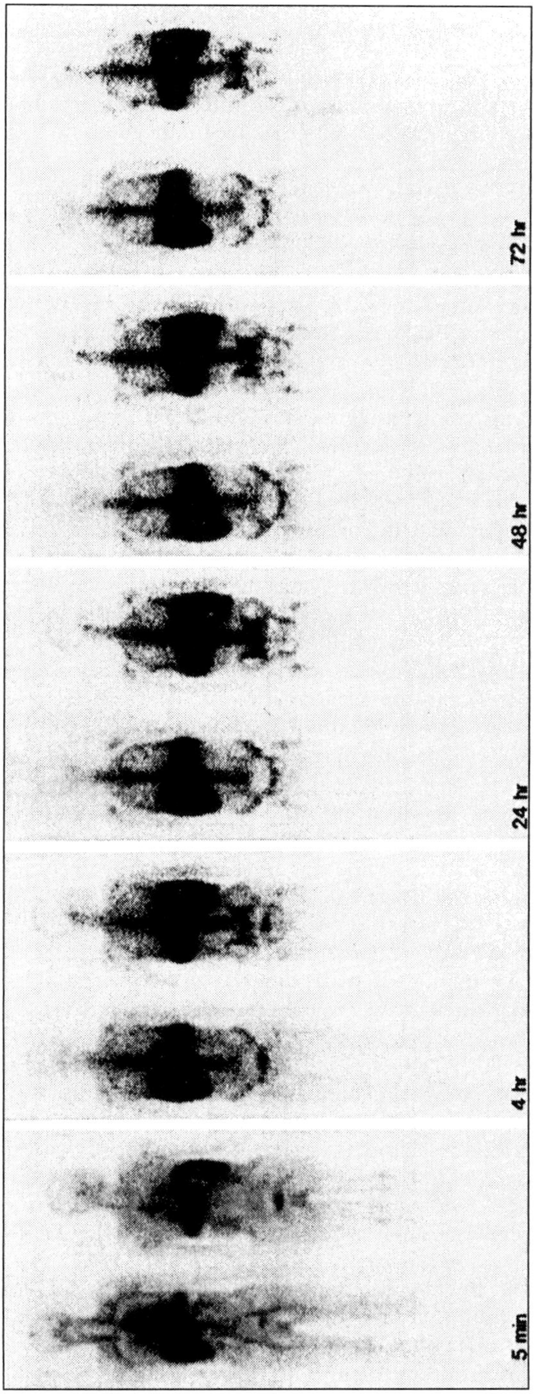

Fig. 18.1. Serial whole-body images with [^{111}In]liposomes. A 59-year-old male was injected with [^{111}In]liposomes composed of DSPC, cholesterol and ionophore A23187 (2:1:0.004 as a molar ratio). The dose of ^{111}In injected was 37MBq in a total lipid dose of 100 mg. Reprinted with permission from Kubo A et al, Eur J Nucl Med 1993; 20:107-113.

lymphoma or LS174T human colon carcinoma. More than 10% of injected dose/g of [^{67}Ga]desferoxamine was observed to have accumulated in tumor. Consequently, effective imaging of the tumors reflecting the biodistribution was also observed. Tumor imaging, however, was successful only because of the site of tumor was implanted at a non-RES site, i.e., a hind leg; radioactivity that accumulated in the RES did not interfere with the tumor imaging.

Tumor imaging mediated by PEG-liposomes containing [^{67}Ga]desferoxamine was proposed by Woodle et al.[31] In that study, liposomes entrapping desferoxamine mesylate were incubated with Ga-oxine for effective labeling of liposomes.

Another type of long circulating liposome, namely palmitylglucuronide (PGlcUA)-modified liposomes, has also been used for tumor imaging. An attractive feature of the use of the long circulating liposome system appears to be that not only tumors at non-RES sites but also tumors in closer proximity of the liver can be efficiently imaged.[32]

Prior to the performance of the tumor imaging study, the biodistribution of PGlcUA-liposomes in tumor-bearing mice was examined by use of liposomes double-labeled with [^{3}H]inulin and cholesteryl[^{14}C]oleate. These liposomes, composed of dipalmitoylphosphatidylcholine (DPPC), cholesterol and PGlcUA (2:2:1 as a molar ratio), were found to effectively accumulate in the tumor tissue.[33] As a reflection of the tumor accumulation of liposomes, the radioactivity in plasma was decreased in tumor-bearing mice compared with that in normal ones. The reason for the accumulation of PGlcUA-liposomes in the tumor tissue might be passive targeting due to partial avoidance of liposomal RES-trapping, although the presence of a specific interaction between tumor tissues and glucuronate on the surface of the liposomes can not be excluded at the present. The tumor accumulation of liposomes is probably not due to an impaired RES-function in the tumor-bearing mice, since the degree of liver accumulation of PGlcUA-liposomes in such animals was similar to that in normal mice. Interestingly, the recovery of radioactivity of ^{14}C was comparable to that of ^{3}H in the tumor tissue, suggesting the localization of intact liposome structures in the tumor. These results are consistent with other findings suggesting that long circulating liposomes are not taken up by the tumor cells, but rather accumulate in the interstitial spaces in the tumor tissue.

When PGlcUA-liposomal [^{99m}Tc]DTPA was administered to sarcoma 180-bearing mice, the time course of radioactivity of liposomal [^{99m}Tc]DTPA remaining in blood was higher than that of the radionuclide encapsulated in conventional liposomes after injection. In this experiment, conventional liposomes were composed of DPPC, cholesterol and dipalmitoylphosphatidylglycerol (DPPG) (further referred to as 'DPPG-liposomes'). When [^{99m}Tc]DTPA was injected in free form instead of as a liposomal formulation, only 0.45% of the injected dose was detected in the kidneys at 2 h after injection, and no tissue was found to have more than 0.1% of this radiopharmaceutical agent at 8 h after injection. On the contrary, radioactivity was detected in various tissues even at 24 h after administration when liposomal formulations were used. A large amount of radioactivity was recovered from tumor tissue when PGlcUA-liposomes were injected, whereas [^{99m}Tc]DTPA encapsulated in DPPG-liposomes preferentially accumulated in the liver.

Tumor imaging using free and liposome-encapsulated [^{99m}Tc]DTPA was performed with a γ-camera. Corresponding to the biodistribution, tumor tissue was mainly and clearly imaged by [^{99m}Tc]DTPA encapsulated in PGlcUA-liposomes, especially at 4 and 8 h after administration of the liposomes. The image at 8 h after injection is shown in Figure 18.2. Some [^{99m}Tc]DTPA in DPPG-liposomes also accumulated in tumor tissue, but most of it accumulated in the RES. It is concluded that PGlcUA-liposomes may be suitable for tumor imaging even when a tumor resides in or near the liver.

Improved Liposomal Tumor Imaging by Reduction of Systemic Background

As discussed above, long circulating liposomes can accumulate in tumors to a considerable degree. However, this long circulation time of radionuclide in the bloodstream also causes a high systemic background. Recently, Ogihara-Umeda and collaborators presented a unique method to reduce the background after accumulation of radionuclide in tumors.[22,34] Such approaches could shorten the lag time preceding the collection of scintigraphic images.

Ogihara-Umeda et al utilized liposomes composed of DSPC, cholesterol and biotinyl-phosphatidylethanolamine, preloaded with desferoxamine. The preformed liposomes were labeled with ^{67}Ga or ^{111}In. By the addition of avidin, aggregation of these liposomes resulted due to specific and strong binding between avidin and biotin. Such liposome aggregation, when induced in vivo, leads to rapid elimination of the circulating liposomes by the RES.

[^{111}In]desferoxamine-encapsulated biotin-modified liposomes were injected intravenously into sarcoma 180-bearing mice. The whole-body imaging was performed before and after avidin treatment at 2 h after the administration of the [^{111}In]liposomes. Before the injection of avidin, the heart was clearly visible, reflecting the high blood level of the radionuclide. The tumor area, however, was not visualized. The avidin treatment caused a dramatic, rapid reduction of the blood background. As a result, radioactivity shifted to the liver. In addition to the liver, the tumor was visualized shortly after injection of avidin. Figure 18.3 shows the time activity curves in the regions of interest, namely, heart, liver, and tumor. The radioactivity in the heart, which indicated the blood pool, was decreased largely within 5 min after avidin injection. Correspondingly, radioactivity in the liver increased rapidly, suggesting liposomes in the blood were trapped in the RES due to their aggregation. Since the activity in the tumor did not change by the addition of avidin, the tumor-versus-blood ratio increased significantly. This methodology for reducing systemic background of radionuclide presented above helps to improve tumor images and enables one to obtain scintigraphy in a short time period, although the method is not suitable for tumors located near or in the RES.

Tumor Imaging by Positron Emission Computed Tomography (PET)

We have developed a method to investigate the real-time liposomal trafficking in vivo using positron emission tomography (PET).[35,36] This technique is also applicable for tumor imaging. The first studies investigated the real-time biodistribution of long circulating liposomes.[35] PGlcUA-liposomes encapsulating [2-^{18}F]-2-fluoro-2-deoxyglucose ([2-^{18}F]FDG) were administrated to mice bearing Meth A sarcoma, and PET scan performed for 120 min. Time activity curves indicate that only the smallest long circulating liposomes, 100 nm, accumulate in the tumor as compared with conventional DPPG-liposomes.

Subsequent studies compared trafficking and tumor imaging achieved with three kinds of long circulating liposomes, ganglioside G_{M1}-, PEG- and PGlcUA-modified liposomes, all having a size of 100 nm.[37] These were investigated in mice bearing

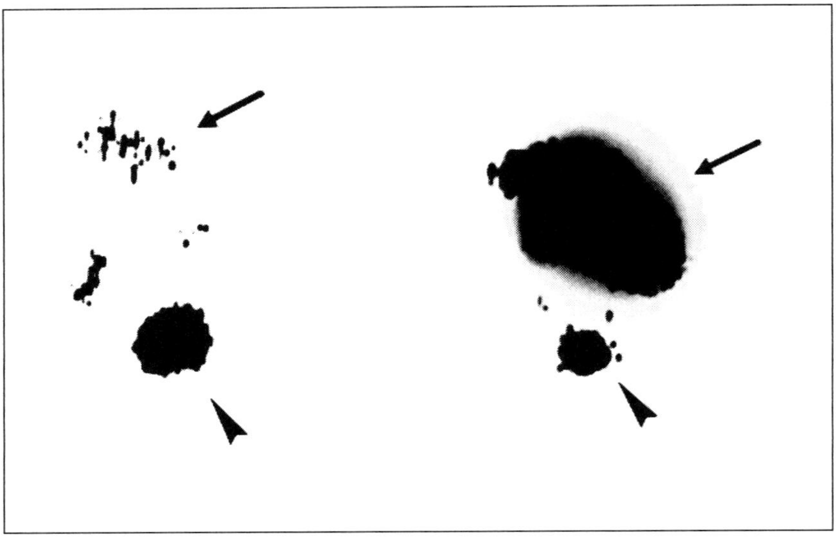

Fig. 18.2. Imaging with [^{99m}Tc]DTPA-labeled PGlcUA-liposomes in sarcoma 180-bearing mice. Sarcoma 180-bearing mice were injected intravenously with [^{99m}Tc]DTPA-containing PGlcUA-liposomes (left image) or DPPG-liposomes (right). Gamma-camera images were obtained at 8 h after injection of the liposomes. The upper part shows the liver accumulation, and the lower part, the tumor accumulation of radionuclide. Arrows indicate liver; and arrowheads, tumor.

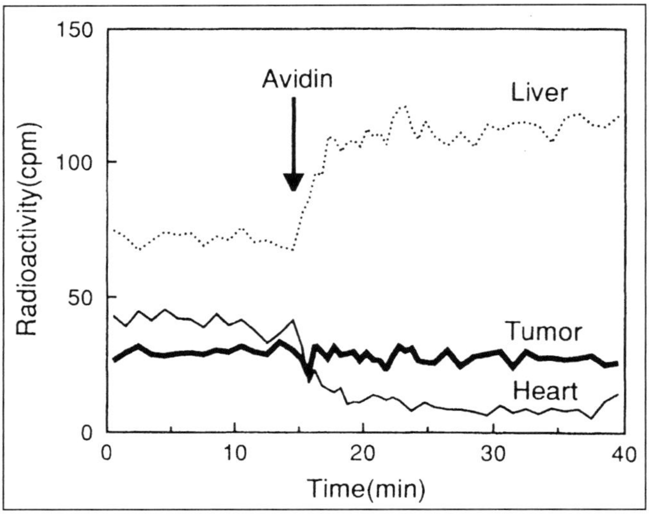

Fig. 18.3. Biodistribution change of biotin-modified liposomes by the addition of avidin. Biotin-modified liposomes with encapsulated [^{111}In]desferoxamine were injected into mice bearing sarcoma 180, and biodistribution of the radionuclide was imaged with a γ-camera. Time-activity curves in the regions of interest over tumor, liver, and heart are shown. Dynamic acquisition was started at 105 min after the administration of the biotin-modified liposomes, and avidin was administered intravenously after an additional 15 min. Reprinted with permission from Ogihara-Umeda I et al, Cancer Res 1994; 54: 463-467.

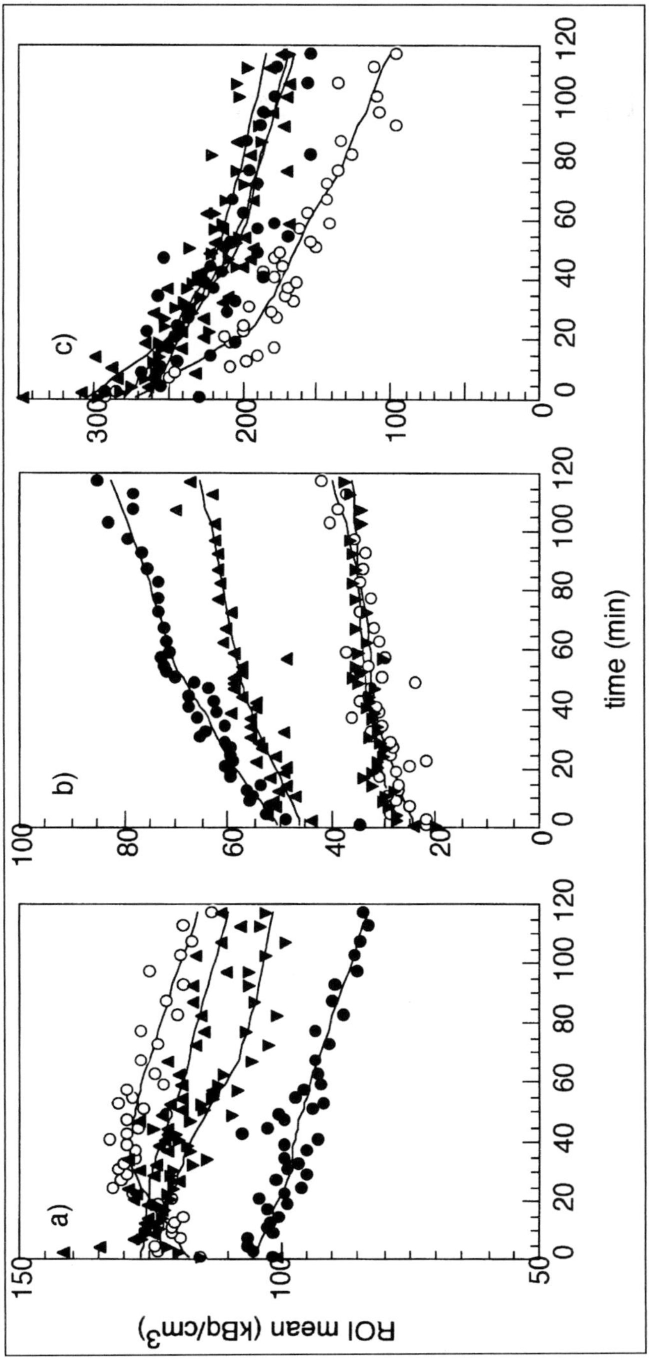

Fig. 18.4. Accumulation of long circulating liposomes in liver and tumor, and retention of them in the bloodstream. Three kinds of long circulating liposomes as well as DPPG-containing liposomes were positron labeled with [2-^{18}F]FDG, and injected into Meth A sarcoma-bearing mice. Then, PET scan was performed. Time-activity curves of 18F in liver (a), tumor (b) and blood pool, namely, heart (c) were obtained. PGlcUA-liposomes (●), PEG-liposomes (▲), G_{M1}-liposomes (▼), and conventional DPPG-liposomes (○).

Meth A sarcoma. Time activity curves indicated that both PEG-liposomes and PGlcUA-liposomes efficiently accumulated in the tumor as compared to both G_{M1}-based long circulating and conventional, DPPG, liposomes. Note that long circulating liposomes using G_{M1}, however, remained in the blood circulation and avoided liver-trapping compared with the conventional DPPG-liposomes. These data suggest that long circulating liposomes may also find application as carrier of PET tumor-imaging agents.

Conclusion

Liposomes can be useful as carriers of radionuclides in the field of diagnostic tumor imaging. They can encapsulate and deliver a large amount of radionuclide to the tumor. One disadvantage of liposomal formulation is that the imaging of tumors located near or in the RES is limited due to strong RES-trapping of liposomes. Long circulating liposomes, however, can overcome this problem and might allow tumor imaging near or even in the RES. In addition, long circulating liposomes localize in tumors to a much higher extent than conventional liposomes.

Although a long circulation time of liposomes is critical for achieving substantial tumor accumulation, it also causes high systemic background. This problem of low and insufficient target-to-background ratios may be solved by utilizing a scavenging method to reduce the background after accumulation of liposomes in the tumor.

Overall, radionuclide-carrying liposomes may be useful for tumor imaging, especially when long circulating liposomes are used. If there is no need to image tumors in the vicinity of the major RES organs, improved tumor imaging (in particular shortened lag time preceding visualization) can be achieved by approaches based on reduction of systemic background.

References

1. Rolland A, ed. Pharmaceutical Particulate Carriers: Therapeutic Applications. New York: Marcel Dekker, 1993.
2. Namba Y, Oku N. Liposomal applications to cancer therapy. J Bioact Compat Polymers 1993; 8:158-177.
3. Gregoriadis G, ed. Liposome Technology. 2nd ed. Boca Raton, FL: CRC Press, 1992.
4. Roerdink FH, Kroon AM, eds. Drug Carrier Systems. New York: John Wiley and Sons, 1989.
5. Hagiwara A, Takahashi T, Oku N. Cancer chemotherapy administered by activated carbon particles and liposomes. CRC Crit Rev Oncol/Hematol 1989; 9:319-350.
6. Gregoriadis G. Liposomes as Drug Carriers. Recent Trends and Progress. New York: John Wiley and Sons, 1988.
7. Seltzer SE. Liposomes in diagnostic imaging. In: Gregoriadis G, ed. Liposomes as Drug Carriers. New York: John Wiley and Sons, 1988:509-525.
8. Perez-Soler R. Liposomes as carriers of antitumor agents: toward a clinical reality. Cancer Treatment Rev 1989; 16:67-82.
9. Senior J. Fate and behavior of liposomes in vivo: a review of controlling factors. Crit Rev Ther Drug Carrier Syst 1987; 3:123-139.
10. Allen TM, Chonn A. Large unilamellar liposomes with low uptake into the reticuloendothelial system. FEBS Lett 1987; 223:42-46.
11. Gabizon A, Papahadjopoulos D. Liposome formulations with prolonged circulation time in blood and enhanced uptake by tumors. Proc Natl Acad Sci USA 1988; 85:6949-6953.

12. Allen TM, Hansen C, Rutledge J. Liposomes with prolonged circulation times: factors affecting uptake by reticuloendothelial and other tissues. Bichim Biophys Acta 1989; 981:27-35.
13. Namba Y, Sakakibara T, Masada M et al. 1-O-Palmityl-D-glucuronate endows liposomes with long half-life in Vivo. Chem Lett 1989; 1989:2145-2148.
14. Namba Y, Sakakibara T, Masada M et al. Glucuronate-modified liposomes with prolonged circulation time. Chem Pharm Bull 1990; 38:1663-1666.
15. Blume G, Cevc G. Liposomes for the sustained drug release in vivo. Biochim Biophys Acta 1990; 1029:91-97.
16. Klibanov AL, Maruyama K, Torchilin VP et al. Amphiphathic polyethyleneglycols effectively prolong the circulation time of liposomes. FEBS Lett 1990; 268:235-237.
17. Allen TM, Austin GA, Chonn A et al. Uptake of liposomes by cultured mouse bone marrow macrophages: influence of liposome composition and size. Biochim Biophys Acta 1991; 1061:56-64.
18. Senior J, Delgado C, Fisher D et al. Influence of surface hydrophilicity of liposomes on their interaction with plasma protein and clearance from the circulation: studies with poly(ethylene glycol)-coated vesicles. Biochim Biophys Acta 1991; 1062:77-82.
19. Klibanov AL, Maruyama K, Beckerleg AM et al. Activity of amphipathic poly-(ethylene glycol)5000 to prolong the circulation time of liposomes depends on the liposomal size and is unfavorable for immunoliposome binding to target. Biochim Biophys Acta 1991; 1062:142-148.
20. Yamauchi H, Yano T, Kato T et al. Effects of sialic acid derivative on long circulation time and tumor concentration of liposomes. Int J Pharm 1995; 113:141-148.
21. Tilcock C, Ahkong QF, Fisher D et al. Polymer-derivatized technetium ^{99m}Tc-labeled liposomal blood pool agents for nuclear medicine applications. Biochim Biophys Acta 1993; 1148:77-84.
22. Ogihara-Umeda I, Sasaki T, Toyama H et al. Rapid tumor imaging by active background reduction using biotin-bearing liposomes and avidin. Cancer Res 1994; 54:463-467.
23. Tilcock C. Liposomal paramagnetic magnetic resonance contrast agents. In: Gregoriadis G, ed. Liposomal Technology. 2nd ed. Vol 2. Boca Raton, FL: CRC Press, 1992:65-87.
24. Unger E, Fritz T, Wu G et al. Liposomal MR contrast agents. J Liposome Res 1994; 4:811-834.
25. Unger E, Fritz T, Shen D-K et al. Gas filled lipid bilayers as imaging contrast agents. J Liposome Res 1994; 4:861-874.
26. Nakamura K, Kubo A, Katayama M et al. Indium-111-labeled liposomes: quality and biodistribution in mice bearing human tumor xenografts. Drug Delivery 1993; 1:69-73.
27. Kubo A, Nakamura K, Sammiya T et al. Indium-111-labelled liposomes: dosimetry and tumour detection in patients with cancer. Eur J Nucl Med 1993; 20:107-113.
28. Presant CA, Turner AF, Proffitt RT. Potential for improvement in clinical decision-making: tumor imaging with In-111 labeled liposomes. Results of a Phase II-III study. J Liposome Res 1994; 4:985-1008.
29. Goins B, Klipper R, Rudolph AS, Phillips WT. Use of technetium-99m-liposomes in Tumor Imaging. J Nucl Med 1994; 35:1491-1498.
30. Gabizon A, Pric, DC, Huberty J et al. Effect of liposome composition and other factors on the targeting of liposomes to experimental tumors: biodistribution and imaging studies. Cancer Res 1990; 50:6371-6378.
31. Woodle MC. ^{67}Gallim-labeled liposomes with prolonged circulation: preparation and potential as nuclear imaging agnets. Nucl Med Biol 1993; 20:149-155.

32. Oku N, Namba Y, Takeda A et al. Tumor imaging with technetium-99m-DTPA encapsulated in RES-avoiding liposomes. Nucl Med Biol 1993; 20:407-412.
33. Oku N, Namba Y, Okada S. Tumor accumulation of novel RES-avoiding liposomes. Biochim Biophys Acta 1992; 1126:255-260.
34. Ogihara-Umeda I, Sasaki T, Nishigori H. Active removal of radioactivity in the blood circulation using biotin-bearing liposomes and avidin for rapid tumour imaging. Eur J Nucl Med 1993; 20:170-172.
35. Oku N, Tokudome Y, Tsukada H et al. Real-time analysis of liposomal trafficking in tumor-bearing mice by use of positron emission tomography. Biochim Biophys Acta 1995; 1238:86-90.
36. Oku N, Tokudome Y, Namba Y et al. Effect of serum protein binding on real-time trafficking of liposomes with different charges analyzed by positron emission tomography. Biochim Biophys Acta 1996; 1280:149-154.
37. Oku N, Tokudome Y, Tsukada H et al. Trafficking of various long circulating liposomes in tumor-bearing mice determined by positron emission tomography. Biopharm Drug Disposition 1996; 17:435-441.

Section III

Future Medical Applications: Surface Ligands and Targeting

CHAPTER 19

Antibody-Mediated Targeting of PEG-Coated Liposomes

Alexander L. Klibanov

Introduction

For two decades the liposome has been portrayed as a classic example of a "magic bullet" concept. A targeting ligand, preferably a target-specific antibody, would be immobilized at the liposome surface, so it would specifically bind only to the target tissue.[1] After iv administration, this immunoliposome would carry a drug for selective delivery to the disease site. Neither drug delivery to normal tissues nor harmful side effects would occur. However, numerous technical problems had to be solved before this concept became feasible.

Liposomes may be degraded by serum lipoproteins and/or rapidly taken up by the liver and spleen. The target may not be readily accessible to liposomes if it is not in direct contact with the bloodstream. Liposomes may not bind to the target if the antibody affinity is not high enough. Selective binding to the target surface does not guarantee that a liposome-associated drug will be delivered into the target cells to act there. To date, liposome-mediated drug targeting research has been focused on resolving such problems. Simply adding cholesterol as a component of the liposome membrane can greatly improve liposome stability in serum.[2] Use of high-affinity antibodies and careful choice of target antigens may improve affinity, specificity and uptake into the target cells.[3] However, reducing the rate of liposome uptake by the liver and spleen has taken years.[4,5]

Red blood cells possess a thick glycocalyx containing sialoglycolipids and sialoglycoproteins, and thereby circulate in blood for many weeks.[6] Sialoglycoprotein fetuin, immobilized on the liposome surface, somewhat improves the circulation time of antibody-coated liposomes.[7] Incorporation of 5-10 mol% of monosialoganglioside G_{M1} or phosphatidylinositol into the liposome membrane increases the residence time in the bloodstream several-fold.[4,8] Consequently, addition of G_{M1} improves targeting of immunoliposomes to mouse lung endothelium in vivo.[9]

Polyethylene glycol attached to proteins[10] and particles[11] allows them to stay in the bloodstream for extended periods of time and to avoid rapid uptake by the liver and spleen. The same is true for liposomes of ~100-200 nm size (see this volume and ref. 5 for review). In this chapter we will discuss antibody immobilization on these PEG-coated immunoliposomes, immunoliposome stability, in vitro and in vivo targeting and circulation.

Long Circulating Liposomes: Old Drugs, New Therapeutics, edited by Martin C. Woodle and Gerrit Storm.

Successful immunoliposome targeting includes:

1) stable, uniform drug-loaded liposomes with a tightly bound, nondenatured antibody possessing high affinity to the target;
2) these immunoliposomes should circulate in the organism long enough to reach the target, bind to it firmly and allow the drug to act;
3) liposome delivery to the target tissue should be much higher than to normal tissues to minimize the harmful side effects of toxic drugs such as anticancer agents.

In this chapter we will discuss some of these points in more detail.

Antibody Immobilization on PEG-Coated Liposomes: An Overview

Currently, there are three basic strategies for attachment of antibodies or other proteins to PEG-coated liposomes (Fig. 19.1A,B; Fig. 19.2). The first, most obvious method attaches the antibody to the liposome membrane via a lipid anchor, as was done earlier for plain liposomes[9,12-14] (Fig. 19.1A). We have mixed the antibody possessing a covalently attached phosphatidylethanolamine derivative with lipids and PEG5000-DOPE in octylglucoside aqueous solution, and prepared liposomes by a detergent dialysis technique. The resulting long circulating liposomes demonstrated specific binding to the target in vitro and in vivo.[15-18] However, while these liposomes stayed in the bloodstream longer than PEG-free immunoliposomes, their binding to the target in vitro and in vivo was not improved, and in some cases was reduced significantly.[15,17]

A second immobilization procedure was based on the avidin-biotin bridge method originally developed for protein immobilization on regular PEG-free liposomes.[19,20] Biotinylated phosphatidylethanolamine is incorporated into the liposome membrane, then avidin is added and attached to biotin on liposomes. Free binding sites on avidin are used to bind biotinylated antibodies which are added later. The presence of PEG5000-PE in the composition of biotinylated liposomes does not completely block avidin binding.[15] Therefore, the same sandwich-type avidin bridge antibody immobilization (Fig. 19.2) is possible for PEG-coated liposomes; this was developed by Allen et al.[21-23]

We have proposed a third approach to take advantage of bifunctional polyethylene glycol, which possesses reactive groups on both ends of the polymer chain[24] (Fig. 19.1B). This permits the attachment of antibodies or other protein molecules to liposomes using the same PEG residues that are used to prolong liposome circulation in vivo. Usually, PEG-PE with an active group on the outer tip of PEG is synthesized, purified and incorporated in liposomes. Various chemistries can be applied for this purpose.[23,25]

We will now consider advantages and disadvantages of these three approaches from the point of view of antibody immobilization and targeting of the liposomes formed.

In order to achieve successful targeting of immunoliposomes, the following requirements must be fulfilled. The antibody should be tightly attached to liposome; PEG-coated immunoliposomes should be stable in circulation. The immunoliposome preparation should be homogeneous in size, affinity and content of antibody, lipids and PEG derivatives. The presence of antibody molecules on the liposomes should not interfere with the PEG-liposome circulation time in the bloodstream. The presence of PEG on the liposome surface should not interfere with antibody access and thereby binding to or affinity for the target antigen.

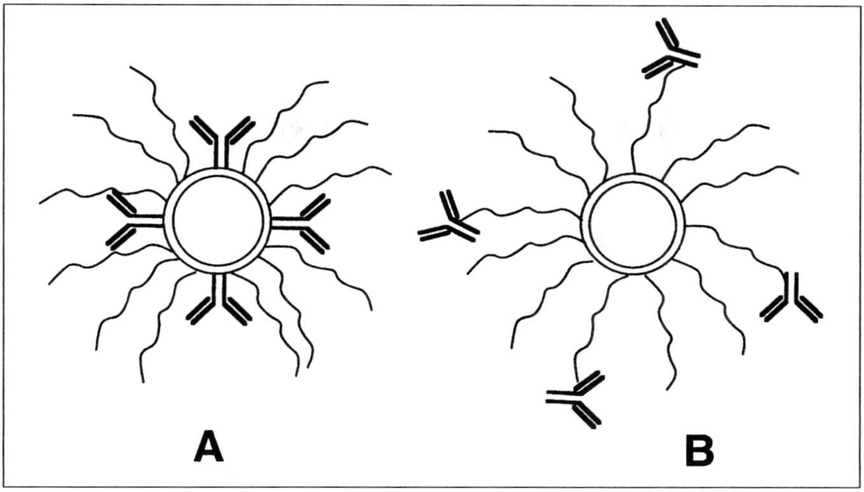

Fig. 19.1. Immobilization of antibody on PEG-PE liposomes: A) direct; B) via PEG-PE spacer arms. The situation in A provides steric hindrance for antigen binding. This is not the case in B. Reprinted from Klibanov AL et al, J Liposome Res 1992; 2:321-334 by permission of Marcel Dekker Inc.

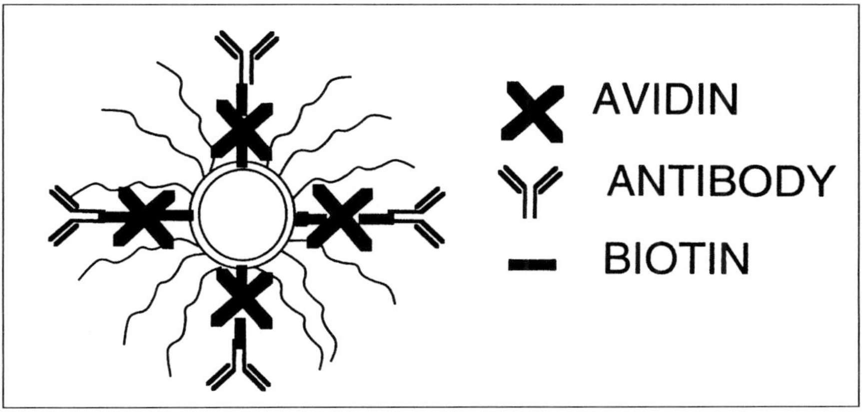

Fig. 19.2. Immobilization of antibody on PEG-PE liposomes via avidin bridge.

PEG-Liposomes with Antibody Attached Directly to the Bilayer: Preparation, Targeting In Vitro and In Vivo

The presence of PEG on the surface of a liposome creates a barrier against the molecules that float in the surrounding medium.[15,26,27] This barrier consists mostly of water with only a fraction of the volume occupied at any given moment by rapidly moving PEG chains. This PEG "cloud" appears to be penetrable by water and small molecules. Larger molecules, e.g., proteins like avidin, are less likely to penetrate this barrier, especially at higher PEG content.[15] PEG-PE, even at low concentrations

(1 mol%) significantly reduces the binding of antifluorescein antibodies to liposome membrane.[28] PEG5000-coated liposomes, with a lipid anchor MPB-PE in the membrane, were able to bind covalently only 2.6 µg antibody/µmol lipid, an order of magnitude less than for PEG2000-coated liposomes, which, in turn, bound 3 times less antibody than MPB-PE liposomes without a PEG coat.[23] Larger structures, like other liposomes in agglutination assays, cannot penetrate the barrier easily, even at the lowest PEG concentrations (0.72 mol%).[15] This PEG "cloud" steric barrier is the most likely reason why PEG-coated liposomes stay in the bloodstream for long periods of time: PEG apparently repels blood proteins from binding to the liposome surface (see chapter 1). However, this barrier could also present problems for targeting of these liposomes via antibodies immobilized on the membrane, if the targeting ligand (antibody) is buried within the PEG layer.

Our first experiments with the targeting of PEG-immunoliposomes in vitro have shown that the presence of a PEG5000 "cloud" on the immunoliposome membrane interferes with its binding to the antigen.[15-17] The antibody was immobilized directly on the liposome membrane via NGPE lipid anchor. While specific targeting of PEG5000-PE containing immunoliposomes to a myosin layer was demonstrated, the amount of liposomal lipid attached to the plate was nearly an order of magnitude less than observed for PEG-free immunoliposomes. The size of PEG5000 molecules in aqueous buffer solution is relatively close to the size of immunoglobulin G, as estimated by gel filtration elution profiles. A simple mechanistic view suggests that small ligands or even small proteins directly attached to the liposome membrane will be incapable of targeting these liposomes. Indeed, a "surrogate targeting experiment" with biotin-PE-containing liposomes shows that biotin on one PEG-liposome cannot be reached by avidin already attached to another liposome, so no aggregation of PEG-liposomes takes place.[15] Plasminogen, a protein twice smaller than an IgG antibody, attached directly to the lipid membrane of PEG5000-liposomes, does not bind to a fibrin-coated surface.[25] Whole antibody molecules, however, allow some specific targeting of PEG5000-PE containing immunoliposomes in vitro and in vivo[15-17,23] despite the presence of the PEG coating.

To obtain reasonable targeting of PEG-coated liposomes, one must either use larger protein molecules, or diminish the action of the PEG steric barrier. Below we discuss the principle approaches reported: limiting the length of the PEG polymer and reducing the content of PEG-PE in the liposome membrane.

The first approach, varying the length of the PEG chain (Fig. 19.3), was investigated for mouse lung endothelium-specific immunoliposomes.[16] Short-chain PEG750 proved inefficient in preventing uptake of these liposomes by the liver and spleen (Fig. 19.3A). These immunoliposomes did accumulate in the lungs as efficiently as PEG-free immunoliposomes. On the other hand, PEG5000-coated immunoliposomes circulated effectively (Fig. 19.3C), but accumulated in the lung only half as efficiently as PEG-free immunoliposomes. Optimal results were obtained with PEG2000-PE-immunoliposomes (Fig. 19.3B) which accumulated in the lung up to 52.6 ± 2.5% of the injected dose; only 34.6 ± 1.5% accumulated in the liver and spleen.[16]

The second approach, varying the surface density of the PEG-lipid, was used for the targeting of immunoliposomes to necrotic myocardium with antimyosin antibodies.[18] Reducing the content of PEG5000-PE in the immunoliposome from 10 mol% to 4 mol% improved the antibody-mediated targeting. At the same time, the circulation time of these liposomes did not fall as abruptly. There exists an optimal PEG-PE concentration, at which the PEG cloud surrounding the liposome is dense enough to

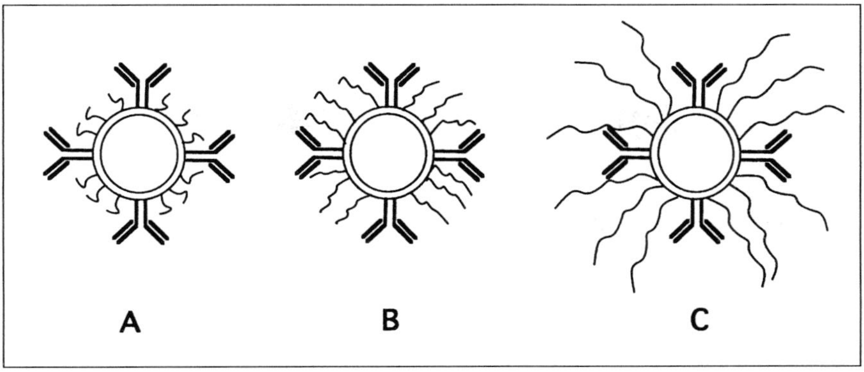

Fig. 19.3. Direct immobilization of antibody on PEG-PE liposomes: A, short PEG polymer (PEG750); B, intermediate length PEG polymer; C, long chain PEG, M = 5000.

allow prolonged circulation, but sparse enough to allow binding of the immobilized antibody to the target. Also, the presence of more antibody molecules on the liposome membrane will increase targetability somewhat, possibly at the expense of circulation time.

The circulation time of liposomes may vary with the type of attachment anchor[29] and the stability and structure of the lipid bilayer. Blume and Cevc argued that use of gel-phase DSPC combined with PEG-DSPE in liposomes does not offer circulation improvement over liquid-crystalline soy PC.[30] However, at 8 h after injection the difference actually was quite clear—a drop from about 60% of the injected dose in blood for DSPC liposomes to approximately 35% for soy PC-liposomes.[30] If the lipid anchor attached to PEG is tightly bound with the membrane, and the liposome itself is composed of long-chain saturated lipids, the liposome appears to circulate longer. Therefore, shorter chain PEG can be used successfully to achieve prolonged circulation of such liposomes.[31] Actually, Maruyama has observed reduced PEG-DSPE-liposome circulation for PEG5000 or higher, and optimal circulation improvement for PEG2000.[31] The use of "solid" liposomes with shorter chain PEGs combined with antibody molecules attached to the membrane might assure antibody accessibility to the target (Fig. 19.3B vs. Fig. 19.3C) without considerable loss in circulation time.

Two observations are possible from both of the approaches described above. Higher content of longer PEG chains at the liposome surface (below a certain threshold) yield longer immunoliposome circulation times, but less efficient binding of these liposomes to the target. A compromise may be found, with only minor reduction in circulation time, while yielding some target-specific binding.[16,18] However, one prefers not to reduce the mol% PEG-PE in the liposome, because this increases uptake of the immunoliposomes by the liver and spleen.

The thickness of the PEG layer as determined by physical methods depends upon both the length of the PEG polymer and PEG-PE surface density.[32] At lower concentrations, PEG chains on liposome surface move freely in the aqueous phase, forming "mushrooms."[32,33] At higher PEG-PE surface densities and/or higher PEG molecular weight, the structure of the PEG layer becomes a thick brush, where PEG molecules extend further from the liposome surface. For instance, 10 mol% PEG2000-PE provides a brush with a 6.6 nm thickness; 5 mol% PEG5000-PE yields 8.8 nm brush thickness.[32] The latter figure is comparable to the size of antibody molecules. Higher PEG-PE concentrations yield even longer brush, potentially shielding the antibody from the target surface (Fig. 19.3C). Binding of the corresponding immunoliposomes to the target indeed becomes less efficient at high concentrations of long chain PEG5000-PE.[15-18]

One can combine a strong PEG steric barrier at the liposome surface with antibody-mediated targeting by employing a longer spacer arm between the antibody and the liposome surface (Fig. 19.1B, Fig. 19.2). This displaces the antibody molecule away from the lipid bilayer. PEG chains remain underneath the antibody binding site, and do not interfere with the antibody binding to the target. In the following sections we will review the methods of preparation of such immunoliposomes, and discuss possible advantages and disadvantages of this approach.

Attachment of Antibody Molecules to Liposomes: Advantages and Disadvantages of a Protein Spacer Arm

PEG-coated liposomes carrying biotin-PE at their membrane can bind some avidin despite the steric hindrance created by PEG chains.[15] The presence of PEG chains on the liposomes limits the binding, though, which thus conveniently inhibits aggregation of these liposomes by avidin.[15] Therefore, avidin may be attached to biotinylated PEG-coated liposome by one of its biotin-binding sites, while other biotin-binding sites are available for coupling with biotinylated antibody (see Fig. 19.2).[21-23]

This approach was used successfully for the first time to target liposomal doxorubicin to lung tumors in mice.[22] A single injection of PEG-coated antibody-avidin-liposomes with doxorubicin cured the majority of tumor-bearing mice (shown in Fig. 19.4);[22] PEG-coated liposomes without an antibody prolonged survival only by a few weeks.

The main advantage of this approach is its simplicity. There is no need for complicated chemical synthesis, since all the components are available commercially. Biotinylation of antibodies is a one-step reaction. It does not cause antibody inactivation.

The disadvantages of this approach for practical applications include the following: first, for every molecule of the antibody attached to a liposome there is a bridge molecule of an immunogenic foreign protein, avidin or streptavidin, attached to this liposome (Fig. 19.2). Secondly, while avidin itself does not aggregate biotinylated PEG-coated liposomes, biotinylated antibodies may aggregate avidin-carrying PEG-coated liposomes. The chemical biotinylation process may yield some of the antibody molecules with several biotin residues attached. Such biotin residues may serve as cross-linkers of liposomes with immobilized avidin. This partial aggregation may be transient and liposomes may dissociate during filtration through Nuclepore filters; however, re-aggregation may occur on storage.[23] This approach was convenient to use in the early stages of targeting research but it may not prove useful for actual targeted drug products.

Instead of using a protein (avidin) spacer to lift the antibody molecule above the PEG "cloud", one can instead attach the antibody molecule to the "cloud" itself, namely to the tip of the PEG chain (Fig. 19.1B). There are two main questions to address here: do the antibodies, attached to the outer tips of PEG chains, shorten the circula-

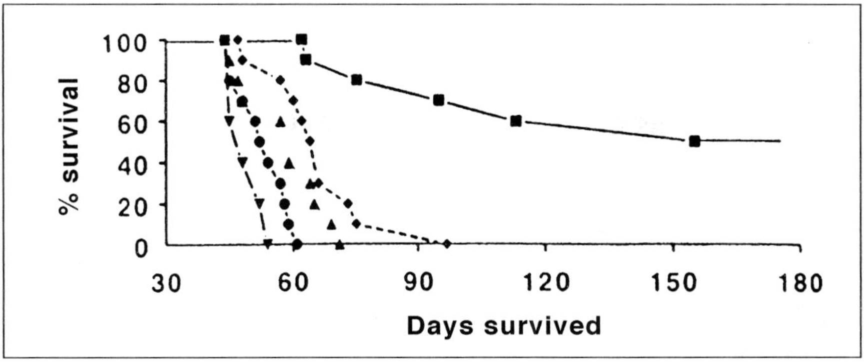

Fig. 19.4. DBA/2 mice (20-25 g; 2 x 5/group) were inoculated iv with 2 x 10^5 KLN-205 cells suspended in 0.2 ml phosphate-buffered saline. Treatment was initiated on the third day with 6 mg/kg of either free DOX (▲), liposomal DOX (◆), mAb liposomal DOX (■), or mAb liposomes (▼) (12 µg Mab, no DOX). Control groups (●) were treated with sterile phosphate-buffered saline. Survival times of the mice were noted and gross pathology was done on all animals. Two long-term survivors were sacrificed on day 170 and showed no evidence of tumor; the remaining animals continue to be monitored. Reprinted with permission from Ahmad I et al, Cancer Res 1993; 53:1484-1488. Copyright, American Association for Cancer Research.

tion time of these liposomes? Does the antibody attachment to the PEG tip (Fig. 19.1B) allow antibody binding and specificity relative to PEG-coated immunoliposomes with antibody attachment to the lipid bilayer (Fig. 19.1A)?

Attachment of Protein Molecules to the Tip of PEG Group on Liposomes

Four types of chemistry were used for the attachment of protein molecule to the outer tip of PEG for liposome binding:

1) Blume and Cevc propose to use PEG with two terminal carboxy residues, one of which is coupled to the primary aminogroup of DSPE.[25] The resulting PE-PEG-COOH is incorporated in the liposome, and the terminus is activated by water-soluble carbodiimide. After the activation the protein is added in mild alkaline medium (see Fig. 19.5A). Up to 2 x 100^{-3} moles plasminogen per mol lipid was bound to liposomes by this method; protein coupling yield was about 50%. The advantages of this method follow: carbodiimide coupling chemistry is well described; the coupling reaction is rapid; the coupling yield is relatively high; and a relatively stable amide bond is generated. The main disadvantage of this approach lies in the possibility that some irreversible protein modification with the carbodiimide reagent may occur. Another known disadvantage of carbodiimide activation is the formation of stable N-acylisourea from the activated O-acylisourea.[34] This N-acylisourea will permanently remain on the tip of PEG attached to the liposome. It is preferable to avoid injection of this by-product in the organism.

2) Allen and Zalipsky[23] propose to use carbohydrate moiety of an antibody for the conjugation. The carbohydrate is activated by periodate with the formation of aldehyde groups. Later these aldehydes are reacted with hydrazide-PEG-DSPE

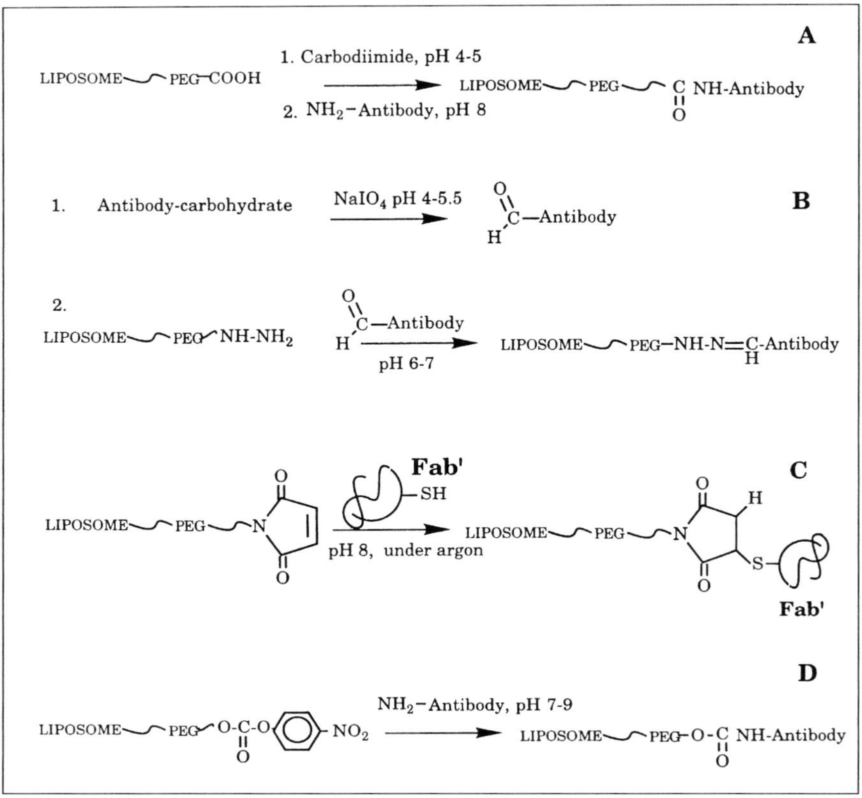

Fig. 19.5. Immobilization of antibody on liposomes via PEG-PE anchors.

incorporated in the liposome membrane (Fig. 19.5B). In order to avoid interantibody crosslinking (aldehyde to the antibody primary aminogroup) the reaction is performed in mild acidic or neutral conditions. This chemistry was used earlier for antibody immobilization on liposomes. At that time, a hydrazide lipid anchor without PEG was used.[35] The coupling yield of the antibody to liposomes with this reaction is usually quite good. Another advantage of this approach is in the high selectivity of binding and orientation of the antibody. It is generally believed that the carbohydrate residue of immunoglobulin is located at the hinge region of the molecule. Selective chemical modification of this region leaves the antigen-binding domains intact, therefore, affinity to the target is not reduced.[36]

After antibody coupling to the liposome is completed, most of the hydrazide PEG tips stay intact, but this is not a major disadvantage. It has been shown that the presence of hydrazide at the PEG tip does not cause accelerated uptake of PEG-coated liposomes by the liver and spleen after intravenous injection.[23] The main problem with this approach is that only some of the antibodies and antibody fragments (mostly IgM) possess extensive carbohydrate moiety suitable for periodate oxidation. For a particular monoclonal antibody of interest, especially IgG, this carbohydrate may be or may not be there. Hydrazone linkages are unstable in an acidic medium,[37] so long-term stability of the immunoliposomes should be studied in detail.

3) Allen[23] has described maleimide-PEG-PE, a reagent proposed by Liposome Technology Inc. (Fig. 19.5C). PEG residue was used as a spacer arm between the lipid anchor and the maleimide group. This chemistry was proposed more than a decade ago for antibody binding to the liposome surface; and at that time, a maleimide bifunctional cross-linking reagent with a short spacer arm was used.[14] A thiol group generated on the antibody molecule or Fab'-fragment, is used to attach to the maleimide residue on tip of the PEG with the formation of a stable thioether bond. The main advantage of this approach is the stability of the conjugate. The thiol group in the Fab'-fragment is located on the opposite site of the molecule from the antigen-binding region, such that a favorably oriented coupling of Fab' with little loss of antigen-binding capacity can be expected.[14] The main disadvantage is the introduction of maleimide derivatives, the toxicity of which must be studied in detail. Impurities should be removed from maleimide derivatives.[38] Reactions should be performed in an oxygen-free medium.[14] As of the writing of this chapter, detailed information on maleimide-PEG-PE immunoliposomes was not yet available.

4) We have proposed p-nitrophenylcarbonyl-PEG-PE (NP-PEG-PE) as an anchor for the coupling of ligands to liposomes (Fig. 19.5D) to take advantage of a chemistry which is widely used for protein modification with PEG.[39,40] Antibodies were simply added to the dispersion of liposomes which contained NP-PEG3350-PE and incubated in a mild alkaline medium. Up to 10^{-3} moles immunoglobulin per mol lipid were attached to liposomes by this technique. This corresponds to about one hundred protein molecules on a 100 nm liposome with about 10^5 lipid molecules. This approach offers several advantages. Nitrophenyl esters are not very reactive, so NP-PEG-PE is quite stable on storage; hydrolysis is also slow. One can easily monitor the progress of the ligand coupling with NP-PEG-PE liposomes and compare it to the background hydrolysis rate in the absence of ligand, simply by monitoring the release of p-nitrophenol spectrophotometrically at 400 nm (yellow color). All the antibody molecules and their fragments possess primary aminogroups, so attachment to a PEG tip via a stable urethane bond is assured for virtually any antibody, fragment or other protein. The yield of this reaction may be adjusted by varying the concentration of liposomes and antibody, pH and reaction time. The main disadvantage of this approach is that antibody molecules typically carry more than a single primary aminogroup, and binding may occur in a nonuniform, statistical fashion. Some of the antibody molecules or fragments may connect to PEG through primary aminogroups near the binding site and lose antigen-binding capacity. However, the majority of antibody molecules may retain antigen-binding capacity and assure targeting of these immunoliposomes.

Targeting and Circulation of Immunoliposomes (Avidin or PEG Spacer Immobilization)

PEG-free immunoliposomes bind to the targets much better than PEG-coated liposomes with the antibody molecules attached directly to the membrane (see above). Use of avidin or PEG as an extended spacer arm for protein immobilization on liposomes drastically improves liposome targeting. When the antibody-mediated binding to the target is not inhibited by the PEG cloud due to steric arrangement (Fig. 19.1B, Fig. 19.2), liposomes bind to the target as well as PEG-free liposomes with the same protein immobilized directly on the membrane.

When the antibody against KLN-205 cells was immobilized on DSPC/Cholesterol/PEG1900-DSPE liposomes via an avidin bridge method (Fig. 19.2), specific binding of these liposomes to cells in vitro was only slightly lower than for PEG-free immunoliposomes.[21] A similar result was obtained when PEG-PE was used as an anchor for protein immobilization: plasminogen-NGPE and plasminogen-PEG-PE liposomes targeting to a fibrin layer in vitro were equally strong.[25] This constitutes a major improvement over direct covalent immobilization of antibody on PEG-liposomes (Fig. 19.1A), where nearly an order of magnitude reduction of in vitro binding was observed.[17]

Allen et al[22] demonstrated that a single injection of doxorubicin-loaded immunoavidin liposomes to lung tumor-bearing mice drastically improves median life span; in the majority of treated animals no signs of tumor could be detected (Fig. 19.4). This is the first report of successful PEG-coated immunoliposome-mediated targeting of an anticancer drug. It is possible that the use of PEG spacer arms instead of avidin for antibody immobilization will further improve targeting efficacy.

In order to study the biodistribution and targeting of antibody-PEG-liposomes, we immobilized lung endothelium-specific antibody 34A or nonspecific immunoglobulin 14 on liposomes via a p-nitrophenyl-PEG3350-PE anchor according to the scheme presented in Fig. 19.5D (Klibanov, Huang, Serbina et al, unpublished). Liposomes coated with the 34A antibody have been targeted to mouse lung endothelium in many experiments.[9,15,16,41,42] PEG5000-coated 34A-immunoliposomes with the antibody attached directly to the membrane did not accumulate in the lung as efficiently as PEG-free immunoliposomes.[15] In contrast, up to 60% of the injected 34A-PEG3350-PE immunoliposomes were targeted to the lung (Fig. 19.6A). Similar results were reported by Maruyama et al, who used PE-PEG-COOH to attach 34A and 14 to liposomes.[43]

The number of protein molecules attached to liposomes plays a crucial role in the biodistribution and targeting of IgG-PEG-liposomes, both for liposomes coated with specific and nonspecific immunoglobulins. While a small amount of immunoglobulin, several molecules per liposome, does not influence the circulation time of nonspecific IgG-coated liposomes very strongly, an increase in the content of IgG does increase the accumulation of liposomes in the liver and spleen and shortens the liposome blood residence time (Fig. 19.6B). The simplest explanation may be that the presence of a large amount of protein material outside of the PEG layer is "tagged" by blood proteins/opsonins, causing uptake of the liposomes by the liver and spleen. Alternatively, Fc-fragments of IgG molecules may bind to the receptors on the macrophage surface. In both cases, it is advisable to use antibody fragments instead of the whole antibody molecules. The use of $F(ab)_2$, Fab, Fv, single-chain Fv or similar molecules may improve the circulation time of ligand-PEG-liposomes even at high ligand-per-liposome ratios. Even if we assume that any part of the protein above the PEG layer, not just the Fc-fragment, is responsible for increased recognition and uptake, the reduction of the attached protein molecular weight from 150,000 (IgG) to 25,000 (Fv) may still result in a 6-fold reduction of protein content and we have observed a corresponding improvement in circulation (Fig. 19.6B). It is likely that the Fc-fragment of the antibody molecule exposed to the outside medium is responsible for accelerated uptake of these liposomes, as shown by Leserman for PEG-free liposomes.[44] Allen et al[23] have observed slightly faster liposome clearance from the bloodstream when 35 μg IgG/μmol lipid was immobilized via DSPE-PEG-hydrazide, as compared with PEG-liposomes without an antibody.

In support of the hypothesis that Fc-fragment exposure above the PEG coating is the most important factor, plasminogen-PEG-PE/DSPC liposomes stayed in the

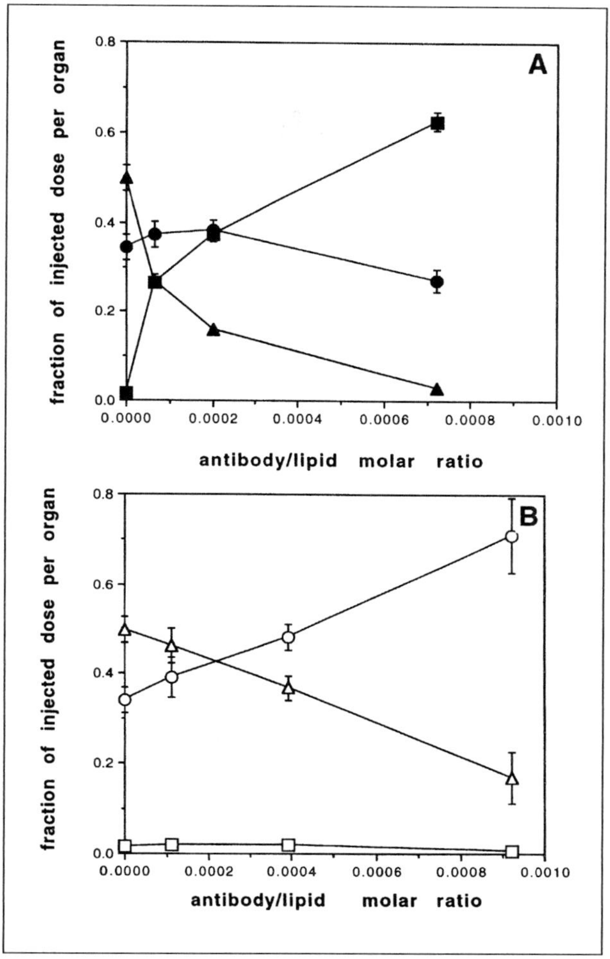

Fig. 19.6. Biodistribution and lung targeting in mice of PEG3350-DOPE-coated liposomes with immobilized 34A antibody (A, closed symbols) or 14-nonspecific IgG (B, open symbols). p-Nitrophenyl-carbonyl-PEG-PE (Fig. 19.5D) was used for the antibody coupling to the tip of the PEG tether. Animal studies were performed as in ref. 15. Animals were sacrificed three hours after the intravenous injection of liposomes. 34A-antibody and 14-IgG were generously provided by Dr. S. Kennel (Oak Ridge National Laboratory). Levels of ^{111}In-labeled liposomes were determined in blood (▲), lungs (■), liver and spleen (●).

bloodstream for hours after intravenous injection.[25] Those liposomes contained 180 µg plasminogen per µmol lipid, which is similar to the amount of antibody at the highest IgG concentration point (Fig. 19.6B), which caused accelerated 14-PEG-PE liposome uptake by the liver.

Targeting and biodistribution of the antibody-PEG-liposomes are governed by several processes. First, like nonspecific IgG-PEG-liposomes, they circulate in the bloodstream for prolonged periods but are eventually captured by the liver and spleen. More antibody molecules per liposome results in higher uptake by the liver and spleen. At the same time, a higher number of antibody molecules per liposome results in the considerable improvement of targeting (Fig. 19.6A).

A search for a workable compromise will be necessary in any immunoliposome targting study. Coverage of the PEG "cloud" with antibody molecules results in faster uptake of these liposomes by the liver and spleen. For targeting to lung endothelium, a target with enormous surface and high blood flow, a large number of antibody molecules on the liposome surface resulted in substantial targeting improvement.

However, if the target is relatively small (e.g., tumor metastases), liposomes must circulate longer so that the higher percentage of liposomes will have a chance to flow through the target area. In this case a large number of whole antibody molecules on the tips of PEG "cloud" will become less advantageous, and a few dozen antibody molecules per liposome may provide better overall targeting than several hundred antibody molecules.

Immunoliposomes: Requirements for Successful Targeting

Immunoliposome Homogeneity

When antibodies were immobilized on liposomes and targeting to the antigen was performed in vitro, it became apparent that heterogeneity with regard to binding affinity and antibody:lipid ratio may occur. A small fraction of the liposomes were the first to bind to the target; these contained more antibody molecules per liposome.[45] Huang et al[46] were able to isolate antibody-rich liposomes by gradient centrifugation. The same phenomenon may take place not just for antibody molecules on liposomes, but also for the PEG-lipid molecules associated with these liposomes. It was already noted that in some cases PEG-DSPE, for instance, tends to form separate micelles which are not incorporated in liposomes in excess of 7 mol% PEG2000-DSPE.[47] We have not observed separate PEG-PE micelles in PEG-DOPE/egg PC/cholesterol liposomes.[15] Blume and Cevc[30] have reported that DSPE-PEG and DSPC form a homogeneous phase up to a 1:1 molar ratio. However, if the multicomponent system includes PEG-lipid, antibody-lipid, phosphatidylcholine, cholesterol and other additives, it will be necessary to insure a lack of heterogeneity such as different liposomes and/or micelles which can form from such a large number of components under some conditions. The structure of a particular system may also depend on the method of its preparation. If an antibody would be added to the preformed liposomes and attached covalently via a PEG spacer group, the chances for the formation of heterogeneous structures may be lower. Alternatively, when all the components are combined with detergents (octyl glycoside or cholate) as mixed micelles, and the detergent is removed by dialysis,[15-18] chances exist that PEG-PE rich and/or antibody-lipid rich particles may form. The size of these PEG-immunoliposomes may also vary, so additional sizing, e.g., Nuclepore filtration, of the liposomes may be required.[15]

Stability

It is important to firmly attach the antibody to the liposome membrane; otherwise, the liposomes will lose the targeting moiety during circulation. Even if a fraction of the antibody molecules detach from the liposome, the free-floating antibodies will compete with immunoliposomes for binding with the antigen molecules at the target site. If the total amount of the antigen available for binding is limited, care should be taken to avoid the presence of the free antibody molecules in the system, before and after injection.

Protein molecules can be attached to liposomes via long-chain fatty acid residues used as hydrophobic anchors.[45,46] Use of the phospholipid molecules which are more compatible with the structure of the bilayer required less chemical modification of the protein globule: a single phospholipid attached to a protein molecule allowed tight protein attachment to the liposome membrane.[12-14] The same is true for PEG attachment to the liposome surface. In our attempt to use PEG stearate, which contains a single fatty anchor, the resulting PEG-coated liposomes circulated longer then PEG-free liposomes but shorter than PEG-PE liposomes. PEG-stearate probably dissociates from liposomes much faster than PEG-PE.[48] The type of the PE in the PEG-PE anchor may also play a role,[29] with DSPE being a better anchor than DOPE. Similar considerations will apply for antibody immobilization at the distal PEG end: DSPE-PEG appears a preferable anchor for antibody immobilization on liposomes.

Aggregation

Aggregation of immunoliposomes occurs when several liposomes react with a single antibody molecule through the multiple functional groups on the antibody. This is a concern with the most convenient conjugation methods using primary amino groups on the protein, mostly derived from the multiple lysine residues exposed on the surface of most antibodies. These amino groups on the antibody surface react with the functional group on the liposome surface. The concern may be even greater with PEG spacers as the functional groups at the end of PEG chains anchored to different liposomes may attach to a single antibody molecule. Aggregation by noncovalent interactions between liposomes and immunoliposomes can also occur. Regardless of mechanism, several unwanted consequences may occur from aggregation:

1) Multiple-point chemical modification of the antibody, with the possibility of denaturation and inactivation.
2) Steric hindrance of the antibody access to the target antigen by several polymer chains attached to it—for instance, one of the chains may attach to the antigen-binding site of the antibody.
3) Attachment of several PEG chains to each other via an antibody molecule may restrict their mobility and decrease their ability to mask the liposome surface from serum proteins.
4) Formation of bridges via PEG and antibody molecules between two or more liposomes, with the consequences ranging from liposome dimer formation to total flocculation and precipitation of the liposomes.
5) Aggregated liposomes are much more rapidly taken up by the liver and spleen.Furthermore, the antibody binding sites are now sandwiched between liposomes, and thus sterically hindered from binding to the target surface. The degree of liposome aggregation should be monitored, and aggregation avoided by adjustment of the reaction conditions and chemistries used.

Immune Response

Liposomes are known to be strong adjuvants.[49] At the same time, chemical modification of proteins with PEG is known to reduce immunogenicity of proteins given by intravenous injection.[50] The use of antibody fragments, humanized antibodies, single-chain antibodies or even smaller peptides has been proposed to reduce anti-antibody immune responses in patients.[51] The type and intensity of immune responses produced against antibodies that are part of immunoliposomes depends on several factors: protein molecular weight, antibody conjugation method, liposome composition, protein to lipid ratio, the dose and injection frequency, and others. For example, a single injection of sub-milligram quantities of Fv-fragments is less immunogenic than multiple injections of milligram quantities of whole mouse immunoglobulin molecules attached to liposomes. Anti-antibody responses represent a significant challenge for use of immunoliposomes in therapeutic situations requiring multiple injections. The use of small protein fragments and conjugation to the surface of a PEG coating on a liposome, such as by use of Fv-PEG-PE type liposomes, to address this problem remains to be seen.

Possible Future Designs

We have discussed the advantages and disadvantages of various strategies to prepare long circulating immunoliposomes. The diversity of possible applications makes a single optimal approach suitable for all the cases unlikely. Possible designs include the following.

1) Immunoliposomes bearing a PEG surface coating and the antibody conjugated to the distal end of the PEG have advantages in minimizing interference with the antibody binding to its target. Moreover, the use of short single-chain Fv antibody fragments conjugated with PEG chains and lipid anchors will be advantageous as compared to use of the whole mouse-derived antibody molecules (see above). Shorter antibody fragments may possess an affinity comparable to that of a whole antibody, but may be less immunogenic and less likely to be recognized by blood opsonins or macrophages.

2) Immunoliposomes with a combination of PEG-PE prepared with both short and long PEG have been proposed to permit optimization of properties.[27] This idea could be adapted for targeting. A PEG cloud attached to the liposome will be used to make these liposomes long circulating. Attachment of the antibody to a tip of a PEG spacer arm on the same liposome may allow us to vary the targetability and circulation time by varying the length of this PEG spacer.

3) Other polymers. Most of the recent work on immunoliposome targeting has been done with PEG-coated liposomes. Still, PEG is only one of the examples of flexible hydrophilic polymers that are available. Other hydrophilic polymers have been proposed as liposome coatings.[52-54] Currently, they are at best comparable to PEG in their efficacy to prolong liposome circulation time. It is possible that new polymers will be found which will provide better shielding at lower polymer concentrations, or with a thinner layer of the polymer "cloud" or "brush". Also, some defined small molecules immobilized on the liposome surface might also improve liposome circulation time even better than ganglioside G_{M1},[9] glucuronic acid[55] or glutaric acid derivatives.[56] In this case, there would be no need for the thick polymer layer, and the ligand immobilized on the liposome surface directly might provide selective binding to the target without the need for a long spacer arm. For some targeting applications long circulating liposomes based on cohesive bilayers or using glycolipids such as ganglioside G_{M1} might prove successful but may require a large target size and substantial blood flow through the target.[9]

Conclusion

Immunoliposomes are not truly "magic bullets"—a bullet usually moves toward the target by itself; whereas liposomes flow with the blood, and in some instances can escape from the bloodstream (e.g., in solid tumors with leaky neovasculature). This may prove advantageous, by providing the liposomes with the opportunity to accumulate at multiple disease sites (e.g., metastases). Vascular delivery might also prove disadvantageous, if the flux of blood and circulating liposomes through the target area is too small. To allow adequate targeting of immunoliposomes, the affinity of the antibody to the target tissue and the density of the antibody target molecules must be high, the 'background' nonspecific binding in normal tissues must be low, and the immunoliposomes must have sufficient opportunity to access the target.

Through gradual understanding of liposomal systems, technological solutions are being developed for the preparation of long circulating targeted immunoliposomes. These immunoliposomes may serve as a basis for the research and development of the next-generation diagnostic and therapeutic agents targeted by specific ligands. To succeed, these liposomal drug targeted agents must demonstrate clear superiority over existing medications in terms of efficacy and lack of side effects. Cost for pharmaceutical production is also becoming a significant factor. Success of doxorubicin-loaded PEG-liposomes in clinical trials has been reported (see other chapters in this volume). Antibodies[57] are currently in use for diagnosis and therapy. While there is no magic bullet, the need for more effective therapeutics continues to motivate studies of targeted pharmaceuticals. The combination of antibodies and PEG-liposomes may prove helpful in this work.

Acknowledgments

Support for the experiments performed in Dr. L. Huang's laboratory at the University of Tennessee and the University of Pittsburgh was provided in part by NIH grant CA 24553. I am grateful to Dr. D. Needham (Duke University) for sharing a preprint of his paper, and to Dr. G. Brandenburger and Dr. K. Adzamli for critically reading the manuscript. 34A-antibody and 14-IgG were generously provided by Dr. S. Kennel (Oak Ridge National Laboratory).

References

1. Gregoriadis G, Neerunjun DE. Homing of liposomes to target cells. Biochem Biophys Res Commun 1975; 65:537-544.
2. Kirby C, Clarke J, Gregoriadis G. Cholesterol content of small unilamellar liposomes controls phospholipid loss to high density lipoproteins in the presence of serum. FEBS Lett 1980; 111:324.
3. Wolff B, Gregoriadis G. The use of monoclonal anti-thy1 IgG1 for the targeting of liposomes to AKR-A cells in vitro and in vivo. Biochim Biophys Acta 1984; 802:259-273.
4. Allen TM, Chonn A. Large unilamellar liposomes with low uptake into the reticuloendothelial system. FEBS Lett 1987; 223:42-46.
5. Forum. Covalently attached polymers and glycans to alter the biodistribution of liposomes. Huang L, ed. J Liposome Res 1992; 2:288-454.
6. Allen TM, Williamson P, Schlegel RA. Phosphatidylserine as a determinant of reticuloendothelial recognition of liposome models of the erythrocyte surface. Proc Natl Acad Sci USA 1988; 85:8067-8071.
7. Torchilin VP, Khaw BA, Berdichevskiy VR, Klibanov AL, Haber E,.Smirnov VN. Complexes of liposomes with immunoglobulins and sialoglycoproteins. Bull Exp Biol Med (Russ) 1983; 95:51-53.
8. Gabizon A, Papahadjopoulos D. Liposome formulations with prolonged circulation time in blood and enhanced uptake by tumors. Proc Natl Acad Sci USA 1988; 85:6949-6953.
9. Maruyama K, Kennel SJ, Huang L. Lipid composition is important for highly efficient target binding and retention of immunoliposomes. Proc Natl Acad Sci USA 1990; 87:5744-5748.

10. Beauchamp C, Gonilas SL, Menapace DP, Pizzo SV. A new procedure for the synthesis of polyethylene glycol-protein adducts; effects on function, receptor recognition, and clearance of superoxide dismutase, lactoferrin, and alpha$_2$-macroglobulin. Anal Biochem 1983:131:25-33.
11. Illum L, Davis SS, Muller RH, Mak E, West P. The organ distribution and circulation time of intraveneously injected colloidal carriers sterically stabilized with a block-copolymer—poloxamine 908. Life Science 1987; 40:367-374.
12. Sinha D, Karush F. Attachment to membranes of exogenous immunoglobulin conjugated to a hydrophobic anchor. Biochem Biophys Res Commun 1979; 90:554-560.
13. Weissig V, Lasch J, Klibanov AL, Torchilin V. P. A new hydrophobic anchor for the attachment of proteins to liposomal membranes. FEBS Lett 1986; 202:86-90.
14. Martin FJ, Papahadjopoulos D. Irreversible coupling of immunoglobulin fragments to preformed vesicles. J Biol Chem 1982; 257:286-288.
15. Klibanov AL, Maruyama K, Beckerleg AM, Torchilin VP, Huang L. Activity of amphipathic polyethylene glycol 5000 to prolong the circulation time of liposomes depends on the liposome size and is unfavorable for immunoliposome binding to target. Biochim Biophys Acta 1991; 1062:142-148.
16. Mori A, Klibanov AL, Torchilin VP, Huang L. Influence of the steric barrier activity of amphipathic poly(ethyleneglycol) and ganglioside GM1 on the circulation time of liposomes and on the target binding of immunoliposomes in vivo. FEBS Lett 1991; 284:263-266.
17. Klibanov AL, Khaw BA, Nossiff N, O'Donnel SM, Huang L, Slinkin MA, Torchilin VP. Targeting of macromolecular carriers and liposomes by monoclonal antibodies against heart myosin heavy chains. Am J Physiol 1991; 261(suppl.):60-65.
18. Torchilin VP, Klibanov AL, Huang L, O'Donnell S, Nossiff ND, Khaw BA. Targeted accumulation of polyethylene glycol-coated immunoliposomes in infarcted rabbit myocardium. FASEB J 1992; 6:2716-2719.
19. Trubetskaya OV, Trubetskoy VS, Domogatsky SP, Rudin AV, Popov NV, Danilov SM, Nikolayeva MN, Klibanov AL, Torchilin VP. Monoclonal antibody to human endothelial cell surface internalization and liposome delivery in cell culture. FEBS Lett 1988; 228:131-134.
20. Loughrey HC, Bally MB, Cullis PR. A non-covalent method of attaching antibodies to liposomes. Biochim Biophys Acta 1987; 901:157-160.
21. Ahmad I, Allen T. M. Antibody-mediated specific binding and cytotoxicity of liposome-entrapped doxorubicin to lung cancer cells in vitro. Cancer Res 1992; 52:4817-4820.
22. Ahmad I, Longenecker M, Samuel J, Allen TM. Antibody-targeted delivery of doxorubicin entrapped in sterically-stabilized liposomes can eradicate lung cancer in mice. Cancer Res 1993; 53:1484-1488.
23. Allen T, Agrawal AK, Ahmad I, Hansen CB, Zalipsky S. Antibody-mediated targeting of long circulating (Stealth^R) liposomes. J Liposome Res 1993; 4:1-25.
24. Klibanov AL, Huang L. Long-circulating liposomes: development and perspectives. J Liposome Res 1992; 2:321-334.
25. Blume G, Cevc G, Crommelin MDJA et al. Specific targeting with polyethylene glycol liposomes: coupling of homing devices to the ends of the polymeric chains combines effective target binding with long circulation times. Biochim Biophys Acta 1993; 1149:180-184.
26. Needham D, Hristova K, McIntosh TJ, Dewhirst M, Wu N, Lasic DD. Polymer-grafted liposomes: physical basis for the "stealth" properties. J Liposome Res 1992; 2:411-430.
27. Lasic D. Sterically stabilized vesicles. Angew Chem Int Ed Engl 1994; 33:1685-1698.

28. Torchilin VP, Omelyamenko VG, Papisov MI, Bogdanov AA, Trubetskoy VS, Herron JN, Gentry CA. Polyethylene glycol on the liposome surface—on the mechanism of polymer-coated liposome longevity. Biochim Biophys Acta 1994; 1195:11-20.
29. Parr MJ, Ansell SM, Choi LS, Cullis PR. Factors influencing the retention and chemical stability of polyethylene glycol conjugates incorporated into large unilamellar vesicles. Biochim Biophys Acta 1994; 1195:21-30.
30. Blume G, Cevc G. Molecular mechanism of the lipid vesicle longevity in vivo. Biochim Biophys Acta 1993; 1146:157-168.
31. Maruyama K, Yuda T, Okamoto A, Ishikura C, Kojima S, Iwatsuru M. Effect of molecular weight in amphipathic polyethyleneglycol on prolonging the circulation time of large unilamellar liposomes. Chem Pharm Bull 1991; 39:1620-1622.
32. Kenworthy AK, Hristova K, Needham D, Macintosh TJ. Range and magnitude of the steric pressure between bilayers containing phospholipids with covalently attached polyethylene glycol. Biophys J 1995; 68:1921-1936.
33. Torchilin VP, Papisov MI. Why do polyethylene glycol-coated liposmes circulate so long? J Lipos Res 1994; 4:725-739.
34. Bogdanov AA, Klibanov AL, Torchilin VP. Protein immobilization on the surface of liposomes via carbodiimide activation in the presence of N-hydroxysulfosuccinimide. FEBS Lett 1988; 231:381-384.
35. Chua MM, Fan ST, Karush F. Attachment of immunoglobulin to liposomal membrane via protein carbohydrate. Biochim Biophys Acta 1984; 800:291-300.
36. Rodwell JD, Alvarez VL, Lee C et al. Site-specific covalent modification of monoclonal antibodies: in vitro and in vivo evaluations. Proc Natl Acad Sci USA 1986; 83:2632-2636.
37. Kaneko T, Willner D, Monkovic I et al. New hydrazone derivatives of adriamycin and their immunoconjugates—a correlation between acid stability and cytotoxicity. Bioconj Chemistry 1991; 2:133-141.
38. Loughrey HC, Choi LS, Cullis PR, Bally MB. Optimized procedures for the coupling of proteins to liposomes. J Immunol Methods 1990; 132:25-35.
39. Klibanov AL. Targeting of immunoliposomes in vivo. Fifth Princeton Liposome Conference, Princeton, NJ, 1993.
40. Klibanov AL, Serbina N, Torchilin VP, Huang L. Attachment of ligands to liposomes via PEG spacer for prolonged liposome circulation and targeting. J Liposome Res 1996; 6:195-196.
41. Maruyama K, Holmberg E, Kennel SJ, Klibanov A, Torchilin VP, Huang L. Characterization of in vivo immunoliposome targeting to pulmonary endothelium. J Pharm Sci 1990; 79(11):978-984.
42. Holmberg E, Maruyama K, Litzinger DC, Wright S, Davis M, Kabalka GW, Kennel SJ, Huang L. Highly efficient immunoliposomes prepared with a method which is compatible with various lipid compositions. Biochem Biophys Res Commun 1989; 165:1272-1278.
43. Maruyama K, Takizawa T, Yuda T, Kennel SJ, Huang L, Iwatsuru M. Targetability of novel immunoliposomes modified with amphipathic polyethylene glycols conjugated at their distal terminals to monoclonal antibodies. Biochim Biophys Acta 1995; 1234:74-80.
44. Aragnol D, Leserman L. Immune clearance of liposomes inhibited by an anti-Fc receptor antibody in vivo. Proc Natl Acad Sci USA 1986; 83:2699-2703.
45. Klibanov AL, Muzykantov VR, Ivanov NN, Torchilin VP. Evaluation of quantitative parameters of the interaction of antibody-bearing liposomes with target antigens. Anal Biochem 1985; 150:251-257.
46. Huang L, Huang A, Kennel SJ. Monoclonal antibody covalently coupled with fatty acid. J Biol Chem 1980; 255:8015-8018.

47. Allen TM, Hansen C, Martin F et al. Liposomes containing synthetic lipid derivatives of poly(ethyleneglycol) show prolonged circulation half-lives in vivo. Biochim Biophys Acta 1991; 1066:29-36.
48. Klibanov AL, Maruyama K, Torchilin VP, Huang L. Amphipathic polyethyleneglycols effectively prolong the circulation time of liposomes. FEBS Lett 1990; 268:235-237.
49. Gregoriadis G. Liposomes as immunological adjuvants for peptide and protein antigens. In: Gregoriadis G, Florence AT, Patel H, eds. Liposomes in Drug Delivery. Langhorn PA USA: Harwood Academic Publishers, 1993:77-94.
50. Katre NV, Immunogenicity of recombinant IL-2 modified by covalent attachment of polyethylene glycol. J Immunol 1990; 144:209-213.
51. Becker W. Goldenberg DM. Wolf F. The use of monoclonal antibodies and antibody fragments in the imaging of infectious lesions. Seminars in Nuclear Medicine 1994; 24:142-53.
52. Torchilin VP, Shtilman MI, Trubetskoy VS, Whiteman K, Milstein A. M. Amphiphilic vinyl polymers effectively prolong liposome circulation time in vivo. Biochim Biophys Acta 1994; 1195:181-184.
53. Maruyama K, Okuizumi S, Ishida O et al. Phosphatidyl polyglycerol prolongs liposome circulation in vivo. International J Pharmaceutics 1994; 111:103-107.
54. Woodle MC, Engbers CM, Zalipsky S. New amphipathic polymer lipid conjugates forming long circulating reticuloendothelial system-evading liposomes. Bioconjugate Chemistry 1994; 5:493-496.
55. Oku N, Namba Y, Okada S. Tumor accumulation of novel RES-avoiding liposomes. Biochim Biophys Acta 1992; 1126:255-260.
56. Park YS, Maruyama K, Huang L. Some negatively charged phospholipid derivatives prolong their liposome circulation in vivo. Biochim Biophys Acta 1992; 1108:257-260.
57. Waldmann TA. Anti-IL-2 receptor monoclonal antibody (anti-Tac) treatment of T-cell lymphoma. Important Advances in Oncology 1994:131-141.

CHAPTER 20

Small Molecule Ligands for Targeting Long Circulating Liposomes

Martin C. Woodle, Danilo D. Lasic, Gerrit Storm

Introduction: Advantages of Targeted Long Circulating Liposomes

Most of the contributions to this volume demonstrate the therapeutic benefits that can be provided by long circulating liposome (LCL) formulations of therapeutic and diagnostic agents. These benefits begin to approach realization of the "magic bullet" drug delivery concept. However, the success of clinically applied LCL formulations is derived from selective but nonspecific localization, i.e., not mediated by binding to a specific receptor moiety, at accessible pathological tissues where leakage occurs from the vascular circulation. This is sometimes referred to as "passive" targeting and can be thought of as simply "leakage in the plumbing". Desires to improve targeting by virtue of binding to receptors by attaching corresponding ligands, i.e., "active" targeting or specific localization, remain largely unfulfilled. An issue for active targeting is the ability of the LCL already localized in the target site to identify and bind to the target cells. Consequently, targeted LCL represent a potentially important means to increase the therapeutic index of encapsulated drugs.[1-7]

An active area of investigation is the use of antibodies with LCL to provide localization specificity, as discussed in chapter 19. However, foremost of the numerous barriers impeding development of antibody-targeted LCL is loss of their long circulation capability due to a reawakening of MPS recognition to exposed protein on the liposome surface. Consequently, small molecular weight ligands, e.g., Fv domains consisting of little more than the antibody binding site and small oligosaccharides, are more likely to be successful.[5]

An alternative application for LCL with ligands coupled to their surface is ligand presentation to blood elements and vascular surfaces.[8-10] This strategy could be used to present decoy ligands which might act as competitive inhibitors of undesired cellular interactions, for example interference with excessive recruitment of neutrophils at sites of inflammation. Alternatively, it could be used to act as a scavenger to bind and inactivate unwanted plasma components such as circulating toxins, viral particles, metastatic tumor cells, etc. This application of ligands on the surface of LCL is just beginning to be investigated but once again may be dependent upon

Long Circulating Liposomes: Old Drugs, New Therapeutics, edited by Martin C. Woodle and Gerrit Storm.

identification of small molecular weight ligands which are compatible with long circulation.

When successful addition of targeting to LCL is accomplished, it will leave one remaining hurdle, intracellular delivery, for the goal of selective intracellular delivery of therapeutic agents, i.e., synthetic virus delivery systems. The original suggestion that liposomes could fuse with cells to deliver all kinds of drugs directly to the cytoplasm has not been realized but cytosolic delivery can be accomplished with reconstituted viral envelopes, often called virosomes, or inclusion of pH-sensitive lipids that destabilize and disrupt the endosome membrane.[11-15] The latter method is based on design of liposomes which undergo a physical change induced by a specific environmental trigger, being in this particular case a change in pH. The changes being considered are mostly bilayer phase transition to a nonbilayer phase, such as a micellar or hexagonal phase. The former simply increases drug leakage while the latter may enable delivery of liposome contents to the cytoplasm.

Many current efforts for intracellular delivery are focused upon liposome and other systems such as cationic lipid complexes of the relatively large nucleic acid genetic agents.[16,17] The ongoing work is pointing out that these systems will also be limited by the same barriers liposome drug delivery has faced, in particular the need to reduce nonspecific recognition and achieve cell-specific localization. To be successful, intracellular delivery will depend on some of the same achievements made with LCL, i.e., blood persistence and extravasation through the endothelial barriers to reach the pathological tissue, but will demand additional capabilities, i.e., specific cellular interactions leading to intracellular delivery to those cells responsible for controlling the pathological condition. Approaches to accommodate these somewhat mutually exclusive properties include coating DNA-lipid complexes with sterically stabilizing polymers and appropriate ligands but in a manner that the coating is shed, exposing fusogenic components driving intracellular delivery after arrival at the target site.[18,19] Therefore, the development of synthetic virus-like delivery systems will depend on separation and control of mutually exclusive requirements: inert surface with ligand targeting properties facilitating distribution to the intended cells on one hand and interactive properties facilitating intracellular delivery of the agent to the target cells on the other hand.

The development of long circulating liposomes for gene delivery has only just begun.[20] As with ligand targeting, an adverse consequence of steric stabilization is inhibition of liposome surface interactions and thereby cellular uptake, although evidence already exists that the barriers can be overcome.[21] Note that the much smaller molecular weight antisense oligonucleotides may be less dependent upon delivery systems designed to provide intracellular delivery and thus existing LCL may prove beneficial.[22] Much more work is needed to obtain effective delivery for the large DNA required for introducing new gene expression. Systemic efficacy will probably require both the prolonged circulation and ligand specificity thought to be provided by targeted LCL in addition to the as yet unrealized capabilities for intracellular delivery.

Overall, LCL bearing small molecule surface ligands can be expected to have many therapeutic applications. The aim of this chapter is to consider the current state of development for therapeutic uses of liposomes "decorated" with small molecules.

Ligand Mediated Targeting

The need for drug delivery selective for sites of pathology has inspired considerable efforts, in particular research on the use of cell-specific antibodies. The use of antibodies to target liposomes has been studied with the intention of combining their respective valuable capabilities. Studies of antibody targeted liposomes, or immunoliposomes, have been extended to the use of long circulating liposomes containing surface-grafted PEG, sometimes referred to as sterically stabilized immunoliposomes (SSIL), as discussed in chapter 19.[1-4,6,7,23-27] Unfortunately, addition of antibodies to LCL, or even conventional liposomes, has not always been beneficial in disease models, e.g., see Vingerhoeds et al.[28] The presence of the antibodies on the surface of LCL can rescind prolonged circulation through rapid sequestration of the immunoliposomes in the liver and spleen. Note that a frequently necessary condition for effective drug targeting is cellular internalization upon immunoliposome binding to the target cell.[29] Not all receptors are internalized upon ligand or antibody binding and in this case cell-bound immunoliposomes can only release drug in the close surroundings of the target cell; released drug then can diffuse into or away from the target cell, as was shown for cisplatin and doxorubicin in ovarian carcinoma.[28,30] A problem specific to PEG-grafted liposomes is that the surface PEG can interfere with antibody binding to its target antigen. Most importantly, due to the success of LCL to achieve selective localization into tumors and sites of inflammation, the ability of ligands, and in particular antibodies, to further enhance efficacy of encapsulated drugs at these sites is not clear.[6,31] Increased antitumor activity of SSIL containing doxorubicin (Dox-SSIL) was observed in vivo as compared to other treatment groups in the treatment of early solid tumors. Also, Dox-SSIL appeared advantageous in the treatment of hematological malignancies where the tumor cells were in the blood or easily accessible from it. However, the selective benefit of Dox-SSIL was not seen in the case of more developed solid tumors.[6,31] Thus in the latter case, other ligands or strategies may prove to be of greater value. At the present time, studies are ongoing to identify therapeutic strategies benefited by addition of antibodies to LCL.

A potential solution to some of the problems with SSIL is to use biodegradable linkages for coupling the PEG to the liposome surface.[32-34] If the linkage is sensitive to specific chemical conditions typical of specific biological environments at the target site, this could permit development of methods for intentional loss of the liposome PEG-coating and restoration of antibody-ligand interactions under the specific condition. However, differences in pH, redox conditions, or other "signal" parameters between healthy and target tissues may be small and thus will probably require considerable optimization of the chemical nature of the PEG-lipid conjugate.

A promising approach to specific targeting of LCL is to add ligands to the exterior of a PEG coating.[8] The general approach is shown schematically in Figure 20.1 using a heterobifunctional form of PEG-PE replacing the distal methoxy with a reactive group for antibody conjugation.[35-37] A number of heterobifunctional PEG-lipid conjugates have been reported, permitting conjugation to the distal end of the PEG with many chemistries, including disulfide, thioether, amide, carbamate, and hydrazone bonds. One particular conjugation method may have considerable advantages: the addition of a hydrazide to the end of PEG-PE (Hz-PEG-PE). This method has proven especially useful for conjugation of proteins and peptides to the surface of PEG-grafted liposomes over methods based on the maleimide reaction with free sulfhydryls.[35] One interesting feature of the method is orientation of the antibody binding domains away from the liposome surface. An alternative approach to obtain specific orientations is the use of this and other conjugation methods with antibody fragments containing a single reactive site and therefore a single orientation.[38] Some orientations may prove preferable to others. The ligand-PEG-lipid conjugates permit preparation and evaluation of many different forms of targeted LCL where the ligand is presented on the outer surface of PEG-grafted LCL.

Ligand Presentation

Given the persistence of LCL in the blood, ligands on their surface can be presented for prolonged periods of time to the blood and tissues where the liposomes localize.[8,9,39-41] Therapeutic applications of ligand presentation by "decorated" LCL are only just beginning to be considered.

One of the first studies of this approach utilized PEG-grafted LCL to deliver an enzymatic activity on their surface to convert a widely distributed small molecule prodrug substrate to the active therapeutic agent, but only where the liposome had distributed.[8] This concept has been extended to combination with antibody targeting, but as yet untested with LCL.[41,42] Other potential applications include the use of prolonged circulation of ligands to act as scavengers for toxins or any undesirable components circulating in the blood. Given the narrow balance between circulation and MPS uptake of LCL with antibodies on their surface (see chapter 19) it should be possible to alter the ligand-LCL system such that binding of the circulating target to the ligand-bearing LCL induces recognition, uptake, and destruction by cells of the MPS. Promising scavenging applications include binding of endotoxins released from bacterial infections or IgE responsible for allergies.

Yet another use of ligand presentation by LCL is to provide a decoy to competitively inhibit cellular interactions. One example of this concept is work with LCL containing sialyl Lewis^{X} glycolipid analogs on their surface, acting as decoy ligands for neutrophils and inhibiting initiation of inflammation associated with "reprofusion injury" and organ transplant rejection.[10] Another similar application is recognition and inhibition of metastatic tumor cells invading other tissues from the blood.[9]

In summary, a number of therapeutic applications may be devised relying on prolonged presentation of cell surface receptors, ligands, inhibitors, or other molecules on LCL to the blood stream. However, this application of LCL is just beginning to be investigated.

Preference for Small Molecule Ligands

The established procedures for coupling ligands to the distal end of the PEG coating of LCL can be applied to many different ligand types and for many different therapeutic applications involving ligand targeting and ligand presentation. However, this approach for attaching ligands to LCL places the ligand in a position for maximal exposure and thus maximal interactions, both desirable and undesirable, e.g., the unwanted human antimouse antibody (HAMA) response. This argues for a preference for small molecular weight ligands with less potential for undesirable interactions.

As pointed out above, antibodies and other large proteins in particular may not be the best ligand for LCL. In particular, the large proteins provide a great chance for induction of immune responses to them. Also, they mostly lack a defined orientation on the liposome surface.[43] Ultimately, successful development of antibody-targeted LCL will require adjustment of many parameters to address these and other issues.

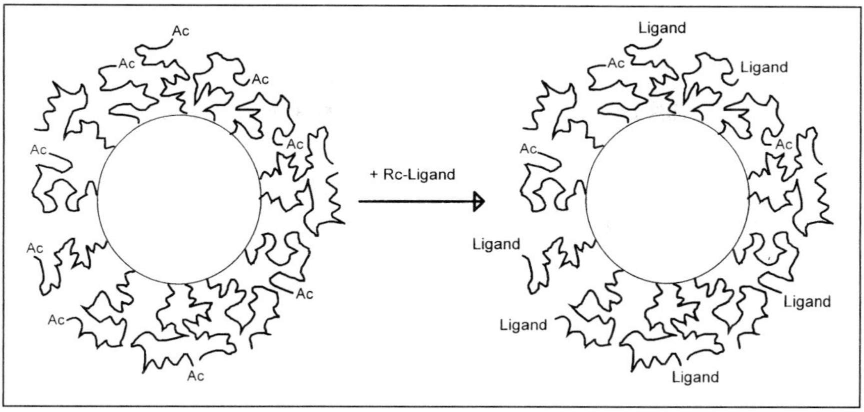

Fig. 20.1. Scheme for ligand presentation on the external surface of PEG-grafted liposomes by conjugation of reactive ligands, designated Rc-Ligand, to an activated group, designated Ac, on the distal end of the PEG-PE containing liposomes. Numerous chemistries have been adapted and proven successful for this method of placing ligands on the surface of PEG-grafted liposomes.

The use of small molecules such as peptides and oligosaccharides as ligands on the surface of PEG-grafted liposomes may prove more useful than antibodies, their fragments, and other large proteins. For example, small peptides or oligosaccharides can provide similar binding constants for cell targeting, but may require a specific conjugation method and thus will tend to have an orientation defined by the conjugation method. Perhaps most importantly, small ligands provide better opportunities to minimize MPS recognition. A study with amino-PEG-PE in liposomes demonstrated that low levels of cationic charge can be accommodated at the end of the polymer.[44] This indicates that positively charged ligands can be accommodated at the surface of the polymer without induction of MPS recognition. Also, a range of conjugation methods applicable to many small molecules have now been developed and are intended either for targeting or for other purposes. Unfortunately, active small molecules are more difficult to identify and in some cases more difficult to prepare, e.g., many oligosaccharides. When they are identified and produced, though, they should help to overcome the disadvantages of antibodies and large proteins. In particular, the extent of immune responses generated by antibodies on PEG-grafted liposomes may be reduced or perhaps completely eliminated with a small peptide. Clearly, the end of the PEG coating is the preferred location for such targeting, as demonstrated using folate to provide selective recognition and uptake by cells expressing its receptor.[45] Also, efforts have been reported using a small peptide fragment of laminin, YIGSR, with strong binding to laminin receptor for the purpose of inhibiting metastasis formation.[9,46] Likewise, similar activities have been reported with small oligosaccharide ligands.[10,47] While this area of investigation is only just beginning, with the availability of many coupling methods for the distal end of the PEG coating, it may be expected to rapidly expand with a large number of potential ligands.

Conclusions

Many directions for further developments in long circulating liposome drug formulations can be envisioned. A clearly important area—as exemplified in the previous chapters of this book—is application of the passive targeting strategy to the many existing drugs and drug candidates for treatment of tumors, infections, and inflammation, all of which represent areas of significant unmet clinical need. This passive targeting also results in a large proportion of these liposomes localizing in the skin and other tissues besides the liver and spleen which may lead to treatment of diseases associated with these sites or potential toxicities.

The next expansion in capabilities, discussed in this chapter, is the promise for addition of decorations to long circulating liposomes which can function as homing or targeting devices, scavenging systems, etc. Efforts are being made to use surface ligands to target LCL, i.e., active targeting. Murine antibodies derived from hybridoma technology have proven to be valuable homing devices in drug targeting systems in animal models. In some instances, however, patients develop human antimouse antibodies (HAMA) upon prolonged parenteral administration of murine antibodies. Occasionally this results in serum sickness, but more frequently reduces the efficacy of the antibody homing capacity by rendering it ineffective. Therefore, alternative homing devices for immunoglobulins are under investigation (e.g. plasminogen, adhesion molecules). "Miniaturization" of the immunoglobulins is studied to avoid unwanted immune responses. The utilization of single-chain antibodies represents an attempt to work with smaller homing devices.[48] Another strategy to reduce the HAMA response is the use of humanized antibodies in which the antigen-combining site of a murine antibody is inserted into the framework of a human antibody (or its fragments) using recombinant DNA techniques.[49]

Another use of decorated LCL is to present ligands to various blood elements. This aspect of LCL development and application to therapeutics is just beginning. Again, small molecule ligands will be instrumental to the success of such systems but will take time to identify and utilize. Nonetheless, initial results with peptides and oligosaccharides are encouraging.

Important further developments in liposome drug delivery aim to increase cellular internalization and cytoplasmic delivery of encapsulated agents such as nucleic acids. This capability is being pursued through the design of liposomes to undergo a physical change upon a specific environmental trigger or after a certain time interval. Ultimately, such capabilities will almost certainly need to be incorporated into liposomes capable of persistent blood circulation as a means to increase and regulate the biodistribution of the gene expression modulation therapeutics.[17] Until such capabilities are attained with LCL, local routes of administration are being used as a stepping stone to identify successful strategies for intracellular delivery in vivo.[50] In order to make the next promise in therapeutic advances a reality, parallel developments are needed in:

1) Understanding disease pathology sufficiently to identify specific ways to interfere.
2) Chemistry and biology of candidate therapeutics.
3) Appropriate drug delivery vehicles including long circulating liposomes.

References

1. Klibanov AL, Maruyama K, Beckerleg AM et al. Activity of amphipathic poly-(ethylene glycol) 5000 to prolong the circulation time of liposomes depends on the

liposome size and is unfavorable for immunliposome binding to target. Biochim Biophys Acta 1991; 1062:142-148.
2. Mori A, Huang L. Immunoliposome targeting in a mouse model: optimization and therapeutic application. In: Gregoriadis G, ed. Liposome Technology. Boca Raton: CRC Press 1993:153-162.
3. Torchilin VP, Klibanov AL, Huang L et al. Targeted accumulation of polyethylene glycol-coated immunoliposomes in infarcted rabbit myocardium. FASEB J 1992; 6:2716-2719.
4. Allen TM, Agrawal AK, Ahmad I et al. Antibody-mediated targeting of long circulating (StealthTM) liposomes. J Lipo Res 1994; 4:1-25.
5. Woodle MC. Sterically stabilized liposome therapeutics. Adv Drug Del Rev 1995; 16:249-265.
6. Goren D, Horowitz AT, Zalipsky S et al. Targeting of Stealth liposomes to erbB2 (Her2) Receptor: in vitro and in vivo studies. Br J Cancer 1996; 74:1749-1756.
7. Emanuel N, Kedar E, Bolotin E et al. Targeted delivery of doxorubicin via sterically stabilized immunoliposomes: pharmacokinetics and biodistribution in tumor-bearing mice. Pharm Res 1996; 13:861-868.
8. Blume G, Cevc G, Crommelin MDJA et al. Specific targeting with poly(ethylene glycol)-modified liposomes: coupling of homing devices to the ends of the polymeric chains combines effective target binding with long circulation times. Biochim Biophys Acta 1993; 1149:180-184.
9. Zalipsky S, Puntambekar B, Boulikas P et al. Peptide attachment to extremities of liposomal surface grafted PEG chains: preparation oth the long circulating form of laminin peptapeptide, YIGSR. Bioconj Chem 1995; 6:705-708.
10. DeFrees SA, Phillips L, Guo L et al. Sialyl lewis x liposomes as a multivalent ligand and inhibitor of e-selectin mediated cellular adhesion. J Am Chem Soc 1996; 118:6101-6104.
11. Wang CY, Huang L. Highly efficient DNA delivery mediated by pH-sensitive immunoliposomes. Biochem 1989; 28:9508-9514.
12. Wilschut J, Hoekstra D. Membrane Fusion. New York: Marcel Dekker, Inc., 1991.
13. Lasic DD. Liposomes: from Physics to Applications. Amsterdam: Elsevier, 1993:575.
14. de Lima MCP, Hoekstra D. Liposomes, viruses, and membrane fusion. In: Philippot JR, Schuber F, eds. Liposomes as Tools in Basic Research and Industry. Boca Raton: CRC Press, 1994:137-156.
15. Duzgunes N, Nir S. Liposomes as tools for elucidating the mechanism of membrane fusion. In: Philippot JR, Schuber F, eds. Liposomes as Tools in Basic Research and Industry. Boca Raton: CRC Press, 1994:103-136.
16. Remy J-S, Sirlin C, Behr J-P. Gene transfer with cationic amphiphiles. In: Philippot JR, Schuber F, eds. Liposomes as Tools in Basic Research and Industry. Boca Raton: CRC Press, 1994:159-170.
17. Lasic DD. Liposomes in Gene Delivery. Boca Raton: CRC Press, 1997:320.
18. Holland JW, Cullis PR, Madden TD. Poly(ethylene glycol)-lipid conjugates promote bilayer formation in mixtures of non-bilayer-forming lipids. Biochem 1996; 35:2610-2617.
19. Lasic DD. Liposomes within liposomes. Nature 1997; 387:26-27.
20. Slepushkin VA, Simoes S, Dazin P et al. Sterically stabilized pH-sensitive liposomes—intracellular delivery of aqueous contents and prolonged circulation in vivo. J Biol Chem 1997; 272:2382-2388.
21. Kirpotin D, Park JW, Hong K et al. Sterically stabilized anti-HER2 immunoliposomes: design and targeting to human breast cancer cells in vitro. Biochem 1997; 36:66-75.

22. Woodle MC, Raynaud FI, Dizik M et al. Oligonucleotide pharmacology and formulation: G3139 anti-BCL2 phosphorothioate in Stealth liposomes and gel implants. Nucleosides Nucleotides 1997; (in press).
23. Maruyama K, Kennel SJ, Huang L. Lipid composition is important for highly efficient target binding and retention of immunoliposomes. Proc Natl Acad Sci USA 1990; 87:5744-5748.
24. Mori A, Klibanov AL, Torchilin VP et al. Influence of the steric barrier of amphipathic poly)ethyleneglycol) and ganglioside GM1 on the circulation time of liposomes and on the target binding of immunoliposomes in vivo. FEBS Lett 1991; 284:263-266.
25. Mori A, Kennel SJ, Huang L. Immunotargeting of liposomes containing lipophilic antitumor prodrugs. Pharm Res 1993; 10:507-514.
26. Ahmad I, Longenecker M, Samuel J et al. Antibody-targeted dilivery of doxorubicin entrapped in sterically stabilized liposomes can eradicate lung cancer in mice. Cancer Res 1993; 53:1484-1488.
27. Zalipsky S, Newman MS, Puntambekar B et al. Model ligands linked to polymer chains on liposomal surfaces: application of a new functionalized polyethylene glycol lipid conjugate. Polym Mater Sci Eng 1993; 67:519-520.
28. Vingerhoeds MH, Steerenberg PA, Hendriks JJGW et al. Immunoliposome-mediated targeting of doxorubicin to human ovarian carcinoma in vitro and in vivo. Br J Cancer 1996; 74:1023-1029.
29. Matthay KK, Abai AM, Cobb S et al. Role of ligand in antibody-directed endocytosis of liposomes by human T-leukemia cells. Cancer Res 1989; 49:4879-4886.
30. Storm G, Nassander UK, Vingerhoeds MH et al. Antibody-targeted liposomes to deliver doxorubicin to ovarian cancer cells. J Lipo Res 1994; 4:641-666.
31. Allen T, Hansen CB, Kao GY et al. Therapeutic opportunities for targeted liposomal drug delivery. Adv Drug Del Rev 1996; 21:117-133.
32. Parr MJ, Ansell SM, Choi LS et al. Factors influenceing the retention and chemical stability of poly(ethylene glycol)-lipid conjugates incorporated into large unilamellar vesicles. Biochim Biophys Acta 1994; 1195:21-30.
33. Holland JW, Hui C, Cullis PR et al. Poly(ethylene glycol)-lipid conjugates regulate the calcium-induced fusion of liposomes composed of phosphatidylethanolamine and phosphatidylserine. Biochem 1996; 35:2618-2624.
34. Kirpotin D, Hong K, Mullah N et al. Liposomes with detachable polymer coating: destabilization and fusion of dioleoylphophatidylethanolamine vesicles triggered by cleavage of surface-grafted poly(ethylene glycol). FEBS Lett 1996; 388:115-118.
35. Zalipsky S. Synthesis of an end-group functionalized polyethylene glycol-lipid conjugate for preparation of polymer-grafted liposomes. Bioconj Chem 1993; 4:296-299.
36. Haselgrubler T, Amerstorfer A, Schindler H et al. Synthesis and applications of a new poly(ethylene glycol) derivative for the crosslinking of amines with thiols. Bioconj Chem 1995; 6:242-248.
37. Zalipsky S. Functionalized poly(ethylene glycol)s for preparation of biologically relevant conjugates. Bioconj Chem 1995; 6:150-165.
38. Park JW, Hong K, Carter P et al. Development of anti-p185HER2 immunoliposomes for cancer therapy. Proc Natl Acad Sci USA 1995; 92:1327-1331.
39. Vingerhoeds MH, Haisma HJ, van Muigen M et al. A new application for liposomes in cancer therapy: immunoliposomes bearing enzymes (immuno-enzymosomes) for site-specific activation of prodrugs. FEBS Lett 1993; 336:485-490.
40. Oku N, Tokudome Y, Koike C et al. Liposomal arg-gly-asp analogs effectively inhibit metastatic B16 melanoma colonization in murine lungs. Life Sci 1996; 58:2263-2270.

41. Storm G, Vingerhoeds MH, Crommelin DJA et al. Immunoliposomes bearing enzymes (immuno-enzymosomes) for site-specific activation of anticancer prodrugs. Adv Drug Del Rev 1997; 24:225-231.
42. Vingerhoeds MH, Haisma HJ, Belliot SO et al. Immunoliposomes as enzyme-carriers (immuno-enzymosomes) for antibody-directed enzyme prodrug therapy (ADEPT): optimization of prodrug activating capacity. Pharm Res 1996; 13:604-610.
43. Trubetskoy VS, Narula J, Khaw BA et al. Chemically optimized antimyosin Fab conjuates with chelating polymers: importance of the nature of the protein-polymer single site covalent bond for biodistribution and infarction localization. Bioconj Chem 1993; 4:251-255.
44. Zalipsky S, Brandeis E, Newman MS et al. Long circulating, cationic liposomes containing amino-PEG-phosphatidylethanolamine. FEBS Lett 1994; 353:71-74.
45. Lee RJ, Low PS. Delivery of liposomes into cultured KB cells via folate receptor-mediated endocytosis. J Biol Chem 1994; 269:3198-3204.
46. Yamamura K, Kibbey MC, Jun SH et al. Effect of matrigel and laminin peptide YIGSR on tumor growth and metastasis. Semin Cancer Biol 1993; 4:259-265.
47. Merwin JR, Noell GS, Thomas WL et al. Targeted delivery of DNA using YEE(GalNAcAH)3, a synthetic glycopeptide for the asialogylcoprotein receptor. Bioconj Chem 1994; 5:612-620.
48. De Kruif J, Storm G, van Bloois L et al. Biosynthetically lipid-modified human scFv fragments from phage display libraries as targeting molecules for immunoliposomes. FEBS Lett 1996; 399:232-236.
49. Crommelin DJA, Herron J, Storm G. (Protein)-targeted delivery with particulate systems. In: Lee VHL, Hashida M, Mizushima YM, eds. Trends and Future Perspectives in Peptide and Protein Delivery. Harwood Academic Publishers, GmbH 1994:207-239.
50. Li S, Gao, X, Son K, Sorgi F et al. DC-Chol lipid system in gene transfer. J Controlled Rel 1996; 39:373-381.

Index

D

E

F

G

H

Q

R

S

T